DEVELOPMENTS IN CONCRETE TECHNOLOGY—1

THE DEVELOPMENTS SERIES

Developments in many fields of science and technology occur at such a pace that frequently there is a long delay before information about them becomes available and usually it is inconveniently scattered among several journals.

Developments Series books overcome these disadvantages by bringing together within one cover papers dealing with the latest trends and developments in a specific field of study and publishing them within *six months* of their being written.

Many subjects are covered by the series, including food science and technology, polymer science, civil and public health engineering, pressure vessels, composite materials, concrete, building science, petroleum technology, geology, etc.

Information on other titles in the series will gladly be sent on application to the publishers.

DEVELOPMENTS IN CONCRETE TECHNOLOGY—1

Edited by

F. D. LYDON
B.E., M.Phil.(LOND.),
C. Eng., M.I.C.E., M.I.C.T.

Senior Lecturer, Department of Civil Engineering and Building Technology, University of Wales Institute of Science and Technology, Cardiff, UK

APPLIED SCIENCE PUBLISHERS LTD
LONDON

APPLIED SCIENCE PUBLISHERS LTD
RIPPLE ROAD, BARKING, ESSEX, ENGLAND

British Library Cataloguing in Publication Data

Developments in concrete technology—(Developments series)
1.
1. Concrete
I. Lydon, Francis Dominick II. Series
620.1'36 TA439

ISBN 0-85334-855-3

WITH 18 TABLES AND 145 ILLUSTRATIONS

Printed in Great Britain by Bell and Bain Ltd, Glasgow

PREFACE

Even a cursory look at the literature on concrete will demonstrate the breadth of attention given to this material. It is of the essence of the construction industry. Its versatility and usefulness are almost unlimited and demands of it continue into new combinations of difficult and even severe circumstances. Such a material, it will be argued, should be sufficiently understood to allow its behaviour to be predicted so that interaction between it and its environment can safely be characterised and, where necessary, extrapolated from current knowledge.

But, as all concrete technologists know, this is not the case; prediction, although much better than it used to be, is virtually impossible in the true scientific sense. The essential inevitability associated with, for example, the classical laws of physics is absent (and, indeed, may never be possible) and predictability, such as it is, stems from empirical studies, from a very imperfect picture of the material and from the intuitive sense that a sound study brings to the perceptive student of concrete. Fortunately, there are several students of the latter ilk whose work and perception have led to better, though still imperfect, understanding of the nature of concrete from which genuine improvements have developed. This is, self evidently, a slow, faltering process; the multiphase composition and innate heterogeneity of the material require cross-linking and inter-communication between several disciplines and, apart from the relatively rare, lucky, fruitful juxtapositions which happily occur between disciplines, sufficient studies are published in each field to slow down even the most committed researcher endeavouring to find a compatible thread to follow. Thus physicists, chemists, materials' scientists, metallurgists, rheologists, mineralogists (and others, no doubt) have all got specialist roles in this progress and the concrete technologist struggles to remain reasonably informed and to abstract and use information which will be helpful in the gradually unfolding behavioural picture. A 'prepared mind' is therefore essential.

A book on developments in concrete technology must reflect the preference of the editor; from a quite large list of topics and trends a very limited number must be chosen which represents, *in toto*, at least a fair consensus of agreement as to aptness and importance. Of course there will be some reservations, perhaps as to emphasis or omission, but taken overall the book should be a significant statement on, and a real, rather than superficial, contribution to, the subject.

Subject matter and authorship, on such occasions, are closely related, so that once the general scheme of the book was set out it was easy to select appropriate contributors. All are authorities, of considerable standing, recognised by their peers for their work in the chosen fields, completely familiar with the background and with current research, ideally placed to provide a sound overview yet not to miss or exclude important major detail. Very busy people all, it is to their considerable credit that contributors undertook the demanding tasks requested of them and were prepared to commit so much precious time and energy to writing their chapters.

It is hoped that the excellent work done by this team of contributors will very much help the 'prepared mind' by providing a really sound, rationally based platform from which further progress can be effected, and by stimulating the imaginative insight so necessary for filling in some of the many blanks in the behavioural rules of concrete.

F. D. LYDON

CONTENTS

This book is dedicated to Bob Elvery whose sad demise has robbed us of even more significant contributions to concrete technology.

LIST OF CONTRIBUTORS

A. F. BAKER, *Research Engineer, Research and Development Department, Taylor Woodrow Construction Ltd., 345 Ruislip Road, Southall, Middlesex UB1 2QX, UK.*

R. D. BROWNE, *Manager, Research and Development Department, Taylor Woodrow Construction Ltd., 345 Ruislip Road, Southall, Middlesex UB1 2QX, UK.*

N. CLAYTON, *Higher Scientific Officer, Building Research Establishment, Building Research Station, Garston, Watford WD2 7JR, UK.*

F. J. GRIMER, *Principal Scientific Officer, Building Research Establishment, Building Research Station, Garston, Watford WD2 7JR, UK.*

J. M. ILLSTON, *Director of Studies in Civil Engineering, The Hatfield Polytechnic, P.O. Box 109, College Lane, Hatfield, Herts AL10 9AB, UK.*

J. B. NEWMAN, *Head, Concrete Materials Research Group, Department of Civil Engineering, Imperial College of Science and Technology, Imperial Institute Road, London SW7 2AZ, UK.*

K. D. RAITHBY, *Principal Scientific Officer, Structures Department, Transport and Road Research Laboratory, Department of the Environment, Crowthorne, Berks RG11 6AU, UK.*

P. J. E. SULLIVAN, *Department of Civil Engineering, Imperial College of Science and Technology, Imperial Institute Road, London SW7 2AZ, UK.*

R. NARAYAN SWAMY, *Reader, Department of Civil and Structural Engineering, University of Sheffield, Mappin Street, Sheffield S1 3JD, UK.*

Chapter 1

THE EFFECTS OF TEMPERATURE ON CONCRETE

P. J. E. Sullivan

Department of Civil Engineering, Imperial College of Science and Technology, London UK

SUMMARY

The properties of different types of concrete change as the temperature increases. Below 100 °*C the migration of water within the concrete affects the concrete properties, producing increased deterioration which appears to reverse at higher temperatures. At a certain temperature limit, permanent degradation of concrete takes place which is irreversible. The use of reinforcement tends to improve the performance of concrete particularly at temperatures below* 100 °*C.*

Tests carried out at temperature transient conditions, i.e. properties measured while the temperature is varying, can give results which are divergent from those carried out when the properties are measured at a steady state of temperature. Tests carried out using a factorial method of experimentation indicate that parameters which appear to be independent of temperature when concrete is tested under steady state are, in fact, temperature dependent when tested under transient conditions.

The chapter also indicates how high temperature resistant concrete is produced and how it behaves at elevated temperatures.

The effects of fire temperature on concrete are described and a comparison of the effects on concrete when subjected to a standard furnace and experimental fires is presented.

The chapter concludes with anomalous deformation behaviour of members heated on one surface when their boundary condition is varied.

1.1 INTRODUCTION

The thermal properties of concrete are dependent on the properties of its constituents, particularly the aggregate type, and also on the volume fraction of aggregate the concrete contains. Since concrete is a porous medium the internal pore structure and the moisture content also influence the thermal flow characteristic which is coupled with the moisture and vapour diffusion rate in the concrete. Mass and heat flow create internal strains within the body and therefore affect the structural characteristics of the concrete.

In general, lighter weight aggregates with low thermal conductivity and coefficient of linear expansion perform better as insulants and also structurally, especially at high temperatures. The lower expansion of lightweight aggregates tends to reduce the stresses developed between the aggregate and its cement matrix at elevated temperatures. However, there is experimental evidence to suggest[1] that vapour expansions, due to heating of the free water within the concrete, can cause internal stresses which can reduce the strength of concretes containing aggregates with a high moisture absorption. At temperatures between 60° and 90 °C the stresses within the body of the lightweight concrete appear to reach a maximum as the structural performance reduces to a minimum. Beyond this temperature the free water evaporates escaping from the surface and relieving the internal stresses, resulting in an improved performance (and for some lightweight aggregate concretes to beyond that at 20 °C).

Normal weight concretes are less efficient insulants than lightweight aggregate concretes. However, because of the higher density, the thermal inertia of a normal weight concrete wall of a given thickness allows a slower rate of temperature change, due to extreme changes of outside temperature, giving lower inside temperature fluctuation.

Normal low porosity aggregates do not appear to deteriorate sharply between 60 and 90 °C, although a reduction of strength and an increase in creep strains have been observed by many research workers. A recovery in strength occurs at about 100 °C when the evaporable water is lost. At much higher temperatures, however, normal concretes deteriorate by varying amounts depending on the type of aggregates used.

The performance of concrete can be measured by the change of its stiffness, strength or other property which would affect its main function in service. As concrete has a low tensile strength it is normally relied on to take compressive forces, the tensile forces being taken by reinforcing steel. A great deal of research work on concrete at elevated temperatures, therefore, has concentrated on compressive strength as a fundamental property in examining its deterioration. Compressive strength is, however, not so sensitive a property to deterioration as tensile or flexural strength under short-term loading at high temperatures, as induced stresses under these conditions are more akin to those imposed by tests in compression. Short-term creep strains under compression[2] are equally sensitive to deterioration as the temperature-induced microcracks soon become superimposed on the applied deformations.

As heat is applied to concrete changes within it begin to take place and, when properties are investigated at elevated temperatures, it is important to recognise that the behaviour of the material during initial heating to a given temperature, *T*, can be considerably different to that during subsequent cycles up to temperature *T*. Typical expansions over one heating cycle of river sand mortar beams and 10-mm gravel aggregate concrete beams[3] are shown in Fig. 1. In these tests the sand/cement ratio for the gravel concrete and mortar beam specimens was identical. The irregular expansion of the mortar during initial heating to just over 400 °C (curve (A)) has been restrained by the

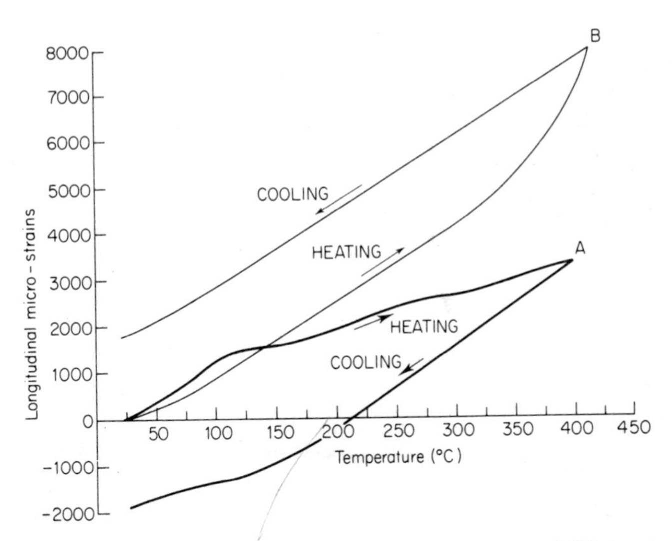

FIG. 1. Effect of temperature on longitudinal movements of (A) mortar (river sand) and (B) concrete (Thames river gravel).

coarser aggregate (curve (B)). On cooling the mortar follows a more uniform contraction ending with an irreversible shrinkage at 20 °C. Subsequent heating cycles would follow this cooling curve. It will be observed that, although the cooling curve for gravel concrete shows an irreversible expansion at 20 °C, indicating disruption of the material, further temperature cycles would also follow the cooling curve.

The difference in behaviour during initial heating from that during subsequent cycles is more pronounced when the movements induced by water migration are less restrained by rigid normal weight aggregates. Thus, cement pastes show an even greater divergence than mortars on first heating and, as indicated in Fig. 2, show a net contraction after

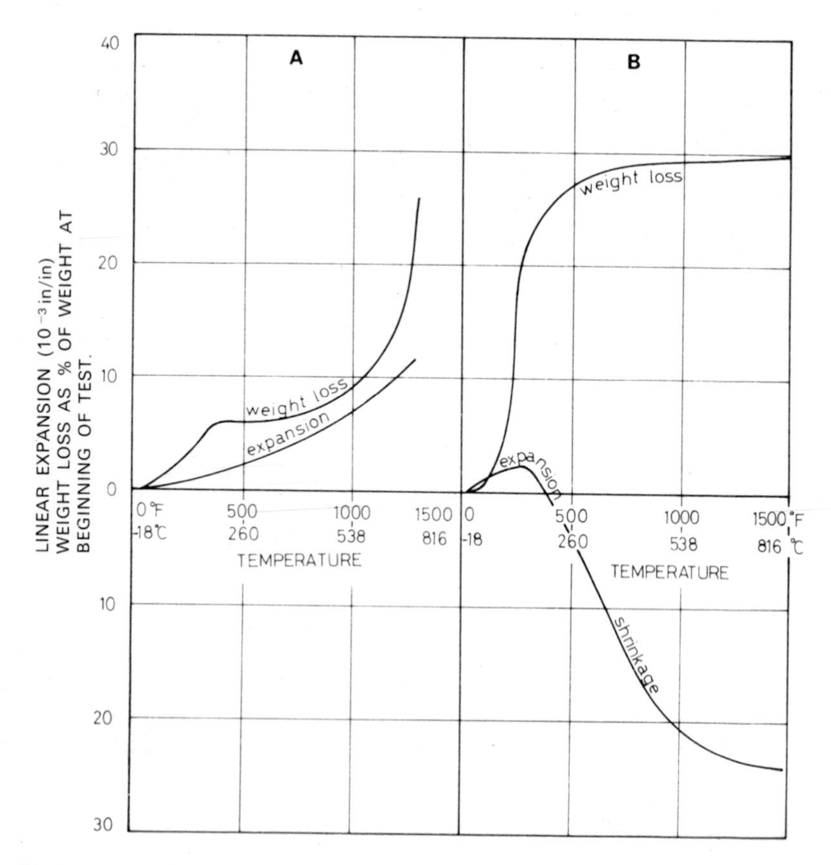

FIG. 2. Expansions and weight loss of (A) gravel concrete and (B) neat cement paste.

heating to temperatures beyond 400 °C. Lightweight aggregates, which impose less restraint on temperature-induced movements than normal aggregates when used in concrete, also exhibit initial heating movements which are very different from subsequent temperature cycles. This reduction of internal restraints in lightweight aggregate concrete reduces the risk of early disruption of the material at elevated temperatures.

The performance of concrete as an insulating material can be measured in relation to its thermal flow properties with increasing temperature. The thermal flow characteristics can give rise to internal stresses and would obviously influence the material's structural performance. However, the stress, strain and strength properties which change at elevated temperatures play a more important role in the performance of concrete used as a structural material at high temperatures.

1.2 THERMAL PROPERTIES

The important properties that determine the insulating qualities of a material are:

1. Thermal conductivity (K) which is used to derive the temperature gradients ($\partial\theta/\partial x$) under steady states of heat flow from the solution of the equation

$$\text{div}\, K\, \text{grad.}\, \theta = 0 \tag{1}$$

or more simply for constant K

$$K\nabla^2\theta = 0 \tag{1a}$$

2. Thermal inertia (i.e. the product of density (ρ) and specific heat (c)) which controls the thermal lag between the outside air temperature and the temperature within a building under fluctuating or transient temperatures.
3. Thermal diffusivity ($D = K/\rho c$) which can be used to derive the changing temperature distributions ($\partial\theta/\partial x$) per unit time ($t$) under transient states from the solution of the equation

$$\nabla^2 K\theta = \rho c \frac{\partial\theta}{\partial t} \tag{2}$$

for K independent of temperature

$$D\nabla^2\theta = \frac{\partial \theta}{\partial t} \tag{2a}$$

4. Radiation and convection properties which control the temperature rise at the boundaries, i.e. surfaces exposed to and remote from the heat source. These two properties are conveniently combined as follows to derive the rate of quantity of heat, dQ/dt, radiated or absorbed from a given surface area A of a body:

$$\frac{dQ}{dt} = A(Eh_r + h_c)\frac{d\theta}{dx} \tag{3}$$

or, more simply,

$$\frac{dQ}{dt} = A(Eh_r + h_c)\frac{(\theta_{air} - \theta_{surface})}{\Delta x} \tag{3a}$$

where $(Eh_r + h_c)$ is the surface conductance of the body, and is a combination of the convection conductance h_c and the radiation (or absorption) conductance Eh_r. This latter term is the product of the black body radiation coefficient h_r and the emissivity factor E, typical values of which are given in Table 1 for typical constructional materials.

TABLE 1

TYPICAL VALUES OF EMISSIVE (E) AND ABSORPTIVE (α) POWER

	E (%)	α (%)
Road surfacing (black top)	85–98	85–98
Concrete	85–98	85–98
Steel	85–98	65–80

The above properties also indirectly influence the thermal structural performance of a material, as thermal gradients may cause thermal stresses.

1.3 THERMAL FLOW

Equations (1), (2) and (3) can be used for the analytical solution of thermal flow through solid bodies for both steady and transient states of heating. For calculation of temperature gradients on bridge decks[4,5] it is convenient to modify the quantity of heat per unit area $(1/A)(\mathrm{d}Q/\mathrm{d}t)$ in eqn. (3) into two parts:

(1) absorption of incoming solar radiation during summer day time, αI, where I is the intensity of solar radiation on a horizontal surface, less heat absorbed by top flange, or
(2) emission of radiation during winter night time, EQ', where Q' is outgoing longwave radiation, less heat lost by top flange.

Thus eqn. (3) expands to

$$\frac{\alpha I}{A} = h_s \frac{\mathrm{d}\theta}{\mathrm{d}x} + \frac{\rho c}{K} x \frac{\partial \theta}{\partial t} \quad \cdots \qquad (4)$$

or

$$\frac{EQ'}{A} = h_s \frac{\mathrm{d}\theta}{\mathrm{d}x} + \frac{\rho c}{K} x \frac{\partial \theta}{\partial t} \quad \cdots \qquad (4a)$$

where $h_s = (Eh_r + h_c)$ and the values of E, Q, α and I are obtainable from meteorological data for the locality of the bridge. If a steel deck rather than concrete is considered K can be taken as 1.

Equations (4) and (4a) transformed into their finite differences can then be used to determine the temperatures at selected finite distances within the bridge at discrete time intervals calculated to ensure that instability of the solution does not occur. Calculated values from this solution agree very closely with measured values of temperature on actual bridges in the UK.[6]

The temperature distribution can be used not only to give data for the realistic assessment of deformations of bridges, but also for the additional stress distributions arising from these transient temperature effects.

For domestic design purposes however, it is more convenient to use the concept of 'standard' thermal transmittance or U-values which are tabulated in the IHVE Guide.[7] This is given by U in eqn. (5),

$$U = \frac{1}{\sum(R_s) + \sum_i (R_i)} \ \mathrm{W/m^2\,{}^\circ C} \qquad (5)$$

where, $R_s = (1/Eh_r + h_c)\mathrm{m}^2\,°\mathrm{C/W}$ (surface thermal resistance), $h_r = 4{\cdot}6$–$5{\cdot}7\,\mathrm{m}^2\,°\mathrm{C/W}$ between 0 and 20 °C, $E = 0{\cdot}9$ for ordinary building materials, $h_c = 5{\cdot}8 + 4{\cdot}1 \times$ (velocity of wind in m/s) $= 3{\cdot}0$ for walls, 4·3 for upward flow to ceilings, and 1·5 for downward flow to floors, and

$\sum_i (R_i) = \sum_i \left(\frac{l}{K}\right)_i$ (the sum of thermal resistances of building components where

l = thickness, and K = 'standard' thermal conductivity of each component forming wall or floor).

The standard thermal conductivities for different building materials which are given in the IHVE Guide[7] are the conductivities of materials containing a given moisture content (1 %–5 %) dependent on the exposure of the building component. Correction factors are also given for components containing different amounts of moisture.

A number of test methods for determing the conductivity, thermal transmittance, conductance, emissivity and heat capacity, applicable for different types of insulating materials, is given in BS 874.[8] However, difficulties arise in the measurement of the thermal conductivity of porous building materials owing to the moisture flow taking place during heating and also owing to the temperature dependence of conductivity. For these reasons BS 874 specifies appropriate conductivity to a constant weight of the specimens prior to testing, and the test temperatures of the hot and cold faces. Thus, for materials which are not normally exposed to temperatures in excess of 50 °C, the conditioning environment is specified at 20 ± 2 °C at a relative humidity of 65 ± 5 %. At the end of test the average moisture content of the specimen is determined in accordance with BS 2972.[9]

Thermal conductivity values of typical concretes together with the thermal conductivity of the aggregate between 5 and 25 °C are shown in Table 2.

The thermal conductivity of a large range of concretes reduces to approximately 0·6 of the saturated values when oven-dried, and intermediate values of saturation can be linearly interpolated.

The thermal conductivity of concrete which is initially wet (see Fig. 3) increases by about 10 % at approximately 60 °C and decreases thereafter. For a predried concrete, conductivity decreases with an increase in temperature as shown also in Fig. 3.

Under transient heating situations, it is more appropriate to utilise diffusivity ($D=K/\rho c$ as in eqn. (2a)) rather than conductivity, which is only suitable when steady states of heating have been attained. Diffusivity which is again closely allied with the diffusivity of the

TABLE 2

THERMAL CONDUCTIVITY OF DIFFERENT TYPES OF CONCRETE

Type of aggregate	*Thermal conductivity (W/m °C)* Rock	Concrete (saturated)
Siliceous rock—quartzite, sandstone, gravel	3·3–4·2	2·4–3·6
Igneous crystalline—granite, sedimentary carbonates (dolomites)	2·3–2·8	1·9–2.8
Igneous amorphous—basalts, dolerites, artificial lightweight	0·7–2·8	1·0–1·6

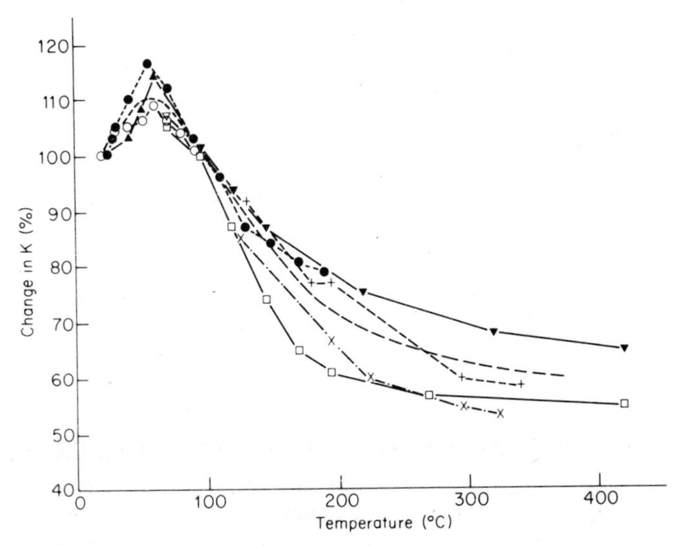

FIG. 3. Effect of temperature on thermal conductivity of initially saturated concrete.[10] ○, ▲, river gravel; ●, limestone; ×, +, serpentine granite; □, ▼, quartzite.

aggregates reduces considerably with temperature as indicated in Fig. 4. The curve gives the results of tests of thermal diffusivity carried out on concrete specimens using calcareous and siliceous aggregate. The average thermal diffusivity at 20 °C of six types of different aggregate concretes can be taken as approximately $1\,m^2/s$ with basalt aggregate concrete being as low as 0·69 and quartz aggregate concrete having an average value of 1·89.[10]

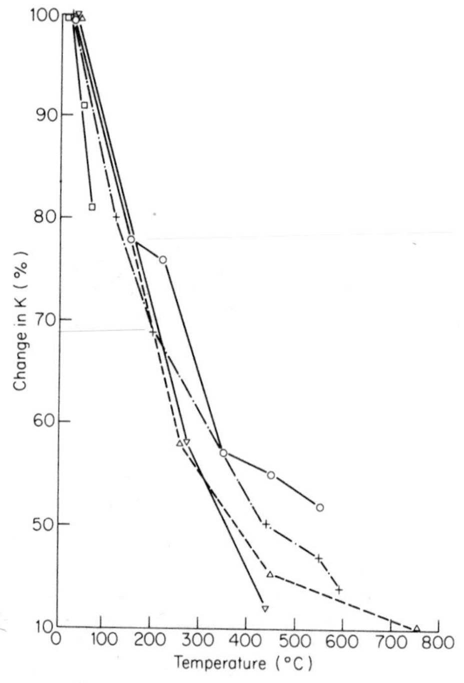

FIG. 4. Effect of temperature on the thermal diffusivity of concrete.[10] +, calcareous aggregate; ○, siliceous aggregate; △, ▽, □, silica.

1.4 PHYSICAL PROPERTIES AT ELEVATED TEMPERATURES

Most of the physical properties of concrete deteriorate with increasing temperatures. It has already been stated that the type of aggregate and moisture content play important roles in the manner that concrete is affected. However, size and shape of aggregate, type of cement, admixtures and water–cement ratio also influence the results during heating. The introduction of reinforcing steel can assist in reducing the sharp drop in strength that can sometimes occur. Other external factors such as form of heating, rate of heating, applied external loads and restraints during heating also influence the ultimate behaviour of the concrete.

The properties most commonly studied have been those associated with durability, including compressive strength, tensile and flexural strength, and modulus of elasticity or rigidity at steady states of temperature, i.e. the specimens, normally small scale, are heated to test temperatures, maintained at these temperatures for a given time to allow conditions (such as moisture) to stabilise and then tested while hot. The lateral expansions and contractions have, generally, been measured initially during heating but the reversible or the thermal coefficient of expansion is normally quoted after heat cycling to test temperature until conditions stabilise. The fire resistance of a material such as concrete is normally carried out on full-scale elements which have been conditioned in a standard manner. They are then heated at a standard rate until collapse or failure of the elements, to satisfy defined serviceability criteria.[11]

Many research workers realising that testing small-scale specimens at steady state temperatures may not, in certain cases, simulate realistic conditions while the specimens are being heated, have undertaken studies into the change in physical properties during transient heating. Testing under these conditions is difficult to perform and the results are difficult to interpret, and attempts have been made to correlate properties derived from these tests with those derived from steady state tests. Another novel approach[12] used a fractional factorial method of experimentation to investigate the significance of altering nine different factors at different intervals of time during heating.

1.5 STEADY STATE TEMPERATURE TESTS

The compressive strength of concrete tested at a steady state of temperature reduces to a minimum value at temperatures between 60 and 80 °C, depending on the moisture content and type of aggregates. The compressive strength then increases, and for a siliceous mortar can reach values 10 % in excess of the initial strength at 20 °C.[3] Figure 5 shows the bounds of the compressive strengths of different types of concrete between 20 and 150 °C. Beyond this temperature there is a gradual decrease in strength. For Thames river gravel concrete the compressive strength drops by approximately 12 % at 400 °C and 25 % at 500 °C. If the specimens are loaded to a constant compressive strength prior to heating, a higher ultimate strength is achieved when tested hot. A further reduction of strength results after cooling to 20 °C.

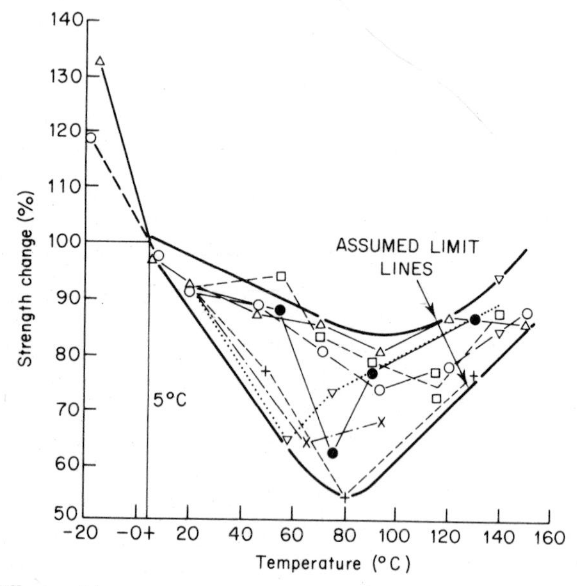

FIG. 5. Effect of temperature on compressive strength of concrete. ●, amphibole snist; ○, △, □, calcareous gravel; +, limestone; × siliceous gravel.

Other testing on river gravel concrete cubes (nominal ratio 1–4–8) which were treated up to 550 °C cooled to 20 °C and subsequently tested in compression after exposing to varying periods of humid atmospheres, showed a further decrease in strength. The older specimens (2 months) reduced in strength from 40 %, after heat treatment, to 20 % after exposure to moist environments up to 6 days, and the younger specimens (2 weeks) reduced to 8 % after exposure in moist environments up to 5 days. This loss was ascribed to the reabsorption of H_2O by the lime, causing expansion and disruption of concrete.

Later work carried out on Ordinary Portland Cement (OPC) and blast furnace slag cement sand mortar (1–3) cylinders heated to 600 °C for 3 h and allowed to cool, indicated that by subsequent curing in water a recovery in strength occurred whereas no recovery was detected if the subsequent curing was in air. The blast furnace cement mortars performed marginally better under the above conditions. Although this work appears to contradict the previous research, it should be borne in mind that mortars are less disrupted than concrete[3] since mortar is more homogeneous. Also, there is more cement available in mortars for rejuvenation than in concrete, which is generally leaner than mortar.

Tests carried out in compression on concrete containing different aggregates showed that the lighter aggregates, including pumice, expanded shale, expanded slag, and sintered pulverised fuel ash (pfa) performed better than normal weight aggregates. Basalts and granites performed better than dolerites, which in their turn behaved better than gravel concretes and sandstones. The upper and lower bounds of the compressive strength for concrete containing different aggregates are shown in Fig. 6.

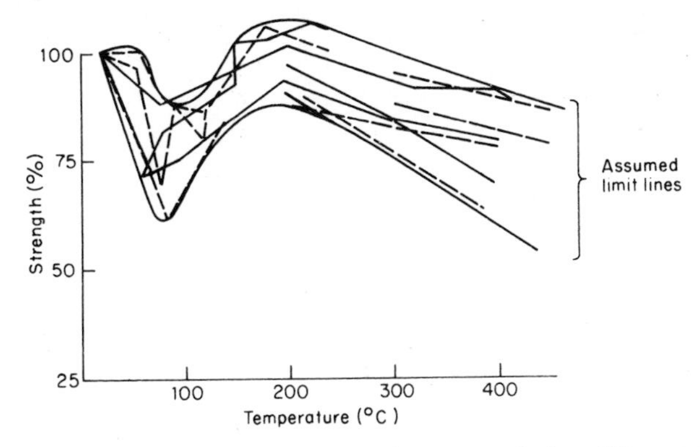

FIG. 6. Effect of temperature on compressive strength for different types of concrete.[10]

1.5.1 Flexural and Tensile Strength

It was early recognised that, although gravel concrete exhibited extensive cracking at temperatures above 300 °C, the carrying capacity of small gravel concrete cylinders in compression had only reduced by 50 % after being subjected to a temperature of 650 °C for periods of 6 h. It was, therefore, proposed that tensile testing would be more sensitive to deterioration at elevated temperatures than testing in compression. This has been borne out by testing in tension and flexure by subsequent research workers.[1,3,13] Figure 7 shows the upper and lower limiting curves of the tensile strength of concrete (containing various types of normal aggregates) plotted against temperature. It will be observed that concrete deteriorates at a faster rate when tested in tension rather than compression but otherwise it shows similar variations with increasing temperatures.

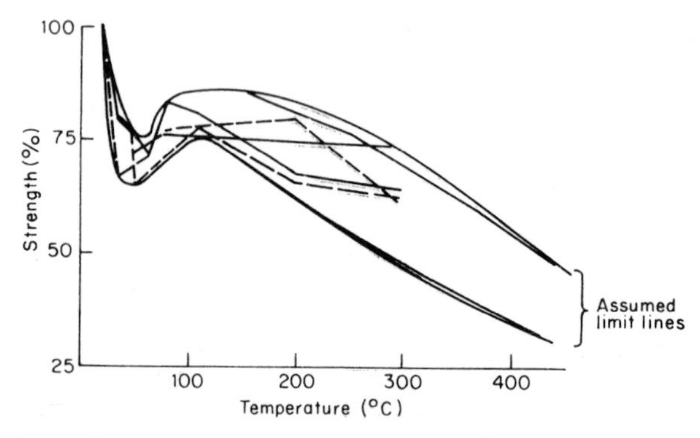

FIG. 7. Effect of temperature on tensile strength for different types of concrete.[10]

The effect of temperature on flexural strength for concrete containing sintered pulverised fuel ash (pfa) is shown in Fig. 8. Similar characteristics are obtained for flexural as for tensile testing, since failure in flexure is controlled by tensile cracking. It will be observed that very low strengths occur at around 60 °C for sintered pfa concrete. The strength of this concrete, however, recovers above 100 °C, and at 500 °C the flexural strength exceeds that at 20 °C by more than 20 %.[14] In practice the tensile zone of concrete beams contains steel reinforcement, and in this instance the failure strength is controlled by the steel. Figure 8 also shows the effect of including mild steel reinforcement in the tensile zone of sintered pfa concrete.[1,14] The sharp drop in strength below 100 °C for the concrete beams is absent for the reinforced concrete specimens and there is only a slight deterioration (10 % drop) up to 400 °C. Beyond 400 °C deterioration increases and at 600 °C the carrying capacities of the reinforced lightweight concrete beams are 30 % of their original capacity. In contrast, the unreinforced material increases in strength beyond 100 °C reaching 20 % above the 20 °C value. This strength is maintained to 500 °C and reduces back to the cold strength at 600 °C. Note also that the Group I beams which contained no shear stirrups failed in shear at a lower load below 300 °C than those in Group II which contained shear stirrups. These latter beams failed in flexure.

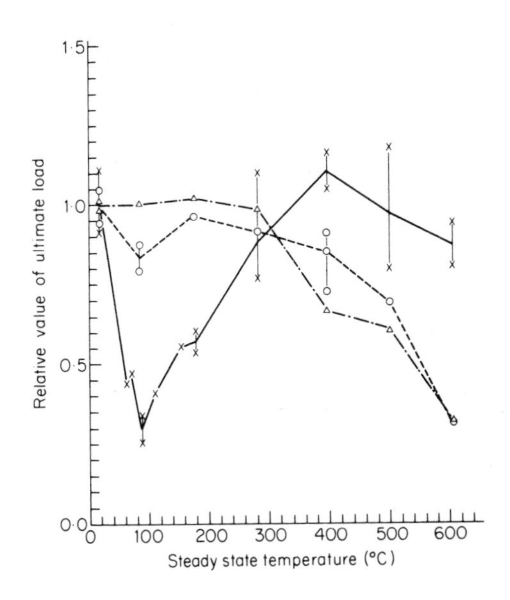

FIG. 8. Effect of temperature on flexural strength for plain and reinforced sintered pfa concrete.[14] ×, plain concrete beams; ○, RC beams, Group 1; △, RC beams, Group II. *Note*: Group I beams contained no shear reinforcement; Group II beams contained shear reinforcement.

Figure 9 compares the difference in load-bearing capacity between reinforced sintered pfa concrete and gravel concrete. Reinforced sintered pfa concrete exhibits a higher ultimate load than does normal reinforced concrete at all temperatures above 20 °C.

Figure 10 compares the drop in strength as a percentage of the cold strength of mortar, concrete and reinforced concrete beams are heated between 100 and 400 °C.[3] The figure illustrates how the addition of a river gravel to a river sand mortar matrix introduces incompatibilities arising from the difference of properties between the aggregate and the matrix; these incompatibilities becoming more severe with increasing temperature. With the introduction of reinforcement to the gravel concrete, the initial percentage drop in strength is higher than that of plain concrete but is not as severe at higher temperatures. When tested

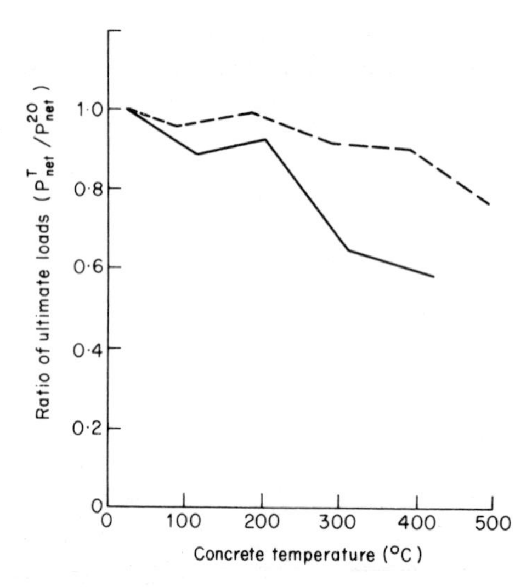

FIG. 9. Effect of temperature on flexural strength of reinforced gravel (—) and sintered pfa (---) concrete beams.

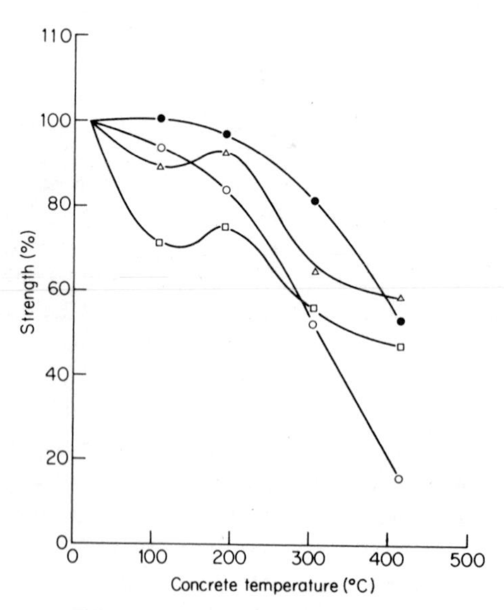

FIG. 10. Per cent strength reduction for mortar, concrete and reinforced concrete with increase in temperature. •, river sand mortar; ○, gravel concrete; □ (hot), △ (cold), reinforced concrete.

cold after heat treatment (see Fig. 18) the percentage residual strength of the RC beams recovers to values higher than the percentage strength of concrete alone tested hot. The steel reinforcement is, of course, not affected when the temperature of the steel is allowed to cool after heating to temperatures up to 400 °C, and hence the drop in residual strength for the RC beams is affected more by the deterioration of the concrete, which had occurred when heated previously. It should be noted that a direct comparison of the load carrying capacity from Fig. 10 can only be made between the mortar beams and the concrete beams as the average failure load values at 20 °C were 591 lb (2·6 kN) for the mortar beams and 569 lb (2·5 kN) for the concrete beams. (See Figs. 16, 17, 18 and 19.) The average load carrying capacity of the RC beams at 20 °C was 2538 lb (11·3 kN), and in absolute terms, therefore, the RC beams had a higher load carrying capacity than the unreinforced beams throughout the temperature range 20–400 °C. The same applies to Fig. 8 which compares the reinforced lightweight concrete with the unreinforced lightweight beams. (See also Figs. 20 and 21).

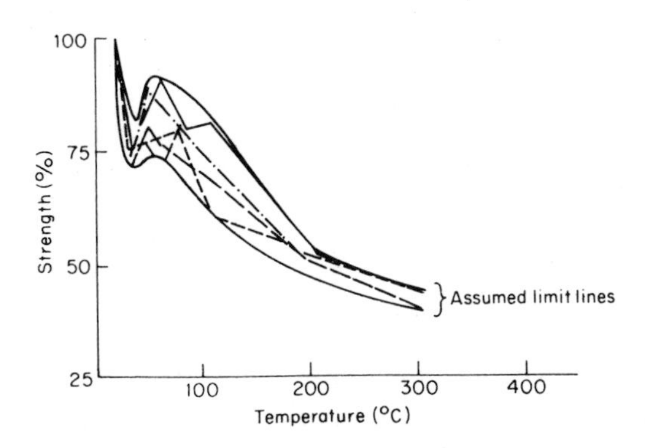

FIG. 11. The effect of temperature on the bond strength of concrete to mild steel.

Figure 11 gives the bounds of the decrease in bond strength of mild steel in concrete. The concrete aggregates used in the tests were sandstone, andesite, serpentine and basalt. Figure 12 shows the variation of bond strength for mild steel in sintered pfa aggregate concrete. It will be observed that after the initial reduction in bond strength at approximately 100 °C, the bond strength recovers and exceeds the initial bond by 40 % at 200 °C. There is then a drop in strength to 90 % of the cold value from 300 to 400 °C and a further reduction to just over 50 % of the cold value at 550 °C.

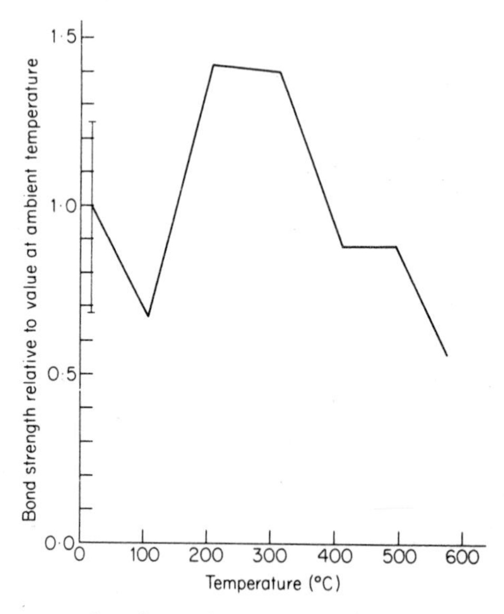

FIG. 12. Bond strength of steel in sintered pfa concrete at elevated temperatures. *Note*: Mean value of bond stress at ambient temperature $= 2{\cdot}36$ N/mm^2.

1.5.2 Elastic Modulus of Elasticity

The modulus of elasticity of concrete decreases with increasing temperatures. As for the strength property, some research workers have detected a steep decrease in elastic modulus just below 100 °C, this decrease being more acute for those concretes containing lightweight aggregates and calcareous aggregates. However, this drop in E-value has also been observed in gravel concrete associated with high thermal creep strains. It is postulated,[14] therefore, that this effect is not necessarily aggregate-induced but is due to a critical relative humidity within the concrete which creates built-in stresses within the material, thus affecting overall behaviour of the material.

A typical decrease of the modulus of elasticity between 100 and 600 °C for a sintered pfa concrete in compression is shown in Fig. 13. The ratio of the compressive stress at test temperatures to that at 20 °C is also shown in this figure, together with a curve representing the ratios of the compressive ultimate strain at test temperatures to that at 20 °C. It will be observed that the drop in E-value is accompanied by a large increase in strain, particularly beyond 400 °C.

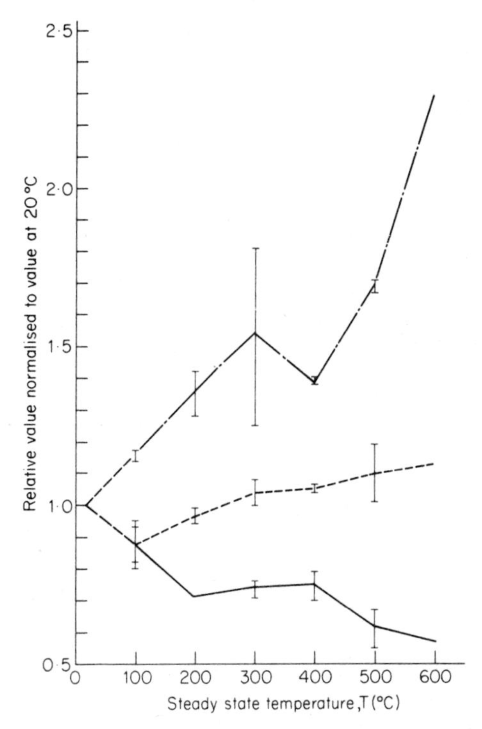

FIG. 13. Variation of stress–strain curve characteristics for LWA concrete in compression. —, ratio of modulus (E_0^T–E_0^{20}); - - -, ratio of ultimate stress (f_{cu}^T –f_{cu}^{20});— - —, ratio of compressive strain due to ultimate stress (e^T–e^{20}).

Figure 14 shows the bounds of the modulus of elasticity as different types of concrete are heated between 20 and 400 °C. The drop in the moduli of elasticity at 100 °C lies between 10 and 32 %, while the moduli of elasticity at 400 °C drop to 30–55 % of their 20 °C value. Those concretes, containing calcareous aggregates, are generally close to the lower boundary, while those containing siliceous aggregates are generally closer to the upper boundary. At approximately 80 °C, the sharp drop in E-value, which occurs particularly with calcareous aggregates has been attributed[10] to greater dislocations which occur in calcareous aggregates, due to their lower thermal expansions, in comparison to those of the cement matrix. Other research workers who have carried out high-pressure porosimetry tests on cement paste heated up to 110 °C, have observed a considerable increase in the pore

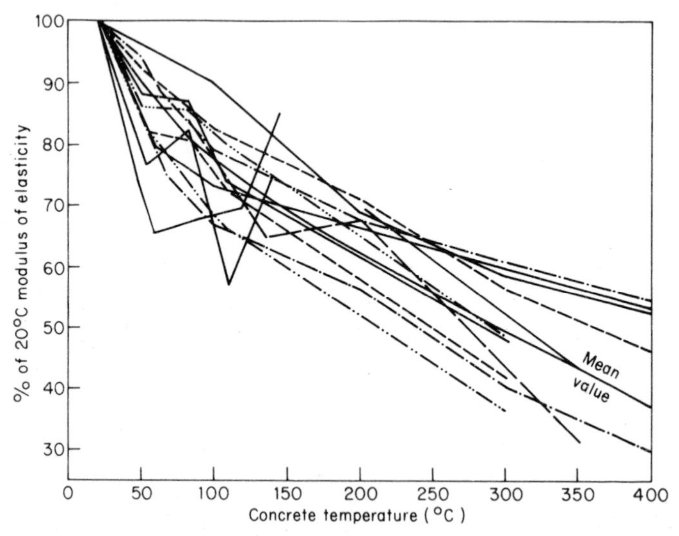

FIG. 14. Effect of temperature on the modulus of elasticity of concrete.[10]

structure of the cement paste. They suggest, therefore, that the unusual deterioration in properties which occurs at 50 °C in concretes with moisture contents between 30 and 40 %, is due to this increase in porosity caused by the evaporation of free water or is due to the expansion of cement paste. The hypothesis of disruptions occurring at temperatures below 100 °C, which has been presented in the introduction of this chapter,[14] where deterioration is ascribed to built-in tensile stresses arising from moisture and vapour expansions, migration, and subsequent loss, is more akin to the latter theory. However, it should be remembered that calcareous aggregates, together with lightweight aggregates, normally have a higher moisture absorption than siliceous aggregates and tend to show a more severe initial deterioration in properties prior to evaporation of the free water.

The effect of the drop in the modulus of elasticity is very clearly seen in Fig. 15 where the variation of initial flexural rigidity of a sintered pfa aggregate concrete is plotted against temperature. The initial flexural rigidity for the unreinforced sintered pfa aggregate concrete beams, which is equivalent to the material's initial tangent modulus of elasticity, shows a severe reduction to 37 % of the original *E*-value between 88 and 110 °C. The modulus then recovers to approximately

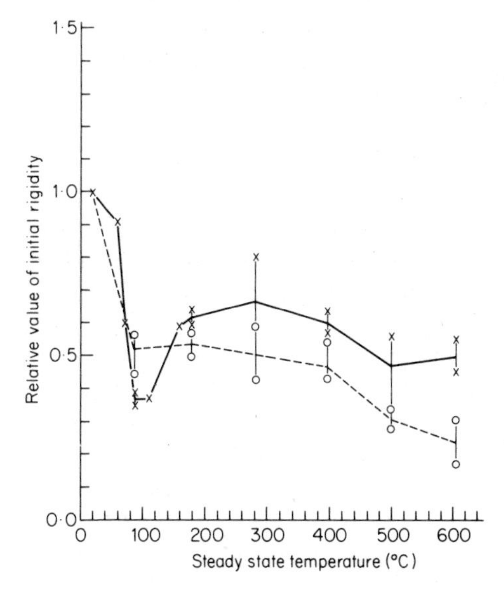

FIG. 15. Variation of initial rigidity of LWA concrete beams.[1] —, plain concrete beams; ---, RC beams.

60 % between 200 and 400 °C, and drops to 50 % at 600 °C. The introduction of reinforcement does not completely eliminate the effect of the initial reduction of the concrete modulus of elasticity, as was the case (see Fig. 8) when the strength properties of reinforced concrete were measured. The rigidity of the reinforced sintered pfa aggregates concrete at 88 °C is approximately 50 % of its original value. This property remains nearly constant up to 400 °C and reduces thereafter to 26 % at 690 °C.

It will be observed that the point at which deterioration is greatest varies, depending on the condition of the specimen and also on the manner of heating. Concrete specimens tend to lose moisture more readily when previously saturated in water and deterioration has been observed to occur at 50 °C. When the sides have been sealed, the temperature at which the maximum deterioration occurs varies between 60 and 80 °C. When properties are measured under a transient temperature state, i.e. under conditions of non-uniform temperature and average concrete temperature generally lower than the furnace temperature, the point at which deterioration occurs is at an average concrete temperature normally higher than when the properties are measured under a steady state of heating, i.e. when the concrete temperature is more uniform and has reached equilibrium with the furnace heating environment prior to testing.

1.5.3 Load and Deformational Characteristics

Typical load-deformational diagrams for river sand mortar, river gravel concrete and reinforced concrete beams[3] are shown in Figs. 16, 17, and 18 and 19 respectively. Figures 16 and 17 show the initial load cycles up to approximately one third of failure load at the test temperature, together with the final load cycle to failure. The slopes of the load deflection curves and the ultimate load at failure decrease with increasing temperatures. As expected, both the deformations prior to loading and at failure load increase with increasing temperature. The load deflection characteristics for the mortar beams indicate that the ductility of the material increases with temperature. This ductility is apparent for the concrete beam only at 304 °C. For the reinforced concrete beams, the deformations prior to loading have been omitted.

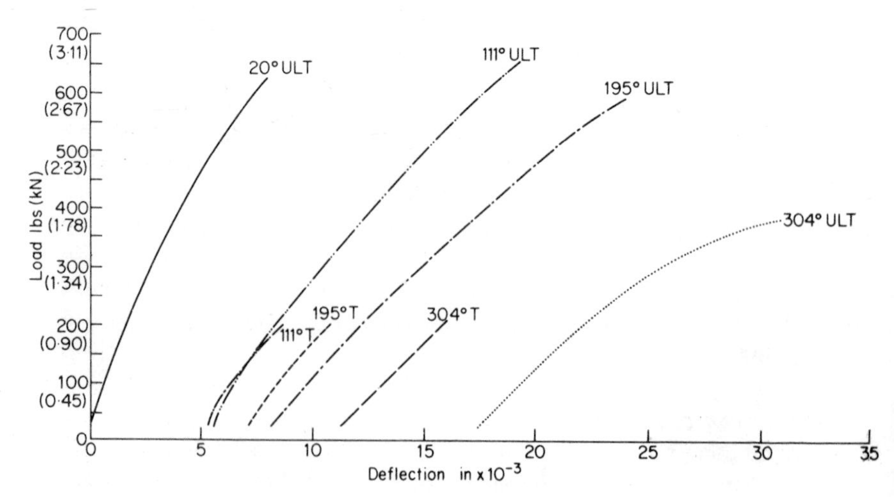

FIG. 16. Load–deflection curves for river sand mortar beams.

It will be observed that the shape of the curves in Fig. 18 are typical of those for under-reinforced beams tested at normal temperature (20 °C). Each curve can be represented by bilinear plots, with their intersection points at the instant of cracking for each beam. The gradients of the

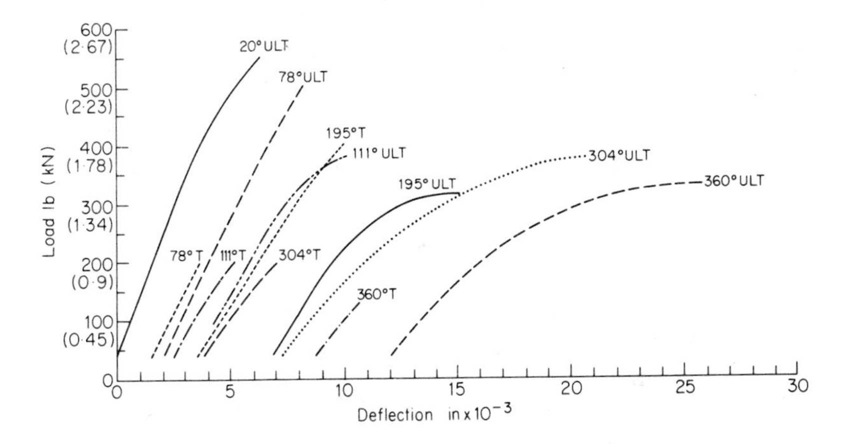

FIG. 17. Load–deflection curves for gravel concrete beams.

curves for the beams tested at 125 and 200 °C are nearly coincident and lower than that for the 20 °C, while the gradients for the beams tested at 300 and 400 °C, which are also nearly coincident, exhibit a still lower gradient. Similar load deflection characteristics to those in Fig. 18 result when reinforced concrete beams are tested hot (see Fig. 19). Comparable failure loads and deflections at failure occur, except for the 400 °C test. In this instance, the plot is more curvilinear and the failure load, 5·6 kN, is lower than the residual failure load of 6·3 kN.

From the initial gradients of the curves for the unreinforced concrete and mortar beams the elastic modulus of the materials can be determined. For the reinforced concrete beams the gradients represent the stiffness (EI) or rigidity (EI/l). The value for effective moment of inertia of the beam (I) changes with the initiation and propagation of cracks caused both by temperature and load effects.

Typical load deflection characteristics up to failure for sintered pfa aggregate concrete beams and reinforced beams are shown in Figs. 20 and 21. For clarity of presentation Fig. 20 illustrating the load–deflection curves for sintered pfa aggregate concrete beams have been displaced horizontally to separate them from the 20 and 88 °C curves, and should strictly all start from the same point (0,0) on the x-axis. Figure 20 shows the severe reduction in the load-carrying capacity

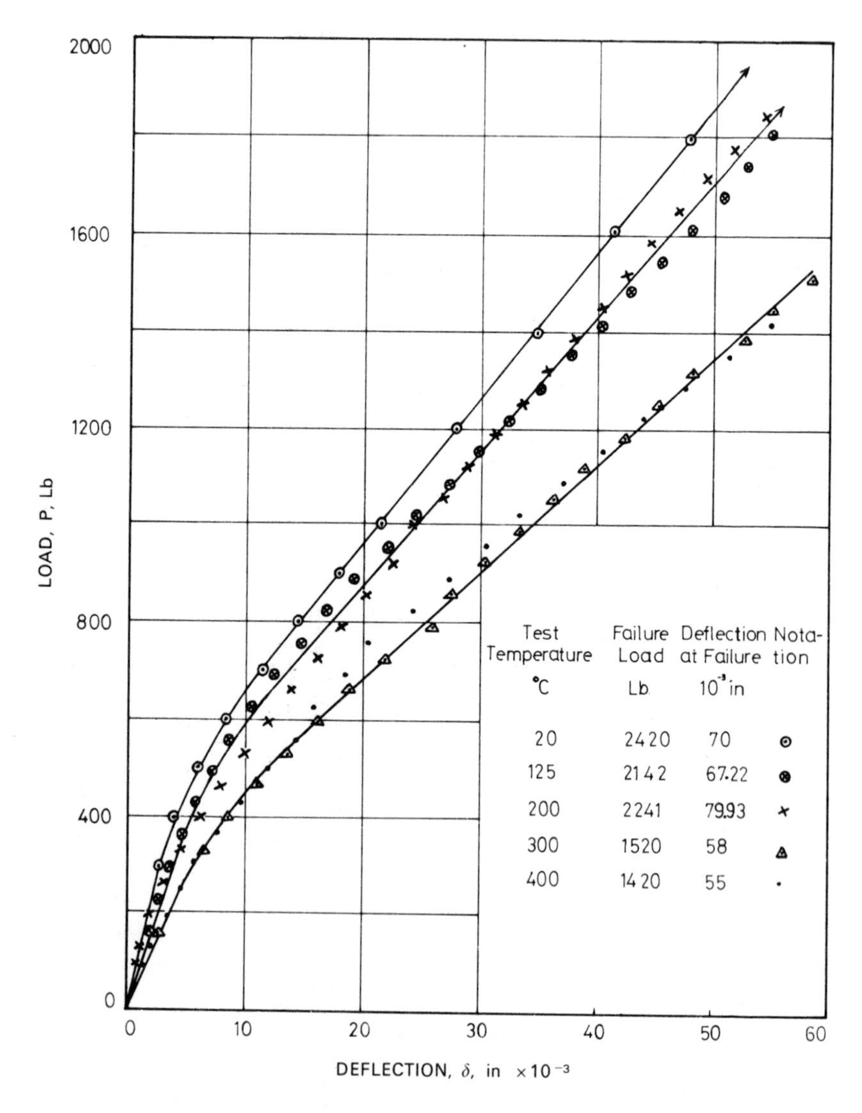

FIG. 18. Residual load–deflection curves for reinforced concrete beams.

which affects the beams as the temperature increases to 88 °C. At 110 °C a slight increase in strength is discernible. In all the beams tested, curvilinear plots result indicating a certain amount of ductility

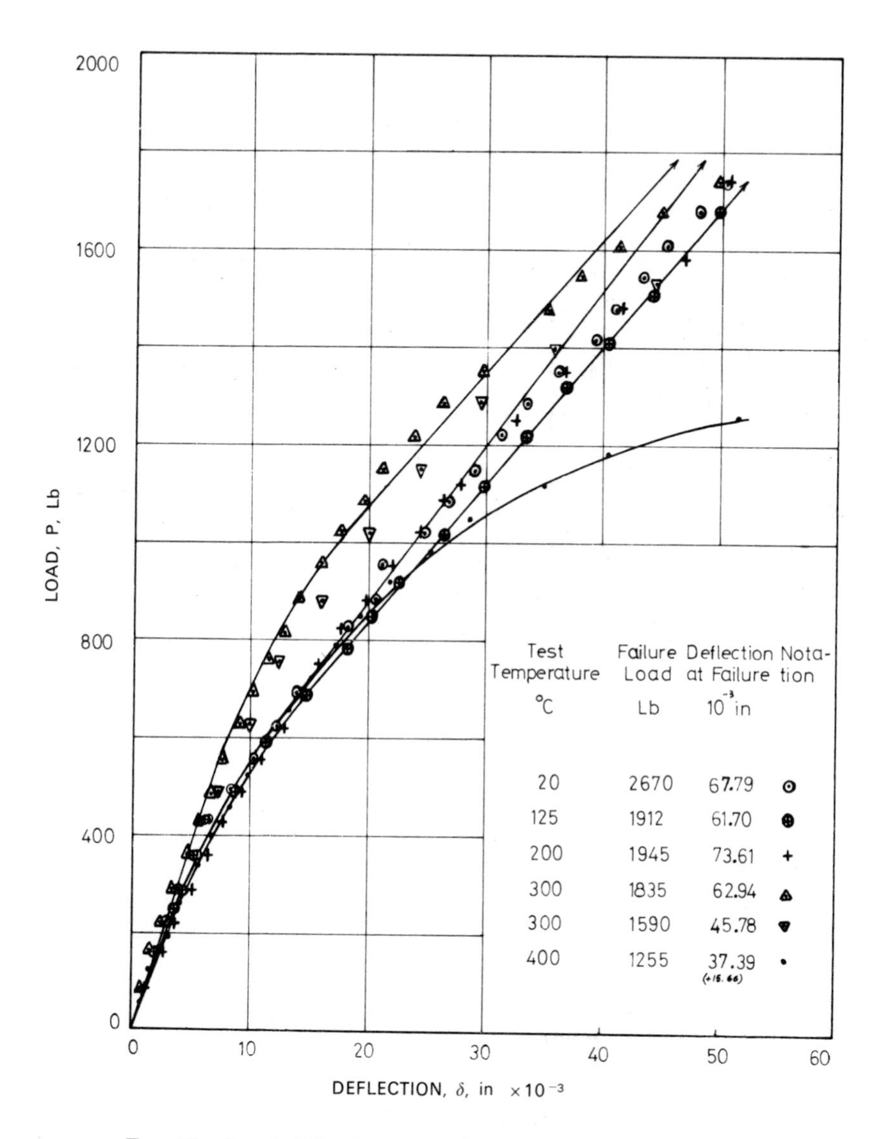

FIG. 19. Load–deflection curves for reinforced concrete beams.

prior to failure. Recovery is observed to take place in the load-carrying capacity of the beams as the temperature is increased to 500 °C, and indicates a levelling off in ultimate load at 605 °C.

Figure 21 showing the load–deflection curves up to failure of reinforced sintered pfa concrete beams at the higher temperatures have

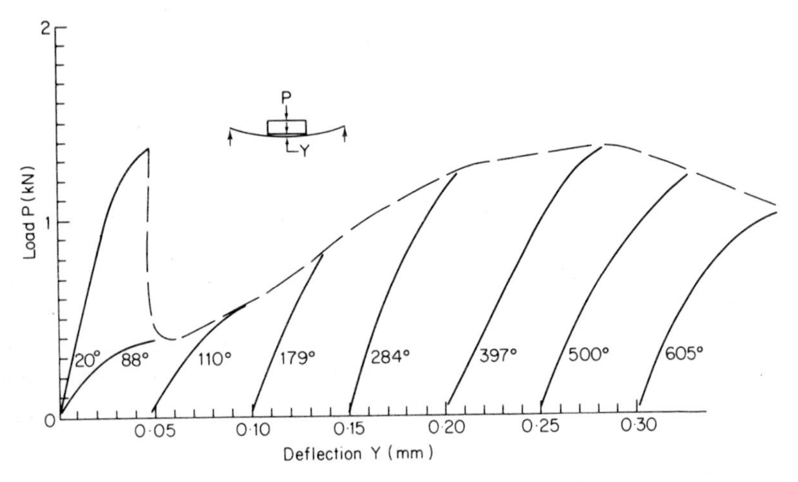

FIG. 20. Load–deflection curves for sintered pfa concrete.

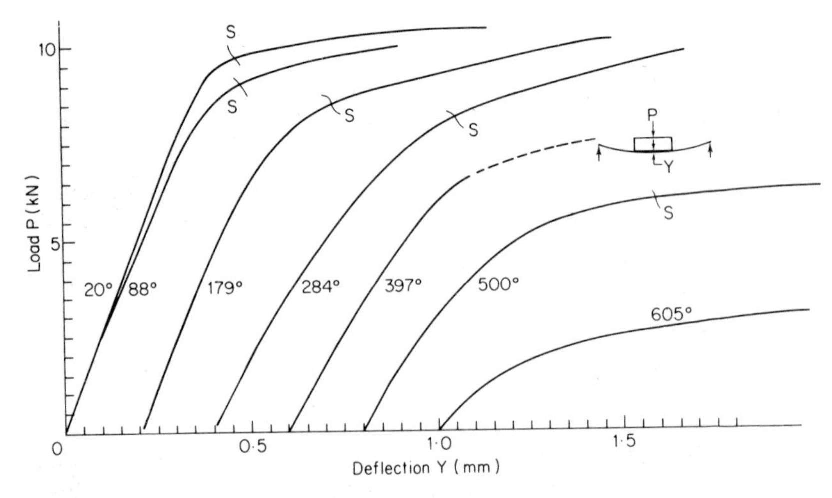

FIG. 21. Load–deflection curves for reinforced sintered pfa concrete.

been displaced horizontally from the 20 and 88 °C curves, for the sake of clarity. Figure 21 shows that the load-carrying capacity of the reinforced beams is only very slightly reduced at 88 °C in comparison to the sharp drop in strength exhibited for the plain beams in Fig. 20. Thereafter, the load-carrying capacity of the reinforced beams remains nearly constant up to 284 °C, the curves becoming more curvilinear indicating increased ductility with temperature. A series of tests which were carried out without shear stirrups failed prematurely at the points,[5] as indicated in Fig. 21.

1.5.4 Stress–Strain Characteristics

Typical stress–strain curves of concrete have been determined from tests carried out on gravel concrete cylinders, load cycled at test temperatures between 20 and 700 °C. Figure 22 shows normalised stress–strain curves[15] derived from these stress–strain relationships. The stress (σ) at any instant has been normalised by the peak stress (σ_m) at each temperature and the strain (ε) has been factored by the corresponding peak strain (ε_m). The global stress–strain relationship between this temperature range can then be represented by $(\sigma/\sigma_m = \varepsilon/\varepsilon_m)e^{1-}(\varepsilon/\varepsilon_m)$ which for small values of $\varepsilon/\varepsilon_m$ can be approximated to

$$\frac{\sigma}{\sigma_m} = K\frac{\varepsilon}{\varepsilon_m}$$

or

$$\frac{\sigma}{\varepsilon} = K\frac{\sigma_m}{\varepsilon_m}$$

This enables the elastic modulus of concrete to be determined at elevated temperatures, given the peak stress and strain values at the temperatures considered.

1.6 TEMPERATURE AND CREEP

Time-dependent deformations due to creep and shrinkage increase considerably with temperature such that, above 100 °C, concrete strains under load can be many times the elastic values at ambient temperatures. The study of creep at elevated temperatures[16–19] is important in that creep can significantly alter the stress distribution within a structure, while creep strains can lead to large structural deformations and loss of prestressing. Concern over the safety aspects of concrete nuclear reactor pressure vessels[20–22] in particular, and structures under fire conditions in general, have recently led to an accelerated research activity into the creep of concrete at elevated temperatures.

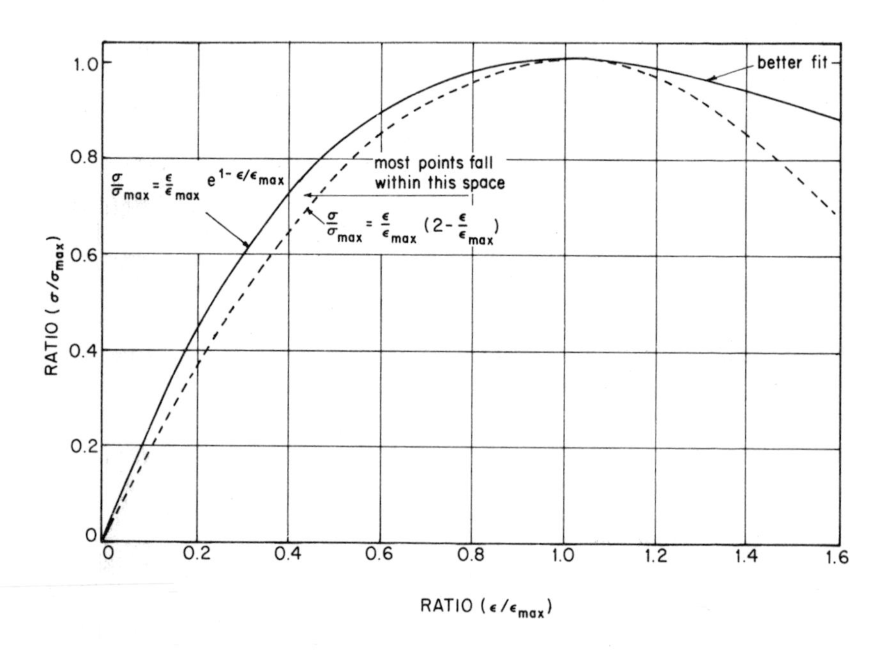

FIG. 22. Normalised stress–strain curves.

Concrete creep curves are normally obtained from laboratory tests conducted on cylindrical specimens which could be sealed[23–27] or unsealed[28–32] and tested at steady state or transient temperature regimes up to approximately 700 °C. Most test specimens to date have been subjected to uniaxial compressive loading[24,29–32] but few have been tested under multiaxial loading conditions.[21,33]

Creep deformations below 100 °C are very sensitive to the moisture condition of the concrete and, generally, increase linearly with temperature and stresses in the working range. The rate of creep of unsealed specimens, however, reaches a maximum at approximately 70 °C after which it declines to a value approximately equal to the 20 °C rate at 100 °C (Fig. 23), whereas already dried concrete shows little creep below 100 °C[18,25,30]

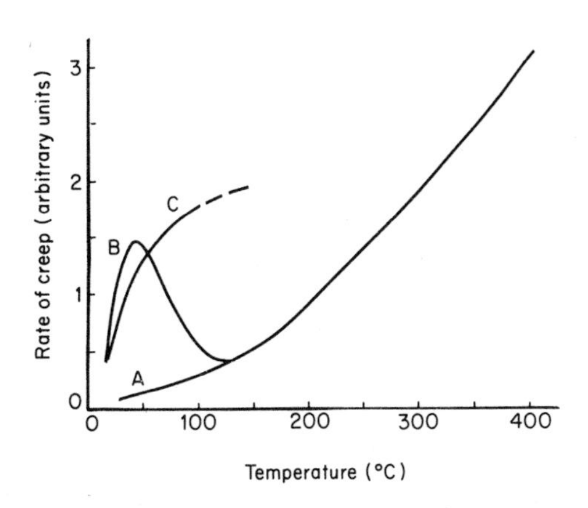

FIG. 23. Variation in the rate of creep related to elevated temperatures for (A) specimens pre-dried at 105°C, (B) saturated specimens tested unsealed, and (C) saturated specimens tested in a sealed condition.[25]

Above 100 °C other chemical and physical factors start to play an increasingly predominant role. At these higher temperatures, a significant increase in creep is related to a weakening of the concrete. Creep strains increase linearly with temperature up to a critical value determined by the thermal stability of its constituents. Large differential expansion between the aggregate and mortar matrix can lead to the disruption of the concrete and hence an increase in creep.[34] Various chemical reactions including dissociation of the $Ca(OH)_2$ lead to similar results.[35] Creep performance is also very dependent on the type of aggregate (Fig. 24) with ordinary gravel showing poorest results, while sintered pfa concrete remains thermally stable to about 550 °C.[32]

1.6.1 Steady State Creep

Until recently most tests were concerned with creep of specimens at steady state temperatures[24,27,30,32,36] Creep strains were shown to increase approximately linearly with temperature (except for the moist unsealed specimens at temperatures below 100 °C mentioned earlier) and stress, up to critical values of both. The creep strains, furthermore, lend themselves to linearity when plotted on a log–log scale.[23,32] We can, therefore, represent the creep strains, ε_c, mathematically by a

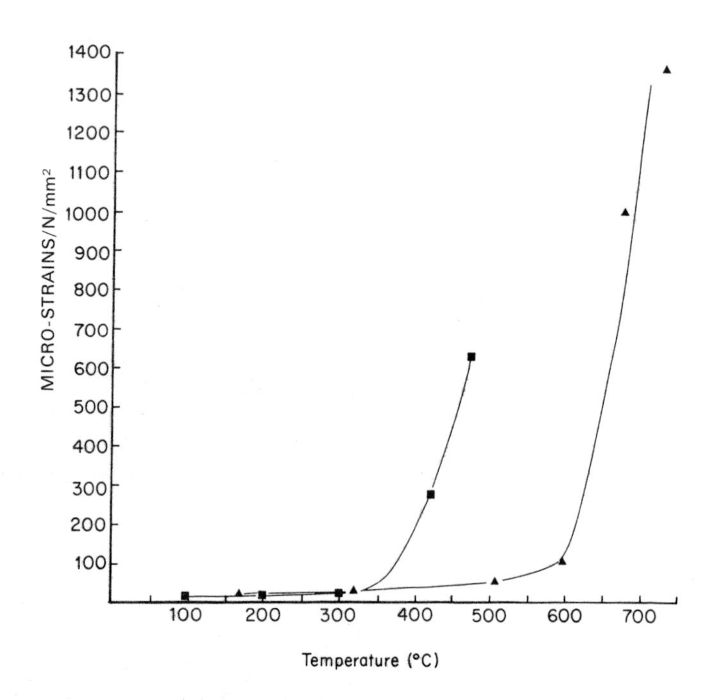

FIG. 24. Specific creep strains 1 day after loading for two types of aggregate concrete: Thames gravel concrete, ■, and sintered pfa (lightweight) aggregate concrete, ▲.[32]

power law confined by a limiting envelope of stress, σ, temperature, T, and time after loading, t, as follows:

$$\varepsilon_c = \sigma(a + b(T - c))t^d$$

It should be noted that these results are all empirical and a change in constituents or even mix proportions can lead to different values of the constants a, b and d.

1.6.2 Transient Creep

Tests performed on specimens subjected to steady loads but increasing temperatures lead to creep strains far in excess of those predicted from the steady state tests.[28,29] It is now accepted that under transient temperatures an additional transient strain component, ε_{tr}, is present and that the total strain, ε_{tot}, becomes

$$\varepsilon_{tot} = \varepsilon_{th}(T) + \varepsilon_{\sigma}(\tilde{\sigma}, \sigma, T) + \varepsilon_{c}(\sigma, T, t) + \varepsilon_{tr}(\sigma, T)$$

where ε_{th} = thermal strain, ε_{σ} = stress-related strain, and $\tilde{\sigma}$ = stress history.

1.7 TRANSIENT TEMPERATURE TESTS

The physical properties of concrete measured during heating (i.e. during transient temperature testing) follow similar patterns to those measured after temperatures have reached a steady state (i.e. under isothermal conditions.[1,40] Figure 25 shows how the relationship for the ultimate strength of gravel concrete beams tested under transient conditions is displaced from that tested under isothermal states.

Figure 26 shows the delay which occurs in the characteristic loss of flexural properties when sintered pfa aggregate concrete beams are tested under transient temperatures. Both the discontinuity points, or the minima, and the maxima points of recovery are shifted along the temperature axis by approximately 100 °C, this delay being dependent on the rate of heating and the thermal inertia of the material of the beams.

As explained previously, thermal properties are very much dependent on the moisture content of the concrete, which during temperature transience is constantly changing. The change in moisture content also influences the deformational characteristics as has been demonstrated earlier (see Figs. 1 and 2). Figure 27 shows longitudinal expansions of a series of tests on mortar beams and concrete beams heated to 125, 200, 300 and 400 °C, approximately. In the case of mortar the initial or instantaneous temperature strains are significantly larger than the strains after heat soaking at the test temperature, particularly at higher temperatures. The differences between the initial and final strains, i.e. after maintaining at test temperature for a long period of time, are due to irreversible shrinkage strains at the temperatures considered. Initial strains for concrete are higher than those for mortar due to the higher expansion of the gravel aggregate. The final strains below 300 °C are lower than the initial strains, showing that the gravel aggregate is restraining the matrix from shrinking, resulting in small irreversible

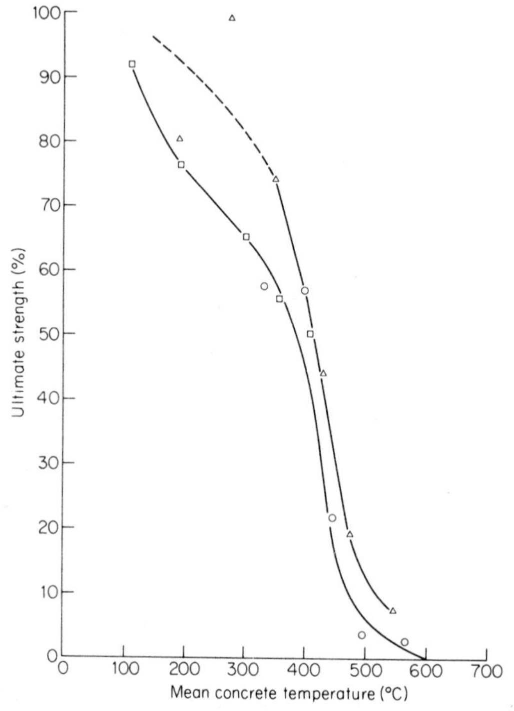

FIG. 25. Per cent ultimate strength at elevated temperature for beams tested under transient and steady state.

shrinkage strains. Beyond 300 °C disruption of the concrete begins to occur since the aggregate begins to fail, and the ultimate tensile strength between the aggregate and matrix is exceeded. It should be noted also that at 415 °C a large, irreversible expansive strain results. Figure 27 shows two of the individual tests for mortar and concrete beams respectively as they are heated slowly to approximately 415 °C and then allowed to cool.

Figure 28 shows the effect of introducing reinforcing steel in concrete on the overall longitudinal strains of the materials combined. In this instance there is a reduction in shrinkage strains below 300 °C since a further restraint to movement of the matrix is imposed by the steel. This steel restraint is also apparent at approximately 400 °C where the irreversible expansions observed for the plain concrete (see Fig. 1) are greatly reduced. Thus, the steel reinforcement assists the aggregate to

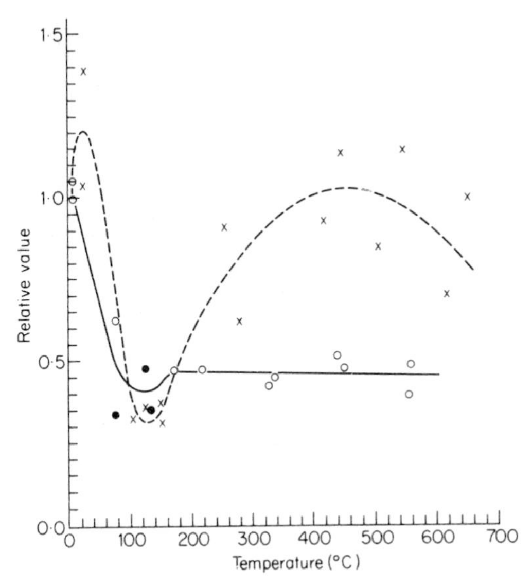

FIG. 26. Variation of flexural properties of plain concrete beams in load–temperature transient tests. —O—, ratio E_0^T–E_0^{20}; – – × – –, ratio P_0^T–P_0^{20}.

reduce shrinkage at temperatures below 300 °C, and beyond this temperature prevents complete disruption which would occur if the concrete was unreinforced.

From the foregoing discussions it is patently clear that many factors influence the properties of concrete to different degrees at different times during heating. For this reason an investigation[12] was carried out to study the effect of varying nine different factors at three different levels on the flexural behaviour of reinforced concrete beams during heating at the standard furnace rate. Because of the large number of tests which would have been required if a full factorial testing programme had been adopted, a fractional factorial design of experimentation was designed. This permitted the minimum number of specimens to be tested to allow a statistical investigation on the experimental results to proceed, and in accordance with Fisher's relationship[41] the required number of tests for the nine variables at three different levels was $(1/3)^6 \times 3^9 = 27$. Using an extension of Fisher's experimental design table[42] it was possible to examine not only the

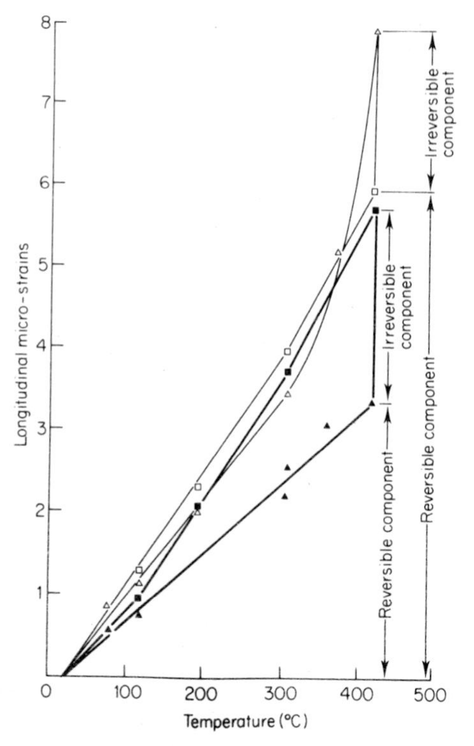

FIG. 27. Reversible and irreversible components of strain for gravel concrete (—) and river sand mortar (━). ▲, △, final strain (adt) ■, □, initial strain.[3]

nine individual factors below but also assess the interaction effects of two of the factors.

The factors, which were assumed to have a major influence on flexural strength and initially assumed to be independent variables, were selected at three different levels as follows:

(1) Scale; the cross sections of beams selected were

$$150 \times 230\,\text{mm},\ 125 \times 200\,\text{mm} \text{ and } 100 \times 150\,\text{mm}.$$

(2) Gravel aggregate proportion with nominal proportions of coarse to fine aggregate and cement of 1–1–2, 1–1·5–2 and 1–2–4.
(3) Water–cement ratios of 0·50, 0·57 and 0·65.
(4) Age of concrete at testing; 6 weeks, 12 weeks and 18 weeks.

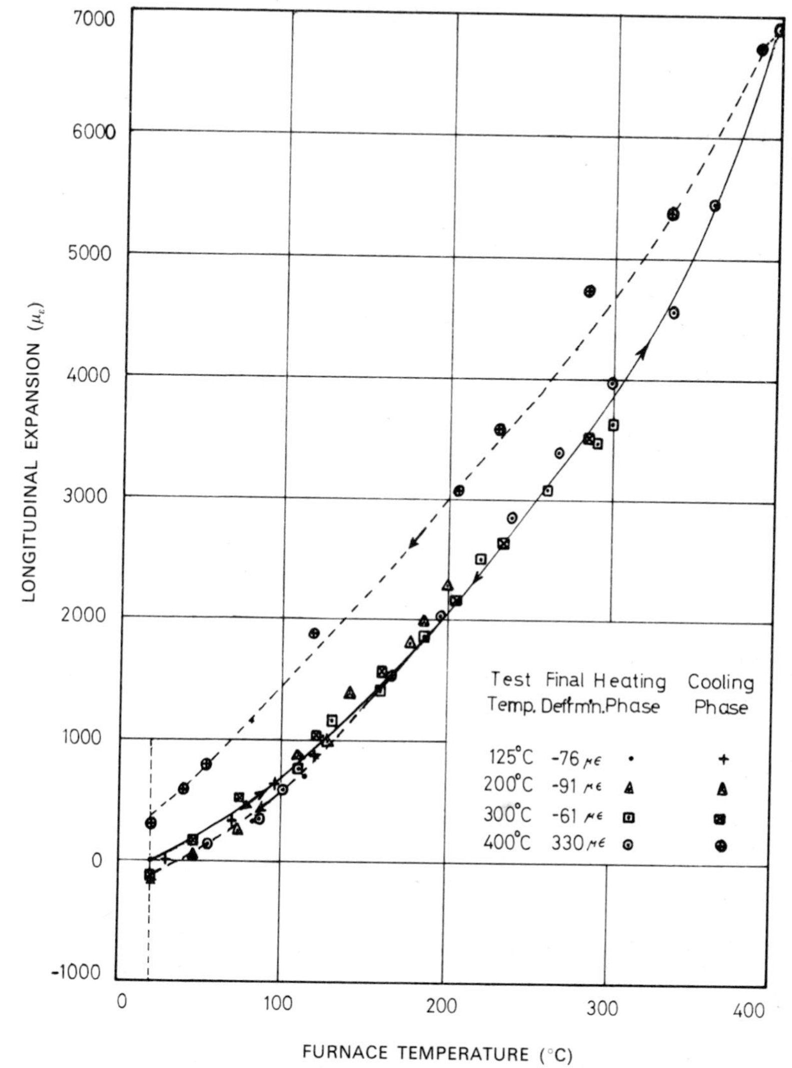

FIG. 28. Effect of temperature on longitudinal movement for reinforced concrete beams.

Prior to heating the beams, design loads, or two overload conditions, were applied. Three different rates of heating were selected with the standard furnace rates as one of the parameters. The period during which the beams were exposed to heating was varied as were the cooling rate and the boundary conditions.

During heating, a record of loads, deflections, temperatures and expansions were taken which allowed the determination of the flexural rigidity of the beams at 5-min intervals of heating. A statistical analysis carried out on the results revealed that certain factors which had a predominant effect initially became less significant with further heating. Table 3 shows the major factors which significantly affect the flexural rigidity of reinforced concrete beams. Thus initially the size, aggregate proportions and the applied bending moment are significant at the 1 % probability whereas the water–cement ratio, age of concrete at loading and rate of heating are significant at the 5 % probability level. After 20 minutes of heating, water–cement ratio, age at testing and rate of heating lose their significance on the deterioration of the beams. At the later stages of heating, heating period becomes significant while the other parameters become less important.

TABLE 3

SIGNIFICANT FACTORS DURING HEATING

Time (min)	*Size*	*Aggregate proportion*	*Water–cement ratio*	*Age of concrete*	*Bending moment*	*Rate of heating*	*Heating period*
5	a	a	b	b	a	b	–
10	a	a	b	–	a	b	–
15	a	a	b	–	a	–	–
20	a	a	–	–	a	–	–
25	a	a	–	–	a	–	–
30	a	a	–	–	a	–	–
35	a	a	–	–	a	–	–
40	a	a	–	–	a	–	–
45	a	b	–	–	a	–	–
50	a	b	–	–	a	–	b
55	b	–	–	–	b	–	a
60	–	–	–	–	–	–	a
65	–	–	–	–	–	–	a

[a]statistically significant at the 1 % probability level.
[b]statistically significant at the 5 % probability level.

This highlights the difficulties which could arise when interpretation is required for transient test conditions, when results are derived from single variable parameter investigations at elevated steady state temperatures. Insufficient or incorrect data may lead to incorrect or misleading conclusions.

1.8 HIGH TEMPERATURE RESISTANT CONCRETE

Concrete can be made to be temperature resistant[43] by (1) mixing finely divided admixtures to OPC concretes, (2) use of special aggregates or (3) special cements. In certain cases special aggregates, cements and admixtures are used in combination to produce the required high temperature concrete. It is common to fire or heat the concrete to working temperatures prior to exposing the concrete to service temperature conditions.

Figure 29[44] shows a diagrammatic representation of the physico-chemical processes which take place as OPC and fire clay aggregate concrete are heated up to 1600 °C. It will be observed that the hydraulic strength (curve 2) reduces with increase in temperature. With

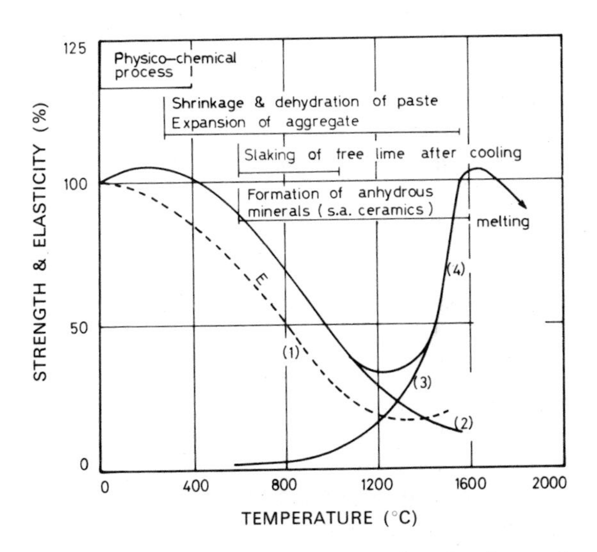

FIG. 29. Change in mechanical properties on first heating and cooling OPC and fire clay aggregate. (1) Modulus of elasticity, E; (2) hydraulic strength; (3) change of ceramic strength; (4) hydraulic plus ceramic strength of concrete.

appropriate firing, ceramic bonds are developed which cause the strength to recover as in curve (4).

When high alumina cement is used as the binder, a more severe loss in strength occurs with temperature but a similar curve as for the OPC results, as shown in Fig. 30. By using a sodium silicate solution in

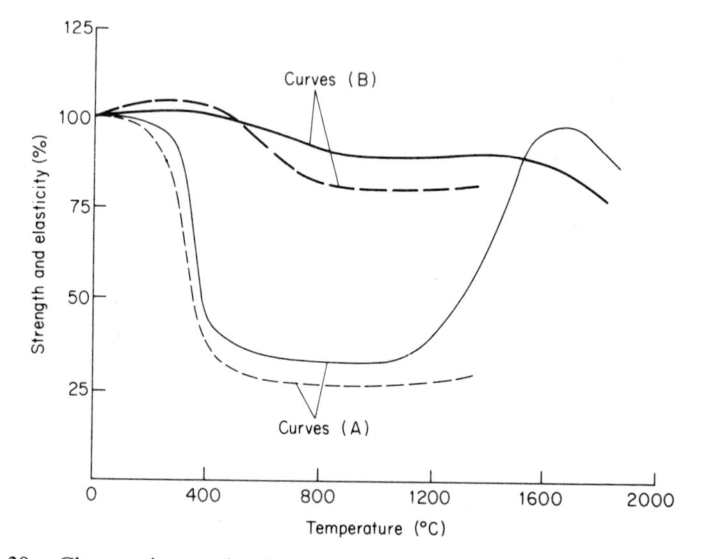

FIG. 30. Change in mechanical properties for: curves (A) high alumina cement, and fire clay aggregate concrete and curves (B) concrete made with sodium silicate solution and fire clay aggregate. —, —, strength; ---, ---, modulus of elasticity.

combination with fire clay aggregate the sharp drop in mechanical property is avoided. The temperature resistance of OPC concrete is improved by the use of different percentages of finely ground additive. These fine grained additives sometimes called ceramic stabilisers can consist of chamottes derived from burning fire resistant clays, clay powder, chrome powder, sintered magnesia powder, quartz dust, fly ash or slag dust.

It is very important with heat resistant concrete to take into account the stresses arising from the difference in the coefficient of thermal expansion of the steel and concrete, the non-reversible shrinkage of the matrix and the change in the mechanical properties of the steel with increasing temperatures. Where the differential stresses developed between the concrete and the steel are such as to destroy the bond between the two materials, precautions should be taken to allow relative movements to take place between the steel and concrete.

1.9 EFFECTS OF FIRE TEMPERATURE ON CONCRETE

It is very difficult to simulate the effect of a fire on a building element, not only because of the severe transient temperature regime imposed on the material but also because of the difficulty of predicting the behaviour and the intensity of an actual fire.

In order to simplify testing and at the same time have a yardstick for comparing fire resistances of different materials, standard furnace tests have been proposed in national and international standards which involve the testing of prototype building elements in standard furnaces heated at the rate similar to that shown in Fig. 31.[11] This curve can also be mathematically represented by the equation $T = 345\log(1+8t)$ where T is temperature in °C and t is time in minutes. The test, which is performance rather than research oriented, is normally carried out by

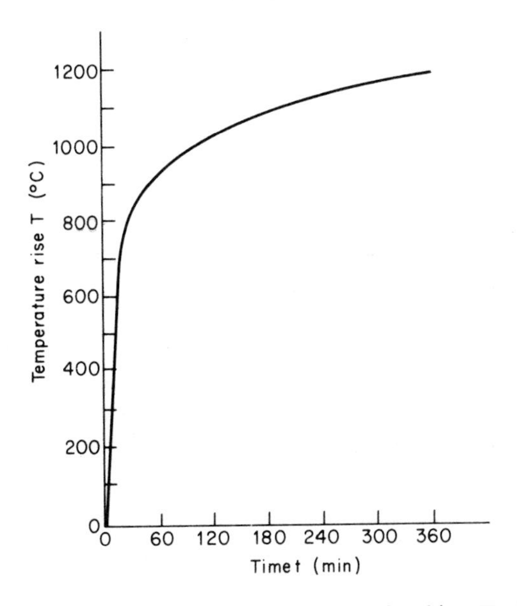

FIG. 31. Standard furnace temperature–time relationship. $T = 345\ \log_{10}(8t+1)$.

commercial and/or government laboratories following the general procedures as follows.

If load bearing, a design load is applied to the building element (i.e. beam, column floor or wall) within the standard type furnace, and temperatures are measured around and, sometimes, on and within the element. The displacements and the structural integrity of the element are observed during heating. The time taken to reach failure is then defined as the fire resistance of the element. Failure is deemed to have taken place on partial or complete collapse, excessive deflections causing unserviceability, or additionally, in the case of separating walls and floors, when the temperature of the cold face has reached a given value.

A further difficulty which may arise when testing concrete is its moisture state, and most codes specify that concrete elements should be conditioned in a defined manner to limit the moisture content of the concrete prior to testing. Also, since the structural behaviour of the reinforced element is very much dependent on the temperature attained by the reinforcement, a limit is sometimes imposed on the steel temperature: the steel temperature is, of course, limited to rise by the concrete cover to the reinforcement. In cases where the cover is thicker than 50 mm, additional mesh reinforcement within the cover is specified to prevent premature spalling and exposure of the main reinforcing bars to extreme temperatures at an early stage of heating.

The moisture condition of concrete is a very important factor in its fire endurance. In certain situations the presence of moisture acts as a deterrent to failure as the latent heat of vaporisation increases the thermal inertia of the material.

Tests[45] carried out on slabs of different thicknesses, where temperature and moisture gradients were measured during heating, indicated that young concrete had a higher fire endurance than mature concrete. Also, these tests generally confirmed the results from small scale tests (Sections 1.5 and 1.6) in that expanded shale aggregate concrete had a higher fire resistance than limestone concrete, which in its turn performed better than siliceous concrete. On the results of these tests an empirical formula was derived which permitted an adjustment to the fire endurance of concrete for specimens which contained more moisture than specified by Standards.

There are many situations, however, when excess moisture in concrete has been the cause of premature failure and, in some cases, explosive spalling. Both these effects may be partly due to the formation of a moisture clog[46] as the moisture is driven from the hotter drier layers to the colder areas. The hotter dry zones increase rapidly in temperature creating steep gradients at the surface while pressure build-ups occur in the cooler areas, particularly if the concrete has a low permeability. Thus in cases of water-curing of specimens, explosive spalling at temperatures between 100 and 200 °C can occur during heating on both young and mature gravel concrete; because most experience of explosive spalling was with gravel concrete, it was thought that the phenomenon was associated with the aggregate. However, a few cases have been reported of limestone concrete and lightweight aggregate concrete being susceptible to this phenomenon, indicating that the moisture condition of the concrete plays the most important role in explosive spalling.

The mechanism producing spalling may be due to the unstable stress situations which arise when small triaxial tensions occur, due to the vapour pressures acting on the gel pores and smaller capillaries, combining with imposed biaxial compressions.[47] Stress theories have also been postulated in an attempt to define the causes of spalling, and to differentiate between the types of spalling occurring under various conditions of heating and stressing.

1.10 ACTUAL FIRE SITUATIONS

It has long been established that, although the standard fire test gives a reasonable comparison between different materials, it does not simulate closely actual fire situations. The effect of a standard experimental fire has, therefore, been compared to the standard fire time–temperature relationship. A standard fire condition can be produced by burning a given weight of wood in the form of a crib in an enclosed volume with a given ventilation. The resulting time–temperature relationships are as given in Fig. 32. The numbers on each fire curve give the fire load density in kilograms of timber per square metre of floor area and the bracketed numbers are the relative ventilated surface area. It should be observed that the range of fire intensities between 7·5 kg/m^2 and 60 kg/m^2 covers those occupancies which have low fire load densities such as hospitals, domestic and some office buildings.

The standard furnace curve to BS 476 is also plotted on the same diagram for the sake of comparison. It will be observed that temperatures in excess of the standard furnace temperatures are quite easily achieved for standard fires in excess of 30 kg/m^2 density.

Thus, for a given duration of an average fire, the temperature can reach higher values than those assessed by the standard furnace tests, resulting in deterioration of the fabric of the material, whereas no deterioration would have been predicted from the lower furnace

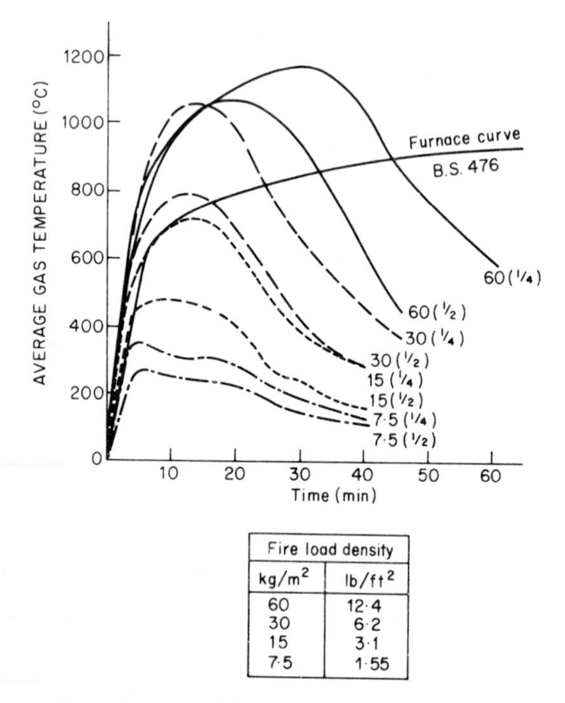

Fire load density	
kg/m²	lb/ft²
60	12·4
30	6·2
15	3·1
7·5	1·55

FIG. 32. Comparison of standard fire tests and standard furnace temperature–time relationships. 60 ($\frac{1}{4}$) denotes a fire load density of 60 kg/m^2 with $\frac{1}{4}$ of the area of a 7·7-m (25 ft 3 in) wall open as ventilation.

temperatures. A relationship has been derived[48] connecting the fire resistance time in a standard furnace (t_{fr}) and the temperature (θ) of a steel column subjected to a standard fire. The equation is

$$t_{fr} = 0{\cdot}14\,\theta\sqrt{RC}$$

where RC is a time constant in seconds for the protected column.

Using the above relationship the temperature of the steel surface can be estimated for cased steel buildings knowing the standard fire-resistance time. The actual fire resistance of the steel building can then more realistically be assessed from this steel temperature.

1.11 ACTUAL STRUCTURES UNDER FIRE SITUATIONS

The building elements which are tested for fire endurance in a standard furnace are normally simply supported. Such conditions are rarely realised in practice since concrete members are generally continuous, advantage being taken of the monolithic form of reinforced concrete construction.

It should be noted that simply supported members heated on one side as in a standard furnace beam test (see Fig. 33) deflect towards the heated zones which would be in compression relative to the colder

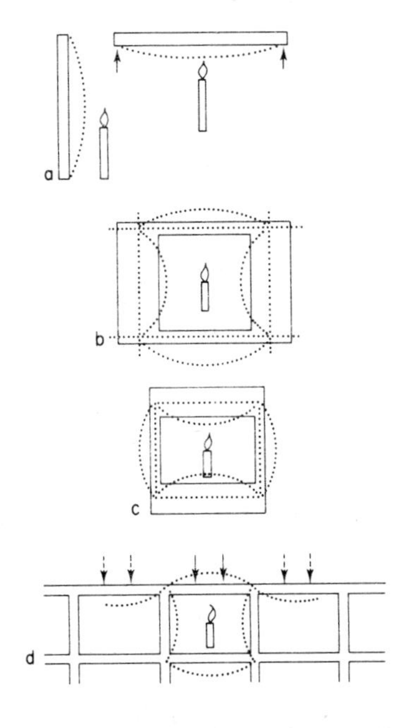

FIG. 33. Deflection of structural members and frames due to heat. (a) Beam and column independent. (b) Interaction of stiff columns with slender beams. (c) Interaction of stiff beams with slender columns. (d) Case when loaded beam over fire may not be critical as (i) unloaded over fire, (ii) loaded beams adjacent to fire. Top beams traded in a framework with stiff columns.

surfaces. Thus the design load would impose a deflection to the beam in the same direction and stresses of opposite signs to those induced by temperature.

However, the deformations induced by a fire within a framework of beams and columns is shown in Fig. 33(b and c). The framework with the stiffer columns (b) deforms such that their hotter faces are convex with compressive thermal stresses and, if the integrity of the connecting joints is not impaired, the beams flex away from the fire with imposed compressive stresses on the hot faces.

The stiffer beams (c) now play the same role as that for the columns (b) and the deformation diagram is reversed. The beams in this instance deform towards the fire and force the more slender columns to deform away from the fire.

Figure 33(d) shows the deformation of part of a multibay, multistorey framework with the upper beams loaded. The upper beams of the compartments adjacent to the fire have load induced deformations in the same direction as those caused by the fire while the load induced deformations on the beam above the fire are in the opposite direction.

The behaviour of real building components in a real fire may, therefore, be quite different to that experienced in a standard furnace. It can be shown that continuity may increase the fire endurance not only of concrete structures but also of steel structures. For this reason, many research workers have been engaged in theoretical and experimental investigations in an attempt to assess realistically the fire resistance capacity of building elements and structures.

Large furnaces now exist which include facilities for controlled end restraints and continuity to building components, but these are suitable mainly for research purposes. The significance of the fire resistance of different building components, therefore, has to be interpreted in the context of the limitations which the standard furnace test imposes, when faced with building elements subjected to actual fires.

Many papers and guides to good practice for fire resistance of concrete and steel structures have been published, for example Refs. 49–52. There has been a tremendous advance in the understanding of materials and structures at elevated temperatures, and rational methods of designing structures for fire endurance have been presented. Although the empirical Building Regulation/Code of Practice requirements offer tables and simple rules which are suitable for everyday construction, the above rational methods proposed should result in savings in cost, without sacrificing safety, for the more sophisticated concrete structures.

1.12 CONCLUDING REMARKS

It has been demonstrated that the performance of concrete as defined by a given property of concrete at elevated temperatures is dependent on a number of factors including:

(1) The type of mix, particularly the species and size of aggregate.
(2) The reinforcement, if any.
(3) The curing conditions prior to testing and the moisture state at testing.
(4) The manner of testing, i.e. whether in compression, tension or flexure, or whether preloaded prior to heating for mechanical properties, or for thermal properties, whether the test is standard or non-standard.
(5) The environment of the furnace at testing, and the manner of determining the property measured (i.e. under a steady or transient state of heating).
(6) Intensity of heating.

When calculating deformations arising from temperatures unexpected curvatures develop which are in some cases quite different to those arising from imposed loads. It should be remembered that the stress states expected from a deformed concrete member due to an imposed load may be in the opposite sense to those which develop when the deformation is temperature-induced.

The moisture condition within the concrete plays a very important role in the behaviour of the material at elevated temperatures. At temperatures experienced in fires, the water in the concrete can, in certain instances, increase the fire endurance, and in others increase its susceptibility to spalling. At temperatures below 100 °C moisture can create a discontinuity of physical properties in concrete which depresses its strength and the modulus of elasticity to low values. The thermal properties are also adversely influenced when moisture within the concrete is present.

It is, therefore, necessary if realistic predictions are required of the behaviour of concrete at elevated temperatures, to select the property appropriate to the moisture condition of the material during its service life.

REFERENCES

1. Labani, J. M. (1977). *The short term structural behaviour of a lightweight aggregate concrete at elevated temperature.* (Ph.D.) London.
2. Sullivan, P. J. E. and Khoury, G. (1978). *Creep of a lightweight aggregate concrete at elevated temperature.* FIP.
3. Sullivan, P. J. E. (1970). *The structural behaviour of concrete at elevated temperature.* (Ph.D.) London.
4. British Standards Institution. (1977) *BS5400. Specifications for the design and specification for steel, concrete and composite bridges.* Part 2. Specification for loads. BSI, London.
5. Emerson, M. (1973). The calculation of the distribution of temperature in bridges. *TRRL Report 561.* Crowthorne, Berks.
6. Emerson, M. (1976). Temperatures in bridges during the hot summer of 1976. *TRRL Report 783.* Crowthorne, Berks.
7. (1970) *IHVE Guide Book A 1970.* Section A3. Thermal and other properties of building structures. Chartered Institute of Building Services, London.
8. British Standards Institution. (1973). *BS 874, Methods for determining thermal insulating properties with definitions of thermal insulating terms.* BSI, London.
9. British Standards Institution. (1975). *Methods of test for inorganic thermal insulating materials.* BSI, London.
10. Blundell, R., Dimond· C. and Browne, R. O. (1976). *Report No. 9, The properties of concrete subjected to elevated temperatures.* CIRIA UEG.
11. British Standards Institution. (1972). *BS 476,* Part 8: *Fire tests on building materials and structures.* BSI, London.
12. Inuzuka, M. (1974). *Flexural behaviour of reinforced concrete members at transient high temperatures.* (Ph.D.) London.
13. Harada, T., Takeda, J., Yaman, E. S. and Furumura, F. (1971). Strength, elasticity and thermal properties of concrete subjected to elevated temperatures, *ACI Special Publication SP 34, Concrete for Nuclear Reactors.* Berlin.
14. Sullivan, P. J. E. and Labani, J. M. (1974). Flexural behaviour of plain and reinforced lightweight aggregate concrete beams up to 600 °C. *Cement and Concrete Research,* **4.**
15. Baldwin R. and North M. A. (1969). The stress curve of concrete at high temperature. A review. *Fire Research Note 785, October.* Fire Research Organisation. Boreham Wood, Herts.
16. ACI (1971). Designing for effects of creep, shrinkage, temperature in concrete structures. *ACI Special Publication SP-27.* ACI, Detroit, Michigan, USA.

17. ACI (1971). Temperature and concrete. *ACI Special Publication SP-25*. ACI, Detroit, Michigan, USA.
18. NEVILLE, A. M. (1970). Creep of concrete: Plain, reinforced and prestressed. North-Holland, Amsterdam.
19. IRVING, J. (1975). The effect of elevated temperatures on concrete and concrete structures. A review of literature. *FIP Publication FIP/3/1*, Slough, UK.
20. ACI (1972). Concrete for Nuclear Reactors (3 Vols). ACI Seminar, Berlin, 1970. *ACI Special Publication SP-34*. Detroit, Michigan, USA.
21. ASCHL, H. (to be published). Concrete for PCRVs: Mechanical properties at elevated temperatures and residual mechanical behaviour after triaxial preloading. *5th SMIRT Conference, Berlin, 1979*. North-Holland Publishing Co., Amsterdam.
22. BROWNE, R. D. (1968). Properties of concrete in reactor vessels. *Proceedings of Conference on Prestressed Concrete Pressure Vessels, 13–17 March, 1967*. The Institution of Civil Engineers, London.
23. BROWNE, R. D. and BLUNDELL, R. (1969). The influence of loading age and temperature on the long term creep behaviour of concrete in a sealed moisture stable state. *RILEM Colloquium on Physical and Chemical Causes of Creep and Shrinkage of Concrete, Munich, April 1968*; also *Materials and Structures* (RILEM), **2,** (8), 1969.
24. NASSER, K. W. (1971). Creep of concrete at low stress–strength ratios and elevated temperatures. *ACI Special Publication SP-25*. ACI, Detroit, Michigan, USA.
25. MARÉCHAL, J. C. (1970). Contribution á l'étude des propriétés thermique et mécaniques due béton en fonction de la température. *Annales de l'Institut Technique du Bâtiment et des Travaux Publics*, Oct.
26. ASCHL, H. (unpublished). Mechanical properties of mass-concrete at elevated temperatures. *3rd International Conference on Mechanical Behaviour of Materials, Cambridge, UK, 1979.*
27. HANNANT, D. J. (1968). Strain behaviour of concrete up to 95 °C under compressive stresses. *Proceedings of Conference on Prestressed Concrete Pressure Vessels, 13–17 March, 1967*. The Institution of Civil Engineers, London.
28. ANDERBERG, Y. and THELANDERSON, S. (1976). Stress and deformation characteristics of concrete at high temperatures. 2. Experimental investigation and material behaviour model. *Lund Institute of Technology, Division of Structural Mechanics and Concrete Construction, Bulletin 54*. Lund, Sweden.
29. WEIGLER, H. and FISCHER, R. (1972). Influence of high temperatures on strength and deformations of concrete. *ACI Special Publication SP-34*, Paper 26. ACI, Detroit, Michigan, USA.
30. GROSS, H. (1975). High temperature creep of concrete. *Nuclear Engineering and Design.*
31. SCHNEIDER, U. and KORDINA K. (1975). On the behaviour of normal concrete under steady and transient temperature conditions. *3rd SMIRT Conference, Vol. 3*, 1975. North-Holland Publishing Co., Amsterdam.

32. P. Sullivan and G. Khoury, (1978). Creep of a lightweight aggregate concrete at elevated temperatures. *FIP Conference, 1978*. Concrete Society, London.
33. Arthanari, S. and Yu, C. (1967). Creep of concrete under uniaxial and biaxial stresses at elevated temperatures. *Mag. Conc. Res.*, **19,** (60).
34. Dougill, J. (1961). The effects of thermal incompatibility and shrinkage on the strength of concrete. *Mag. Conc. Res.*, **13,** (39).
35. Maréchal, J. C. (1972). Creep of concrete as a function of temperature. *ACI Seminar on Concrete for Nuclear Reactors, Berlin, 1970. ACI Special Publication SP-34*. Detroit, Michigan, USA.
36. Geymayer, H. G. (1972). Effect of temperature on creep of concrete. A literature review. *ACI Seminar on Concrete for Nuclear Reactors, Berlin, 1970. ACI Special Publication SP-34*. Detroit, Michigan, USA.
37. Schneider, U. and Weiss, R. (1977). Kinetic considerations on the thermal destruction of cement-bound concrete and its mechanical effects. *Cement and Concrete Research*, **7,** pp. 259–68.
38. Zoldners, N. (1971). Thermal properties of concrete under sustained elevated temperatures. *ACI Special Publication SP-25*, Paper 1. ACI, Detroit, Michigan, USA.
39. Furumura, F. *The Stress–Strain Curve of Concrete at High Temperature.* Tokyo Institute of Technology, Japan.
40 Sullivan, P. J. E. (1974). La tenue du béton de granulats légers aux températures élevées. *Annales de l'Institut Technique du Bâtiment et des Travaux Publics*, No. 321, Octobre.
41. Fisher, R. A. (1957). *Design of Experiments. 11th Ed.* Oliver and Boyd, London.
42. Taguchi, G. (1959). Linear graphs for orthogonal arrays and their applications to experimental design. *Report of Statistical Application Research, Union of Japanese Society of Engineers*, **6** (4) Dec.
43. Petzold A. and Rohrs, M. (1969). *Concrete for High Temperatures.* Applied Science, London.
44. Murasher, V. I. (1954). *Some Characteristics of the Theory of Design of Heat Resistant, Plain and Reinforced Concrete Structures.* State Press for Literature on Building and Architecture, Moscow. (B.R.S. translation).
45. Abrams, M. S. and Gustaferro, A. H. (1968). Fire endurance of concrete slabs as influenced by thickness, type and moisture. *Journal of the PCA*, May.
46. Harmathy, T. S. (1965). Effect of moisture on fire endurance of building elements. *ASTM Special Publication No. 385—Moisture in Materials in Relation to Fire Tests.*
47. Sullivan P. J. E. and Zaman, A. A. A. (1967). Explosive spalling of concrete exposed to high temperatures. *First International Conference on Struct. Mech. in Reactor Technology, Berlin, 1971.* HMSO, London.
48. Law, M. (1967). Analysis of some results of experimental fires. *Proc. Behaviour of Structural Steel in Fire. Symposium, 2, January, 1967.* HMSO, London.
49. Abeles, P. W. and Bobrowski, J. (1972). Fire resistance and limit state design. *Concrete*, **6,** (4), April.

50. GUSTAFERRO, A. H. (1973). Design of prestressed concrete for fire. *PCI Journal*, November/December.
51. FIP (1975). Guides to good practice. *FIP/CEB Recommendations for the Design of Reinforced and Prestressed Concrete Structural Members for Fire Resistance.* FIP, Slough, UK.
52. CONCRETE SOCIETY AND INSTITUTION OF STRUCTURAL ENGINEERS. (1978). *Design and Detailing of Concrete Structures for Fire.* Joint Committee—Concrete Society and Institution of Structural Engineers, London.

Chapter 2

TIME-DEPENDENT DEFORMATIONS OF CONCRETE

J. M. Illston

Director of Studies in Civil Engineering, The Hatfield Polytechnic, Hatfield, UK

SUMMARY

The effects on drying shrinkage of various influential factors are described; the factors include both extrinsic, such as relative humidity, and intrinsic, such as mix ratios. The overall effect is summarised in the form of a simple chart to enable designers to estimate the magnitude of drying shrinkage.

A similar presentation is given for creep, also culminating in a simple design chart. In addition, various methods of evaluating the creep under a history of variable stress are described and compared.

2.1 INTRODUCTION

Concrete exhibits volumetric deformations without being subjected to any form of imposed load. Part of such deformation follows changes in ambient temperature; the flow of heat in concrete is relatively rapid so that, in most structures, the response to temperature change is equally rapid, and thermal movement can be taken as instantaneous rather than time-dependent. Accordingly, it is not considered in this chapter.

The remaining volumetric deformation is largely associated with moisture exchange between the concrete and the environment. The movement of moisture in concrete is much slower than that of heat, and the response of the material to changes in moisture content is time-dependent for all structures. The most common manifestation is of concrete drying from an initial state of near-saturation, and the contraction observed as moisture moves to the outside atmosphere is referred to as drying shrinkage. The consideration of moisture-related deformation forms the first main section of this chapter.

If the concrete is subjected to stress, however induced, there is an immediate response which is associated with an elastic modulus and which results in volumetric and deviatoric strain components which can be considered in accordance with the principles of elastic theory. If the stress is sustained, further strains occur, again with volumetric and deviatoric components, and these are referred to as creep strains. They form the subject of the second main section of this chapter; elastic strains are introduced only in so far as they affect the definition or discussion of creep.

Many hundreds of technical publications have appeared on the topics of creep and shrinkage, and the possible treatments are numerous and diverse. In an overall view of the kind contemplated here a particular standpoint must be adopted in the interests of producing a reasonably comprehensive description. Here it is the needs of the designer of concrete structures that provide the necessary focus, and the intention is to demonstrate and discuss the estimation of the magnitudes of creep and shrinkage for use in design calculations, and to describe techniques for handling special situations, such as the creep occurring under regimes of variable stress.

Further accounts of these and other aspects of time-dependent deformations are available in, for instance, Refs. 1–5.

2.2 DEFINITION AND STRATEGY

The total strain in concrete is made up of the four components introduced above. Omitting the thermal movement, the superposition of components is demonstrated in Fig. 1, which is drawn in terms of linear strains such as would be observed by measurement on concrete test specimens. It is assumed here that the concrete is drying from saturation in an atmosphere of constant temperature and humidity. The first specimen is not subjected to external load and therefore exhibits drying shrinkage only, while the second specimen is loaded during the period t_1 to t_2 and has a total strain consisting of the sum of shrinkage, immediate strain and creep.

In these circumstances shrinkage is defined as the volumetric strain of an unloaded specimen, and it is normal to quote it as a linear strain. Corrections must be made for any concomitant changes in temperature, and it must be realised that the measured shrinkage is a function of the size of the specimen and is not an absolute value for the given circumstances of mix and environment.

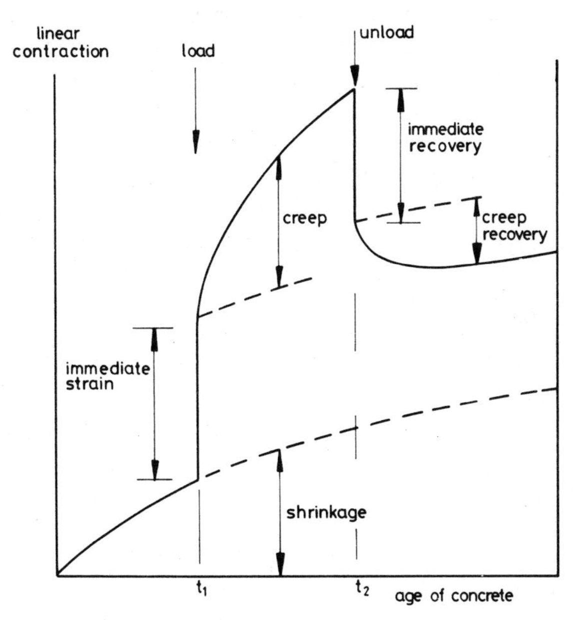

FIG. 1. The time-dependent strains in concrete.

Creep is similarly defined as the time-dependent strain in response to an applied stress. It is thus the strain additional to the shrinkage and immediate strain and it is quoted as the linear strain in the direction of the applied stress. It is found experimentally by deducting the shrinkage and immediate strains from the total strain. As can be seen in Fig. 1, it is partially recovered after the removal of the stress, and, as a general rule, the time-dependent recovery approaches completion in a relatively short time so that the curves of shrinkage and total strain become approximately parallel soon after the removal of the stress.

The determination of creep depends on measurement or estimation of the shrinkage on a companion specimen, and on a similar determination of the immediate strain on application and removal of the stress. The former is straightforward but the latter is complicated by the rapid rate of creep as soon as a change of stress occurs. However, reasonable consistency can be ensured by determining the immediate strain from a modulus of elasticity measured under carefully defined conditions. In brief, this is best done by finding the secant modulus at a stress level of about 40 % of the short-term strength. The measurement should be made after exercising the concrete through several stress cycles to work off the early creep, taking a period of some 5–10 min. The test method suggested in BS 1881[6] is acceptable and it forms the basis of the data given later.

The strain components, as defined above, can be estimated individually in terms of the appropriate variables expressing the mix and the environment. This presents a considerable problem in that each of the components is significantly influenced by a number of variables which may not even have been identified, or may well not be the best representation of the factor that is truly responsible for the behaviour. These matters will be examined in more detail later, but the example of shrinkage can be quoted; shrinkage is a function of humidity and temperature, and also of mix ratios and type of cement. Moisture content would probably be a better variable than humidity, and although it may be the content of tricalcium aluminate in the cement that is responsible for the changes with the type of cement, this has not been proved.

The lesson to be learned at this stage is that estimation is at best imperfect, and that it will be greatly affected both by the choice of variables and by the quality of the information on those variables. It is suggested that, in practice, levels of estimation should be identified relating to the starting data and to the importance of the estimation to the particular structural application. Three levels present themselves as being clearly identifiable.

(1) The lowest level, based on the minimum of initial information and providing the crudest estimate of the deformation, and thus suitable for the most approximate calculations. The simplest form is merely a single strain quoted for normal conditions, but this level also includes estimations based on the design cube strength of the concrete, extended perhaps to include notions of the humidity of storage and of member sizes.
(2) The intermediate level, in which more detailed starting data are

available on, for instance, the concreting materials and mix ratios. This level requires a more comprehensive set of charts providing the appropriate estimation factors; the method given for creep in the CEB/FIP recommendations is typical.[7] In addition, the method will usually include the means to follow the development of strains with time, including any special methods of time-dependent superposition of strains.

(3) The highest level, relevant to structures in which time-dependent movements have a very significant effect. The maximum number of data are required, and they will probably have to be assembled by means of a comprehensive laboratory programme mounted for just that purpose. Special mathematical techniques may also be developed to deal with the more sophisticated constitutive relations that will be required.

In this chapter, some reference will be made to all three levels, but attention will be mostly given to level (1) at which the most recent developments have occurred.

2.3 SHRINKAGE

2.3.1 Pattern of Behaviour

Hardened cement paste (hcp) has a disordered structure containing both crystalline hydration products, such as calcium hydroxide, and microcrystalline products, such as calcium silicate hydrate, thought to be in the form of distorted sheets only a few molecules thick and often referred to as cement gel because of its capacity for taking up and giving out water. Hcp has a variety of pore sizes down to the very smallest capable of containing only a few layers of water molecules, and up to the macropores typified by entrapped air in concrete.

The implication of the structure is that water is held in various states with different bonding energies. Thus reference is made to capillary water which is free from the influence of surface forces, adsorbed water which is held to a solid surface, and interlayer water which penetrates between a pair of solid surfaces. Drying shrinkage is observed as a result of the forces of contraction arising in the hcp as the water is removed by drying. Lower relative humidities are needed to remove the water with higher bonding energies, and, in general, smaller and smaller pores empty as the relative humidity decreases. The wide range of pore sizes in hcp ensures that water is lost continuously as the relative humidity drops, and hence shrinkage also increases continuously with the degree of drying.

Water returns to the hcp as the relative humidity (RH) increases, and swelling is observed. However, irreversible changes occur on first drying, such as the closing of certain pores which do not reopen on wetting; thus not all the original shrinkage is recovered when the material is resaturated. That is, there is a substantial irreversible shrinkage on first drying. The general effect is illustrated in Fig. 2,[8] in which most of the following features are evident:

(1) Continuous saturation in water leads to an overall swelling.
(2) As already described, a substantial proportion of first shrinkage is irrecoverable.
(3) The recoverable shrinkage (or swelling) does not change too greatly in cycles of drying and wetting after the first.
(4) The rate of water absorption (and hence of swelling) is faster than the rate of desorption.
(5) The magnitude of shrinkage of hcp is very large relative to other deformations such as the elastic strain under working stress. As will be seen later this is not the case with normal concretes, in which the magnitudes of shrinkage and especially swelling are much reduced.

It can also be seen in Fig. 2 that in a constant atmosphere shrinkage or swelling continues albeit at a decreasing rate for a very long time. This is a matter of the size of the specimen. Thus, in the typical structural member, water is lost initially from the surface layers of the section thus causing a moisture gradient between the inner core of the section and the outer layers. Water moves down the moisture content gradient to the surrounding atmosphere, and drying becomes a continuous process in which the moisture content always increases with distance in from the surface. The rate of drying clearly depends on the size of the section and the ease with which the moisture moves. The process can be described reasonably well in accordance with a diffusion-type law with a diffusivity which is a function of the moisture content.[9]

The hcp responds immediately to changes in moisture content by shrinking or swelling, and it is possible to determine distributions of free shrinkage by a similar diffusion-type law. Normally the concrete specimen or member is relatively long in the direction orthogonal to

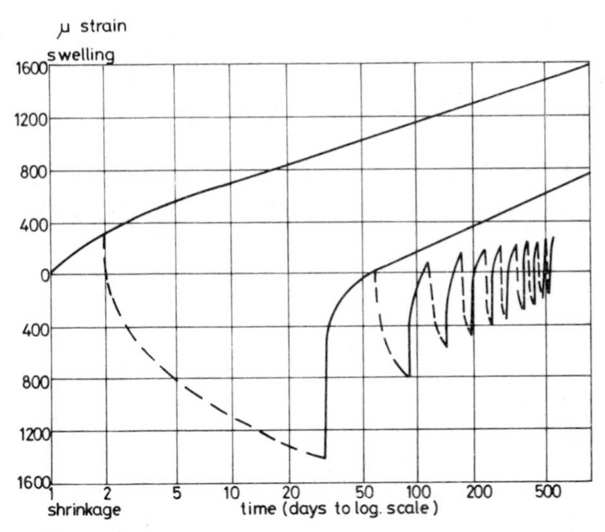

FIG. 2. Characteristics of shrinkage of hcp. Specimens of water/cement ratio of 0·25, alternately in water and dry air at 50 % relative humidity, Water,——; dry air, - - - -.

the moisture movement, thus imposing a continuity on the section which prevents the occurrence of free shrinkage and indeed causes the observed strains (parallel to the long axis) to be virtually identical at all points of the section. Since there is no applied stress, these strains are the observed shrinkage for the specimen or member. The conflict between free shrinkage and continuity means that distributions of self-equilibrating stress are induced through the section, with resulting distributions of creep and elastic strain. In general, the maximum tensile stress occurs on the surface, and the maximum compressive stress at the core of the section. Thus it should always be recognised that what is measured and described as shrinkage is, in reality, a composite strain of free shrinkage, elastic strain and creep. In some circumstances the induced stresses may be of structural significance, but these special cases are excluded from the further discussion, which concentrates on the observed rather than the free shrinkage.

2.3.2 Influential Factors

2.3.2.1 *Environmental Effects*

The effects of long-term storage, from the saturated state, in different relative humidities are shown in the curves of Fig. 3 taken from the results of the classical, long-term tests reported by Troxell, Raphael and Davis.[10] It is evident that the lower humidities lead to a much greater rate of shrinkage over the first year of storage (for this size of specimen). After that the rate decreases but some shrinkage is still

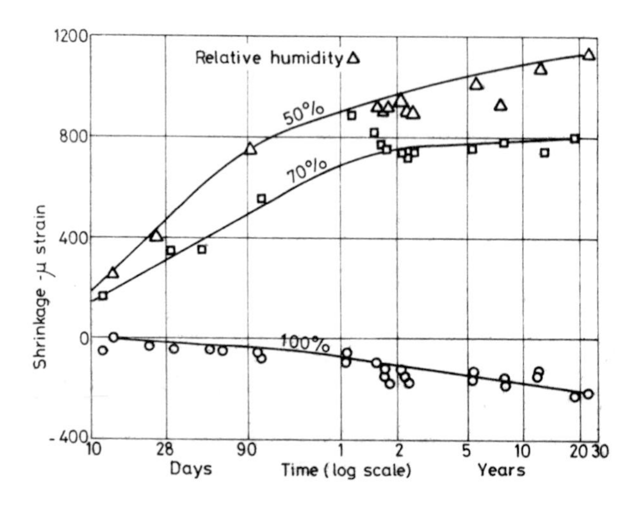

FIG. 3. Long-term shrinkage for various storage humidities.

occurring as long as 20 years after the start of drying. Storage in water leads, as before, to continuous swelling which also continues over the full time-scale.

In accordance with the discussion of the previous section, the rate of shrinkage is highly dependent on the size of the shrinking section. This is conveniently expressed in terms of an effective thickness defined as

$$\frac{\text{area of the section}}{\text{the semi-perimeter exposed to the drying atmosphere}}$$

In the case of a slab, drying from top and bottom faces, this reduces to the actual thickness of the slab. The rate of shrinkage is also affected by temperature, and as indicated by the ideas of increased mobility of water at higher temperature, the shrinkage occurs more rapidly when the temperature is raised.

The size of the section appears to influence the magnitude of shrinkage as well as its rate. This may be just a matter of the rate of shrinkage in larger members becoming so slow that the state of hygral equilibrium with the humidity of the environment cannot be approached within a realistic time-scale. Other contributory phenomena may be:

(1) the greater degree of hydration in the larger section resulting from the longer time that water lingers within the section.
(2) the different induced stress distributions resulting from the bigger areas of the larger sections. It follows that the creep distributions will also be different with a consequent effect on the observed shrinkage.

The former of these two possibilities is supported by the evidence for the effect of age on shrinkage of hcp. The work reported in Ref. 11 shows that younger hcps shrink more than older ones. Surprisingly, this has not been shown to be a significant effect in concretes, and, in most practical circumstances it is possible to ignore any possible effect that the age of the start of drying may have on the subsequent shrinkage.

2.3.2.2 *Materials and their Proportions*

There is no doubt that the proportions of the various constituents in cement influence the shrinkage of the concrete made with the cement. Thus, in a study of this matter Blaine[4] measured the shrinkage of nearly 200 different American cements under identical conditions. The mean shrinkage was 300×10^{-6}, but extreme values reached as high as 420×10^{-6} and as low as 150×10^{-6}. Tests of statistical significance showed that a number of factors might be influential, the most important of which were the content of tricalcium aluminate (C_3A) and the sulphate content (SO_3), deriving (normally) from the addition of gypsum. It has not proved possible to express these (or other) findings in such a quantitative manner that the shrinkage may be estimated from the composition of the cement. It must be concluded that the most appropriate variables to represent the composition of the cement have so far proved elusive.

Admixtures may influence shrinkage because they allow changes to be introduced in the mix itself; this is particularly true of water-reducing agents such as superplasticisers. However, assuming that no such changes are made in the mix, there may still be an effect on shrinkage. It is important that admixtures should be investigated on this count, and it is true that many have been shown to make no significant difference to the shrinkage of the concrete. This is not the case with calcium chloride accelerators, the presence of which does increase the shrinkage. Once again it is not possible to express the change in any orderly quantitative manner.

The addition of aggregate to the hcp has the dual effect of dilution and restraint. This is well demonstrated by the following expression for the shrinkage of concrete, which is founded on theoretical ideas.[12]

$$\varepsilon_{sc} = \varepsilon_{sp} - \frac{(\varepsilon_{sp} - \varepsilon_{sa})2(Ka/Kp)Va}{1 + (Ka/Kp) + Va(Ka/Kp - 1)} \tag{1}$$

Where ε_{sp} is the shrinkage of the hcp, ϵ_{sa} is the shrinkage of the aggregate, Ka/Kp is the ratio of the bulk moduli of aggregate and hcp, and Va is the volume concentration of the aggregate.

Normally the aggregate can be considered as inert, so that its shrinkage, ε_{sa}, is zero. The restraint provided by the aggregate is then a function of its elastic rigidity in relation to that of the paste, represented in the formula by the ratio of the bulk moduli, Ka/Kp. Most normal natural aggregates are considerably stiffer than the hcp, and variations in stiffness between them, although significant in themselves, make little difference to the degree of restraint; it follows that the stiffness of normal aggregates is a secondary influence on shrinkage. This is not the case for softer aggregates, and artificial lightweight aggregates provide comparatively little restraint to the shrinkage of the paste, and it is found that, for the same mix, lightweight concretes shrink distinctly more than normal weight concretes. In addition, the development of shrinkage with time may differ in that lightweight concretes have been shown to have an initially slower rate of shrinkage. Potentially more serious is the effect of a shrinking aggregate. This is demonstrated in eqn. (1), and that it is a practical reality is reported in Ref. 13.

Equation (1) also shows the influence of the proportion of aggregate, which is best represented by the volume concentration. Va. Again common sense is justified in that the more stiff, restraining aggregate there is in the concrete the less it shrinks. The alternative variable, more familiar in concrete technology, is the aggregate–cement ratio, and, other variables remaining unchanged, the shrinkage diminishes with increasing aggregate–cement ratio. The other major mix variable, the water–cement ratio is equally important, and this too is no surprise in view of the well known significance of water–cement ratio to strength. Thus a lower water–cement ratio gives a higher strength and a smaller shrinkage. A possible simplification arises from the combination of the two mix ratios in the single variable, original water content. As shown in Fig. 4,[14] shrinkage is related to original water content over a wide range of values of aggregate–cement and water–cement ratios.

2.3.3 Estimation of Shrinkage

The essence of estimation at level (1)—the lowest—is that it should be based on the minimum number of data and hence be suitable for approximate calculation only. Such a method of estimation is described with supporting evidence in Ref. 15; the working sheet for the method

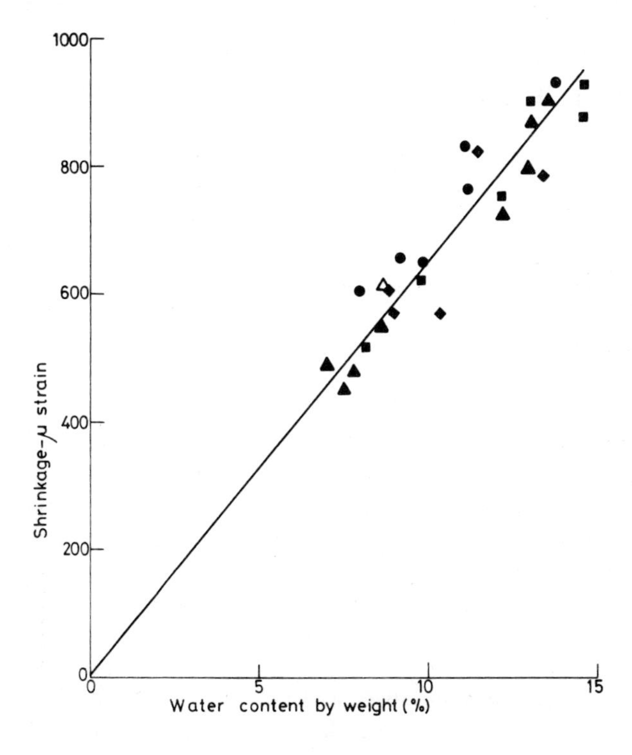

FIG. 4. The linear relation between shrinkage and water content for a variety of mix proportions. Water–cement ratio 0·35— ●; 0·47— ▲; 0·59— ■; 0·71— ◆.

is shown in Fig. 5. The starting data consists merely of the exposure conditions and the size of section. It also introduces an additional factor which is of importance in certain structural members, namely the area of reinforcement as a proportion of the sectional area. The sheet is self-explanatory, and it is necessary just to draw attention to the inevitable simplifications that have been made. Thus:

Drying in a constant relative humidity is assumed. However, an idea of the proportion of the shrinkage that is reversible is given, and the mean relative humidity representing the (variable) outdoor conditions in the UK is also indicated.

Secondary variables are omitted. These include temperature, aggregate stiffness (with the exception of a mention of lightweight aggregate concrete), the effects of cement composition and admixtures, and the detailed development of shrinkage with time of drying.

The chart is prepared for a mix with an original water content of 8 %, and adjustments for other water contents can be made on the basis of direct proportionality between water content and shrinkage.

The worksheet was prepared by Hobbs and Parrott[15] on the initiative of the Movements Working Party of the Concrete Society as a simpler alternative to the method contained in the recommendations of the CEB/FIP.[7] This latter is nearer to level (2), requiring rather more starting data, and providing rather fuller information. The shrinkage of the concrete is expressed by

$$\varepsilon_{sc} = \varepsilon_c k_b k_e k_t k_p \tag{2}$$

where ε_{sc} is the shrinkage of a standard concrete in terms of the relative humidity of storage, k_b is a modifying multiplier representing the mix, using as variables the water–cement ratio and the cement content. k_e similarly is a multiplier for member size in terms of section thickness as described above in Section 2.3.2.1, and k_t allows the shrinkage at any time to be determined with allowance made for member size. k_p allows for the presence of reinforcement. The increased shrinkage of lightweight concrete is mentioned.

Reference 7 appeared in 1970 and is referred to in the British Standard Code, CP 110.[16] A later version has appeared in the CEB Model Code.[17] The strategy adopted is similar to that of the earlier version, but the presentation is somewhat modified. There are simplifications in that the shrinkage is quoted in tabular form for

The drying shrinkage of structural concrete of normal workability made without additives can be estimated from the graph opposite. The required input information is:

(i) Exposure condition.
(ii) Effective section thickness (i.e. twice the volume ÷ exposed surface area).
(iii) Relative area of reinforcing steel.

The graph relates to a concrete with an original water content of 8 % by weight and made with a high quality dense aggregate. The use of less rigid aggregates (i.e. those which cause a low elastic modulus of the concrete) or of shrinkable aggregates will increase the concrete shrinkage. Shrinkage is broadly proportional to the original water content so estimates may be simply adjusted for concretes known to have a water content other than the usual value of 8 %.

Concrete stored outside in the UK under conditions of fluctuating relative humidity will exhibit seasonal, cyclic strains of ±0·4 times the 30-year shrinkage superimposed upon the average strain estimated from the graph opposite; the maximum will occur at the end of each summer.

The reduction in shrinkage caused by symmetrically placed reinforcing steel can be allowed for by the equation:

Shrinkage of symmetrically reinforced concrete

$$= S/(1 + K\rho)$$

where S is the potential shrinkage of the plain concrete, ρ is the area of steel relative to that of concrete and, K is a coefficient = 25 for internal exposure and 15 for external exposure.

It should be remembered that potential movements due to concrete shrinkage may be reduced by external or structural restraint.

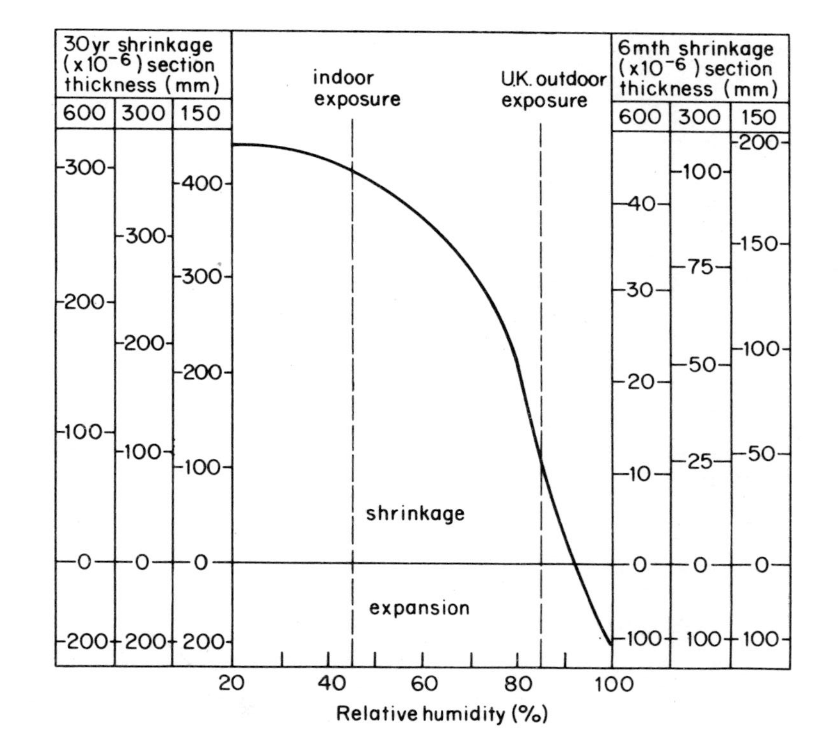

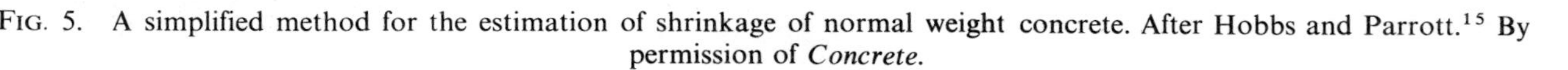
FIG. 5. A simplified method for the estimation of shrinkage of normal weight concrete. After Hobbs and Parrott.[15] By permission of *Concrete*.

particular storage conditions (i.e. in water, at 90, 70 and 40 % RH), and the mix is expressed through three grades of consistency. The size of section and the development of shrinkage with time are dealt with as before, though there is an improvement in that the intervention of storage conditions in the size-development relations is included. Also, some allowance is made for the temperature of curing and storage by adjusting the age in accordance with a maturity function which reflects the difference between the actual storage temperature and the standard value of 20 °C; there is too a correction to the time scale to suit the type of cement.

2.4 CREEP

2.4.1 Pattern of Behaviour

Creep is a function of stress and the interaction between the two is admirably conveyed in Fig. 6.[18] The strains are shown for concretes

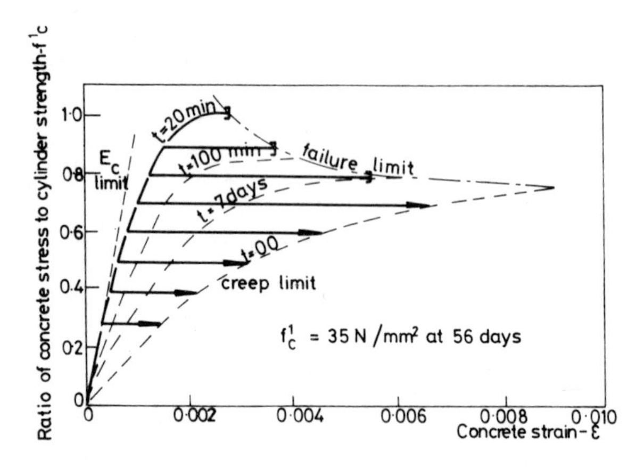

FIG. 6. Stress–strain curves found from tests on concrete subjected to different levels of sustained stress. t = time under load.

under sustained uniaxial compressive stresses set to various proportions of the short-term strength. The results lie within a zone bounded by three limits only one of which (the failure limit) can be readily determined.

The E_c limit is given by the slope at the origin of the typical short-term loading test, the 20-min curve; it thus corresponds to the initial tangent modulus of elasticity. It is apparent that there is no truly linear portion of the short-term curve and that the curvature increases with stress. Clearly it is only in the early stages that the short-term behaviour approximates to the linear elastic.

The creep limit, like the E_c limit, cannot be followed experimentally since creep, like shrinkage, continues indefinitely, albeit at an ever-decreasing rate. It is therefore obtained by extrapolation, but, as can be seen, the forms of the stress–strain curves at any given time have a consistent shape. They are approximately affine to the short-term loading curve and again have no truly linear portion. However, it may reasonably be argued that stresses in concrete in service are usually a relatively small proportion of the short-term strength (say, not more than 40 %), and creep may be treated as a linear phenomenon in the same manner as immediate strain is treated as linearly elastic. On this basis creep, ε_c, may be expressed

$$\varepsilon_c = \sigma c \tag{3}$$

where σ is the applied stress and c is specific creep, that is the creep per unit stress (associated with a particular time). The specific creep is thus a constitutive coefficient and expresses the property of time-dependent deformation under load. It may be quoted alternatively in non-dimensional form as a creep coefficient, φ.

Then

$$\varepsilon_c = \frac{\sigma}{E_c}\varphi \tag{4}$$

where E_c is the modulus of elasticity of the concrete. The creep coefficient is thus the ratio of the creep to the elastic strain.

It follows that

$$cE_c = \varphi \tag{5}$$

The failure limit in Fig. 6 shows that time-dependent rupture can occur in concrete and that the period before failure reduces as the level of sustained stress increases. Time-dependent rupture does not normally occur at stresses below about 75 % of the strength and it is thus not often a feature in the design of concrete structures. It may be of significance in some special applications such as slender columns, but attention here is confined to the predominant structural circumstances in which stresses remain at the lower levels below which linear stress-strain relations are an acceptable practical assumption.

The large majority of experimental investigations into the creep of concrete have, like those in Fig. 6, employed a uniaxial compressive stress, and the general behaviour for such a regime is now well authenticated. A much smaller number of tests have been conducted under the more difficult conditions of uniaxial tension, and, although the evidence is by no means conclusive, it supports the view that the pattern of behaviour is similar to that in compression. Thus tensile creep occurs at an ever-decreasing rate, it is only partially recoverable and there is a possibility of time-dependent failure at higher stress–strength ratios. Some evidence indicates that specific creep in tension is rather greater than that in compression in the period soon after loading, and it may be that the stress–strain curves are nearer to linearity. These are comparatively minor differences and it is certainly a practical proposition to assume identical specific creeps for both senses of stress. It may also be correctly deduced that combinations of stresses of the two senses, as in flexure or torsion, give rise to the same pattern of time-dependent behaviour under load as that already discussed.

2.4.2 Extrinsic or Environmental Influences on Creep

Creep is profoundly affected by temperature and humidity, both while the concrete is curing before loading, and during storage under load.

The effects of curing have been investigated by measuring the creeps of specimens, sealed and at constant temperature, after undergoing various regimes of temperature and humidity before loading. Thus there is no exchange of heat or moisture with the ambient environment during the loading period, and for these reference conditions the creep is referred to as basic.

The curing conditions have a dual effect. Firstly, the progress of hydration of the cement is accelerated by a higher curing temperature, and it may also be inhibited by a lack of water if the concrete dries out during the curing period. The greater the degree of hydration, that is the greater the maturity of the concrete, when the load is applied the less will be the subsequent creep.

Secondly, creep is a function of the moisture content and temperature of the sealed specimens. There is general agreement that a completely dry concrete does not creep, but the evidence is conflicting on the relative magnitudes of creeps at intermediate moisture contents. On balance it seems likely that the greatest creep occurs in specimens sealed in or near to the saturated state, and that there is a comparatively small drop-off in magnitude for predrying down to about 50 % RH.

The (constant) temperature of storage under load is a most

important variable for creep as illustrated in Fig. 7.[19] The consensus view is that creep is approximately linear with temperature (in °C) up to 80 °C. Most creep tests have been conducted at or near to the ambient of 20 °C, and it is of relevance to the designer of special structures (such as nuclear pressure vessels) operating at, say 60 °C, to realise that a creep of three times the normal may be expected in his structure.

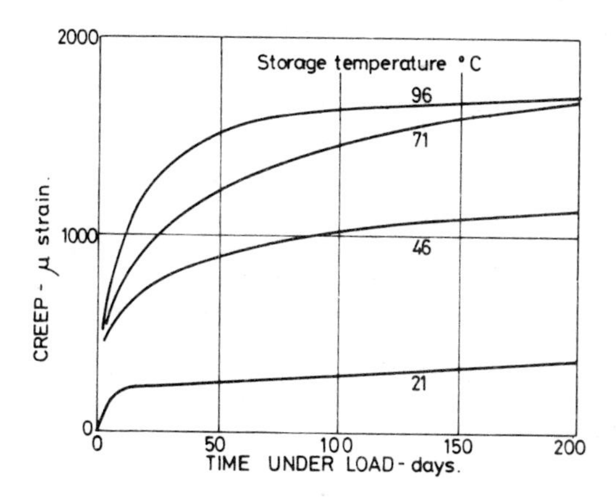

FIG. 7. The effect on creep of the temperature of storage. Concrete heated under water for 1 week before loading at an age of 1 year.

Further enhancement of creep may be observed if the moisture content or temperature changes while the concrete is under load. Thus if the concrete dries out for the first time under load, the total creep is likely to be two or three times the basic creep. The additional, largely irrecoverable, strain is referred to as drying creep, and a similar increase may be observed when a predried concrete is wetted for the first time. Later cycles of drying or wetting do not produce the same large increases as the first change in moisture content. Creep has certain similarities to shrinkage and a particularly good linear correlation has been found between drying creep and the concurrent shrinkage.[20] Additionally, creep under drying conditions is affected by the size of the specimen in much the same way as shrinkage, as described earlier.

A comparable creep component is observed if the temperature is raised for the first time while the concrete is under load. This is referred to as transitional thermal creep[21] and it does not occur on a subsequent drop in temperature, or on further cycles of temperature to the same level. Transitional thermal creep occurs rapidly, continuing for only a short time after equilibrium is attained at the higher temperature; it is almost wholly irrecoverable.

2.4.3 Intrinsic or Mix Influences on Creep

The similarity to shrinkage is evident throughout this section; thus:

(1) Higher water–cement ratios lead to higher porosities and lower strengths. The hcp structure is then less resistant to stresses, whether internal or external, and the creep and shrinkage (and elastic strain) are greater. The common influence of porosity on strength, elastic modulus and creep leads to the possibility of determining creep in terms of either of the other two. Thus either elastic modulus or strength can serve as a single variable representing the influence on creep of the combination of water–cement ratio and maturity. This is of practical significance as either strength or elastic modulus may be known to a designer when data on the mix and curing regime are still uncertain.

(2) The presence of the aggregate again has the dual effects of dilution and restraint, so that the creep of concrete is normally very much less than that of hcp contained in it. It should follow that the creep of lightweight concrete is considerably greater than that of concrete made with a normal hard aggregate. The evidence on this point is not so convincing as it is in regard to shrinkage, and it appears not unreasonable to take the same creep coefficients for both materials.

(3) Variations in cement composition and the effects of additives and admixtures cannot be analysed quantitatively, nor are their chemical and physical implications fully understood. In a number of instances variations in the mix from this source have been shown to have an insignificant influence on creep. However, this is unlikely to be true in all cases, and the absence of any effect needs to be demonstrated whenever new materials or new proportions of compounds are introduced.

2.4.4 Considerations of Age and Time

Assuming that the concrete is cured and stored at normal ambient temperature, creep is a function of two time variables, most conveniently chosen as the age of loading, τ, and the time under load, $t - \tau$, where t is age of the concrete at which the creep is determined. The general creep—age of loading—time under load surface is sketched in Fig. 8 which demonstrates clearly the ever decreasing rate of creep with time under load, and the diminution in its magnitude with age of loading (or with maturity if there are variations in temperature).

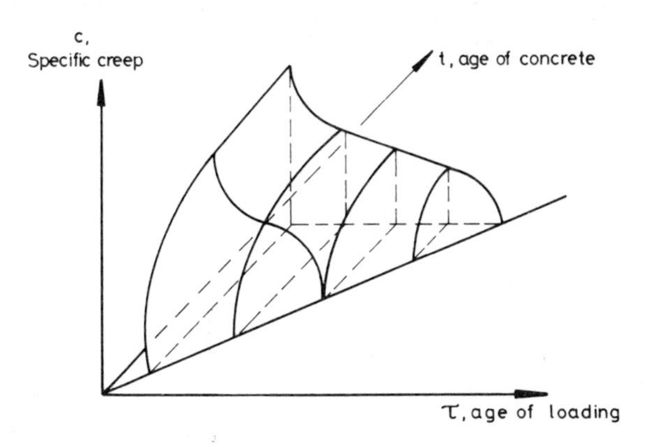

FIG. 8. A typical creep surface.

Empirical expressions have been developed to represent the creep surface and the most common approach is founded on the assumption that the creep curves for different ages of loading have the same shape. Then, accepting also the linearity of creep and stress, the creep, ε_c, is given by the product of two functions:

$$\varepsilon_c = \sigma f_1(\tau) f_2(t-\tau). \tag{6}$$

Various empirical forms of the function $f_2(t-\tau)$ for the development of creep with time under load have been suggested. Two that imply, incorrectly, that creep reaches a limiting value are:

the hyperbolic formula,

$$\varepsilon_c = \frac{a(t-\tau)^n}{b+(t-\tau)^n} \tag{7}$$

where coefficients a, b and power n (often taken as unity) define the curve,

and the exponential formula,

$$\varepsilon_c = c(1-e^{-d(t-\tau)}) \tag{8}$$

where c and d are coefficients.

Reasonable fits with experimental results are achieved in the medium term, but the creep after longer times is then underestimated.

Two other expressions imply no limit on creep and they give satisfactory fits in the short-term, but overestimate the creep for long periods under load. They are:

the logarithmic formula,

$$\varepsilon_c = g + h\log(1 + t - \tau) \tag{9}$$

with coefficients g and h.

and the power formula,

$$\varepsilon_c = j(t - \tau)^k \tag{10}$$

with coefficient j and power k.

The function $f_1(\tau)$ for the age multiplier can also be expressed as a power of the age of loading, τ^{-m}.[22] Furthermore the relationship between creep and strength clearly implies that an inverse function of strength could replace $f_1(\tau)$.

None of these empirical expressions gives satisfaction for the universality of creep curves, and the CEB/FIP recommendations[7] avoid the difficulty by presenting individual creep curves in graphical form for different ages of loading.

In an alternative approach creep is considered as the sum of two components, namely irrecoverable creep (or flow) and recoverable creep (or delayed elastic strain). Characteristics of the components have been determined experimentally[23] and they are shown for two ages of loading, τ_1 and τ_2, in practical form in Fig. 9. Thus the rate of irrecoverable creep is independent of age of loading, and depends (for a given mix, etc.) merely on the age of the concrete. It follows that the irrecoverable creep can be found, for various ages of loading, from a single curve.

The recoverable creep is measured from recovery curves after unloading, and, as previously mentioned, it approaches a limiting value relatively rapidly. The limiting value is approximately constant for different ages and conditions of drying, and even for higher temperatures. For practical purposes it may be taken as 0·4 x (the elastic strain). Although the rate of occurrence is age-dependent (as shown in Fig. 9) it is again a reasonable practical decision to take it as independent of age, or even, where the early creep is not of interest, to include it as an instantaneous addition to the elastic strain.

With reference to Fig. 9, the irrecoverable creep, at age t_1 resulting from a stress σ applied at a loading age of τ_1, is given by:

$$\varepsilon_{cf\,1} = \sigma \frac{\varphi_f}{E_c} [f_f(t_1) - f_f(\tau_1)] \tag{11}$$

where φ_f is the irrecoverable creep coefficient, and the function $f_f(t)$

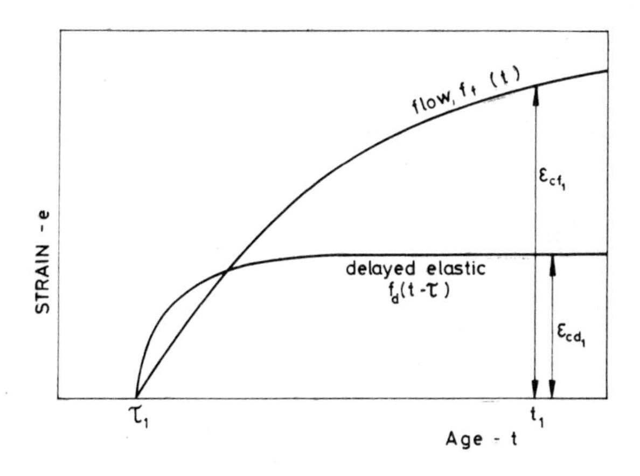

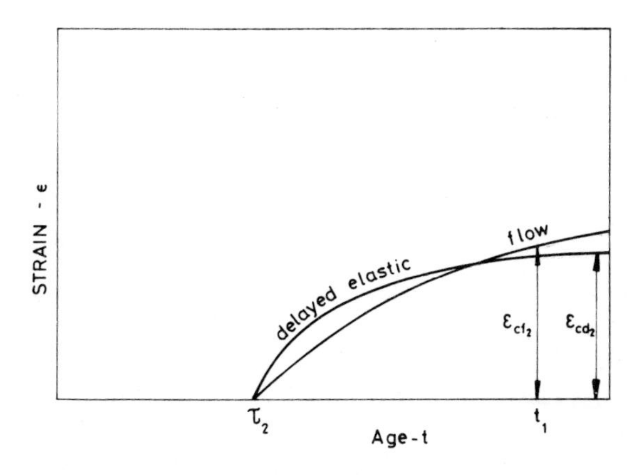

FIG. 9. Recoverable and irrecoverable components of creep (delayed elastic strain and flow), for two ages of loading, τ_1 and τ_2.

expresses its development with time. Similarly, the recoverable creep, ε_{cd_1}, is given by

$$\varepsilon_{cd_1} = \sigma \frac{\varphi_d}{E_c} f_d(t_1 - \tau_1) \tag{12}$$

where φ_d is the limiting recoverable creep coefficient, and function $f_d(t - \tau)$ expresses its development with time under load.

If the total creep of the concrete loaded earliest (i.e. the first curve of Fig. 8) is exactly reproduced by the sum of the two components, then it is an experimental fact that the later curves will be somewhat underestimated in this approach.

2.4.5 Regimes of Variable Stress

Because the relation between creep and stress is linear, the Boltzmann principle of superposition can be applied to concrete. In practical terms this means that the total creep at age t, $\varepsilon_c(t)$, can be found by summing the creeps of the individual stress increments applied during the preceding stress history. Numerically this may be expressed, for n such increments:

$$\varepsilon_c(t) = \Sigma_1^n \Delta\sigma_r c_r = \Sigma_1^n \Delta\sigma_r \frac{\varphi_r}{E_c} \tag{13}$$

where $\Delta\sigma_r$ is a typical stress increment and c_r is the specific creep appropriate to the time under load of $t - \tau_r$.

More generally, for a continuous variation in stress, the creep at time t for a first age of loading of τ_1, is given by:

$$\varepsilon_c(t) = \int_{\tau_1}^{t} \frac{d\sigma(\tau)}{d\tau} \cdot \frac{\varphi(\tau, t - \tau)}{E_c} \cdot d\tau \tag{14}$$

where, as indicated, σ is a function of τ, and φ is a function of both τ and $t - \tau$.

Direct solutions of this integral are only possible for simple stress histories, and even then only if simplified and approximate functions are taken for φ.

Various techniques have become established because of their success in circumventing the difficulties of solution. All four of those given below can be, and often are, employed as the basis of a numerical step-by-step process, and all but one provide the reasonable possibility of closed form solutions to particular structural problems.[24]

(*i*) *Virgin superposition.* Here the numerical procedure of eqn. (13) is applied directly. The stress history is expressed as a step function with time, thus providing the necessary stress increments, and the corresponding curves of specific creep are found from a creep surface. The word virgin is incorporated in the title of the technique to indicate both that the creep surface is determined from tests on virgin (not previously loaded) concrete, and to differentiate between this technique and the generality of the principle of superposition which applies to all four techniques. Virgin superposition is the one technique that is essentially numerical.

The application of virgin superposition (including elastic strain) is demonstrated in Fig. 10 for a positive increment of stress, $\Delta\sigma$, applied

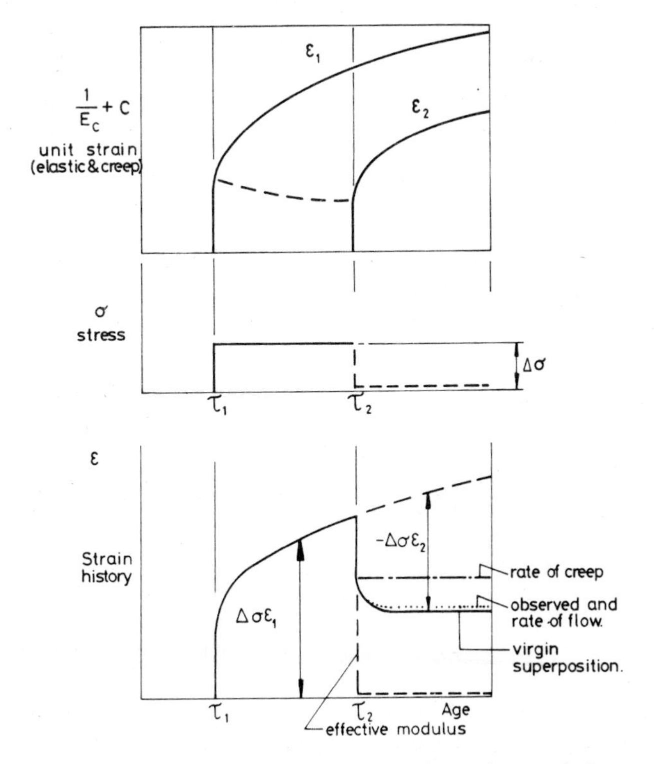

FIG. 10. Estimations of strain history by the various techniques.

at age τ_1, followed by an equal and opposite increment at τ_2. The strain determined by virgin superposition overestimates the observed creep for stress increments after the first; this is shown in the figure by the excessive creep recovery. Clearly the creep rate is affected by the previous stress history so that the principle of superposition (applied in this direct manner) does not apply perfectly to concrete. In other words, for the best estimates some account must be taken of the previous stress history.

(*ii*) *Rate of flow or the two-component technique.* The principle of superposition can again be applied, this time separately to each of the components (recoverable and irrecoverable). The technique is based on the observation of the creep recovery after unloading, and it thus pays some heed to the effects of previous loading on creep. The result is that it can estimate exactly the strain history shown in Fig. 10 and experimental comparisons have indicated that estimates of strain by the two-component technique are rather better than those by virgin superposition for more general regimes of applied stress. As previously mentioned, the reverse side of the coin is that virgin creep curves are underestimated by the two-component approach, and if such estimates are used for variable stress computations, virgin superposition may prove the more effective. Thus, it may be said that both techniques give acceptable engineering answers for even the most severe conditions of applied stress, but which provides the closer estimate of strain will depend on the circumstances.

The two-component approach does have an advantage in requiring data on just one creep curve and one recovery curve instead of the full creep surface needed for virgin superposition. There may also be savings in computation, especially if the recoverable creep is taken as instantaneous and included with the elastic strain. Then, in any time interval, $\tau_r \rightarrow \tau_{r+1}$, during which the total stress remains constant at σ, the irrecoverable creep is given by

$$\Delta\varepsilon_{ci} = \sigma \frac{\Delta\varphi_{ir}}{E_c} \tag{15}$$

where $\Delta\varphi_{ir}$ is the irrecoverable creep coefficient occurring during the time interval.

In contrast, the same computation by virgin superposition requires that the effect of all stress increments be summed $(\Sigma\Delta\sigma_r(\varphi_r/E_c))$ for both τ_r and τ_{r+1}, and the difference taken.

(*iii*) *Rate of creep.* Although the rate of creep technique has been in use for the longer time, it is a simpler version of the rate of flow. The total creep is considered to be irrecoverable, and, as before, its rate is independent of the previous stress history. It follows that the only data required are obtained from a single creep curve, and the computations are correspondingly briefer. This has the further advantage that closed form solutions of a comparatively straightforward algebraic kind are available for a number of practical problems; however, this is only achieved at the expense of some loss of accuracy.

The performance of the rate of creep technique can be assessed in Fig. 10, where it can be seen that the time-dependent recovery after unloading is completely missing in the rate of creep estimate. This kind of underestimate of the effects of later stress increments also means that the creep of virgin concrete at later ages is underestimated significantly more by rate of creep than it is by rate of flow. Nevertheless, many acceptable engineering solutions are achieved using the rate of creep technique, and this is done with an agreeably small degree of difficulty.

(*iv*) *Effective modulus.* This is the simplest technique, and, as might be expected, the least accurate. The creep is included with the elastic strain to give an effective modulus of elasticity, which is sometimes referred to as the reduced modulus, E_c'. Then

$$E_c' = \frac{\text{stress}}{\text{strain}} = \frac{\text{stress}}{(\text{creep} + \text{elastic strain})} = \frac{\sigma}{\sigma c + (\sigma/E_c)} = \frac{E_c}{1 + cE_c} = \frac{E_c}{1 + \varphi} \qquad (16)$$

The considerable advantage is that the storehouse of elastic solutions to structural problems is immediately available; all that is needed is to feed in the (reduced) value of the effective modulus. The estimated strain depends merely on the current stress, and, as shown in Fig. 10, this means that a full and instantaneous recovery is predicted after the load is removed, giving an estimate that is greatly in error. Again, provided that the stress history does not have large fluctuations, the effective modulus technique is capable of providing adequate engineering solutions (with the minimum of difficulty).

2.4.6 Estimation of Creep

In this section the intention is, as for shrinkage, to give most prominence to the introduction of a method of estimating creep at level (1). That is, the estimation of creep from the minimum of initial information is sought, with the realisation that the answer is necessarily of restricted accuracy. The method appears in Fig. 11 in the form of a single sheet. It was prepared by Parrott,[25] again following the initiative of the Movements Working Party of the Concrete Society.

At the basis of the method is the assumption that a designer will always know the strength of the concrete in the structure that he is designing, and cube strength is thus the one piece of firm information required. Additionally, some idea must exist of the age of loading, the humidity of the environment (if only whether indoor or outdoor) and the size of member.

A simple relationship is used to determine the elastic modulus at any given age, from the strength. This elastic modulus can, as discussed earlier, represent the water–cement ratio of the mix, and, at least partially, the age of loading. In the equation for (total) creep the modulus thus acts as a multiplier for these variables. The creep factor (or coefficient) itself is found from the single chart, using the information on age of loading, member size and humidity of storage. The computation yields the final creep resulting from a constant stress, and no more than a notion is provided of its development with time.

Some influential factors are not included. Thus it is assumed that the temperature has a normal ambient value during both curing and storage, and that high quality dense aggregate is used in normal proportions (though there is a mention of lightweight concrete). Further assumptions would be needed in order to make estimates of strain history under variable stress.

The single chart in the simple method just described derives from the data embodied in the charts of the CEB/FIP recommendations,[7] which both requires more initial information and provides fuller answers. It can be reasonably categorised as a level (2) method. The creep is given in the form of a creep coefficient, φ_t. Thus

$$\varepsilon_c = \frac{\sigma}{E_{c28}} \varphi_t \tag{17}$$

where E_{c28} is the reference modulus of elasticity at 28 days of age.

φ_t is given by the product of five partial coefficients, each of which is found from a chart:

$$\varphi_t = k_c k_d k_b k_e k_t \tag{18}$$

k_c depends on the humidity of storage.

The simple design method for predicting the elastic modulus and creep of normal weight concrete presented below is based on the CEB Recommendations but has been formulated in such a way that all of the required input data will be available at an early design stage. The method was developed in conjunction with The Movements Working Party of The Concrete Society. The method has been approved officially by The Concrete Society, and additionally it has been generally approved by a number of practising design engineers.

Information required

The following information is required:

Cube strength at time of loading (σ_t) and/or at 28 days (σ_{28}) in N/mm^2; age of concrete when loaded; exposure conditions; effective section thickness (Twice volume ÷ Exposed surface area).

Elastic modulus

The elastic modulus at the time of loading (E_t) is given by

$$E_t = E_{28}\left(0{\cdot}4 + 0{\cdot}6\frac{\sigma_1}{\sigma_{28}}\right)$$

The strength ratio (σ_t/σ_{28}) is obtained either from measurements or from Table 1.

TABLE 1

STRENGTH RATIOS FOR ORDINARY PORTLAND CEMENT AND RAPID HARDENING CEMENT CONCRETES AT DIFFERENT AGES.

	Age (days)	7	28	90	365
$\frac{\sigma_t}{\sigma_{28}}$	OPC	0·65	1·00	1·20	1·35
	RHC	0·75	1·00	1·15	1·20

The elastic modulus at an age of 28 days is given by

$$E_{28} = 20 + 0{\cdot}2\sigma_{28}(\text{KN/mm}^2).$$

The value 20 in this equation is appropriate to an average, high quality, dense aggregate. Lower values (down to 10 for some aggregates) will be appropriate for less rigid aggregates. Estimates of E_t will have a coefficient of variation of about 15% for the high quality dense aggregates commonly used in the UK.[26]

Total strain and creep

The final (30 year) load-induced strain

$$= \text{elastic strain} + \text{creep strain} = \frac{\text{stress}}{E_t} + \frac{\text{stress} \times \text{creep factor}}{E_t}$$

The creep factor can be determined from the figure. It will have a coefficient of variation of about 15%. The figure is based upon the CEB/FIP Recommendations.[7] About 80 %, 50 % and 30 % of the final creep occurs, after a six-month period under load, for effective thicknesses of <200 mm, 300 mm and >400 mm respectively.

If drying of the concrete is prevented (eg, by immersion in water or by sealing), creep will develop at a rate corresponding to an effective section thickness >400 mm. Creep is partly recoverable after a reduction in stress. The final creep recovery about one year after the reduction in stress

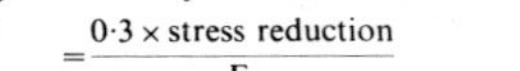

$$= \frac{0{\cdot}3 \times \text{stress reduction}}{E_t}$$

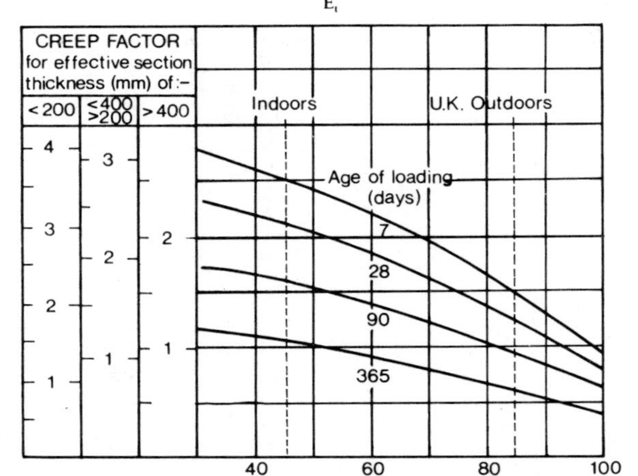

FIG. 11. A simplified method for estimating the elastic modulus and creep of normal weight concrete. After Parrott.[25] By permission of *Concrete*.

k_d gives the effect of maturity of the concrete on loading. Maturity is given in days for the normal ambient temperature of 20 °C, but a second scale is in term of maturity in °C–days, with an origin of −10 °C for temperature. The chart for k_d also provides the differentiation between normal and high early strength cements.

k_b represents the effect of the composition of the concrete and covers the two factors water–cement ratio and cement content. The effect of aggregate type is, as usual, included only in the form of a reference to lightweight concrete.

k_e depends on the size of member, which is, also as usual, expressed as an effective thickness.

k_t covers the development of creep with time under load. Different curves are included for various member sizes, but no account is taken of the additional effect of storage humidity.

The creep history under variable stress is found by the technique of virgin superposition.

In the later version of the recommendations,[17] the main section on creep is drastically revised to adopt the two-component representation of creep, together with a further correction to cover the underestimate of virgin creep discussed earlier.

The limiting recoverable creep is taken as 0·4x (the elastic strain), and its development with time is given in a single plotted curve. The irrecoverable creep (coefficient) is quoted for specific humidities of storage, as for shrinkage, (i.e. in water, 90, 70 and 40 % RH), and a multiplying factor is given graphically for the effect of size of member (again as an effective thickness). A further chart gives the development of irrecoverable creep with time for various sizes of member (with an additional correction for the effect of the humidity of storage). As for shrinkage, the mix proportions are considered through the medium of the consistency of the concrete. Similarly, corrections are made to the time scales in the development of both components, to account for the type of cement and any difference from 20 °C in the temperature curing and storage.

The principle of superposition is, as usual, invoked for computing the strain history under variable stress, after the manner already described earlier. Various possible simplifications are put forward to facilitate the solution of engineering problems; one such is the consideration of the recoverable creep as an instantaneous addition to the elastic strain.

2.4.7 Final details

The main stream aspects of creep have been covered already, but this tailpiece is added in order to recognise two further systems of stress which arise in special circumstances.

(*i*) *Multiaxial stress.* Creep has been measured under a variety of systems of biaxial and triaxial compressive stress, and the findings compared with measurements of lateral and longitudinal creep under uniaxial stress (i.e. transverse to and in the direction of the applied stress respectively). The concept which has received the most support in the collation of such experimental evidence is that of a creep Poisson's ratio. Thus, for uniaxial stress, the lateral creep, ε_{c2}, is given in terms of the longitudinal creep, ε_{c1}, through the creep Poisson's ratio, v_c.

That is,

$$\varepsilon_{c2} = -v_c \varepsilon_{c1}. \tag{19}$$

It is found that, for working stresses, v_c is only somewhat affected by the relative magnitudes of the three stresses, and it does not greatly vary with the age of loading or the time under load. Its value rarely exceeds 0·20, though, at the higher stress levels when internal microcracking contributes to the strain, the magnitudes increase rapidly with stress.

Unlike the elastic Poisson's ratio, v_c is not apparently a function of the mix proportions, nor is it greatly affected by the moisture content or the temperature. However, the evidence indicates that it is significantly lower when drying creep occurs. For basic creep v_c normally lies within the range 0·15–0·20 and it is acceptable to use a constant value regardless of the stress system or of the stress history. Thus, for principal stresses σ_1, σ_2 and σ_3 the creep in the typical direction 1 is given by:

$$\varepsilon_{c1} = [\sigma_1 - v_c(\sigma_2 + \sigma_3)]\frac{\varphi}{E_c} \tag{20}$$

(*ii*) *Cyclic stress.* Experimental evidence on creep under cyclic stress is relatively scarce, but a convincing picture for short periods under load can be obtained from Ref. 27. The most useful approach is to compare the creep under cyclic stress with the creep under a constant stress equal to the mean of the extremes of the cyclic range of stress. Then, it is evident that the cyclic creep is greater than that under constant stress, and the increase is proportional to the range of stress. It is also greater for greater mean stresses. Since the results are based on short times of load application, it may be concluded that the cyclic stress accelerates the creep process, but that it does not necessarily increase the eventual magnitude of creep.

REFERENCES

1. NEVILLE, A. M. (1973). *Properties of Concrete*, 2nd edition. Pitman Publishing, London.
2. NEVILLE, A. M. (1970) (Chapters 17–20 written in collaboration with Dilger, W.). *Creep of Concrete: Plain, Reinforced and Prestressed*. North-Holland Publishing Co., Amsterdam.
3. LEA, F. M. (1970). *The Chemistry of Cement and Concrete*, 3rd edition. Edward Arnold (Publishers) Ltd, London.
4. BLAINE, R. L. (1968). A Statistical Study of the Shrinkage of Neat Cements and Concretes, Paper 1-M. *Shrinkage of Hydraulic Concrete. Proceedings of International Symposium*, RILEM/Cembureau, Madrid.
5. CONCRETE SOCIETY (1974). *The Creep of Structural Concrete*. Technical Paper No. 101.
6. BRITISH STANDARDS INSTITUTION (1970). *Methods of Testing Concrete*. BS 188. Part 5. B.S.I., London.
7. CEB/FIP (1970). *International Recommendations for the Design and Construction of Concrete Structures*. London
8. L'HERMITE, R., CHEFDEVILLE, J., and GRIEU, J. J. (1949). Annales de l'Institut Technique du Bâtiment et de Travaux Publics, No. 106. *Liants Hydrauliques* No. 5.
9. PIHLAJAVAARA, S. E. (1969). *An Approximate Solution of a Quasi-Linear Diffusion Problem*. Report of the State Institute for Technical Research, Helsinki, Finland.
10. TROXELL, G. E., RAPHAEL, J. M. and DAVIS, R. E. (1958). *Proc. ASTM*, **58,** 1101.
11. HELMUTH, R. A. and TURK, D. H. (1967). *J. Res. Develop. Lab., Portland Cement Assoc.*, **9**.
12. HOBBS, D. W. (1971). *Matériaux et Constructions*, **4,** 107.
13. BUILDING RESEARCH STATION (1963). *Shrinkage of Natural Aggregates in Concrete*, Digest No. 35. HMSO, London.
14. HOBBS, D. W. (1974). *Proc. ACI*, **71,** 445.
15. HOBBS, D. W. and PARROTT, L. J. (1979). *Concrete*, **13,** 19.
16. BRITISH STANDARDS INSTITUTION. (1972). *The Structural Use of Concrete*, CP 110. B.S.I., London.
17. CEB/FIP (1978). *Model Code for Concrete Structures*. London.
18. RUSCH, H. (1960–1). *Proc. ACI*, **57,** 1.
19. NASSER, K. W. and NEVILLE, A. M. (1967). *Proc. ACI*, **64,** 97.

20. Gamble, B. R. and Parrott, L. J. (1978). *Mag. Concrete Res.*, **30,** 129.
21. Illston, J. M. and Sanders, P. D. (1974). *Mag. Concrete Res.*, **25,** 169.
22. Bazant, Z. P. and Osman, E. (1976). *Matériaux et Constructions*, **6,** 149.
23. Illston, J. M. (1965). *Mag. Concrete Res.*, **17,** 21.
24. Constantinescu, D. R. and Illston, J. M. (1974). *Matériaux et Constructions*, **7,** 395.
25. Parrott, L. J. (1978). *Concrete*, **12,** 33.
26. Teychenné, D. C., Parrott, L. J. and Pomeroy, C. D. (1978). Estimation of the Elastic Modulus of Concrete for the Design of Structures. *BRE Current Paper*, *CP*23.
27. Whaley, C. P. and Neville, A. M. (1973). *Mag. Concrete Res.*, **25,** 145.

Chapter 3

BEHAVIOUR OF CONCRETE UNDER FATIGUE LOADING

K. D. Raithby

Principal Scientific Officer, Transport and Road Research Laboratory, Department of the Environment, Crowthorne, UK

SUMMARY

The fatigue behaviour of plain concrete is reviewed in the context of limit state design principles for concrete structures, where there is a requirement to ensure freedom from serious cracking due to repeated loading.

Recent work suggests that under constant amplitude loading the effects on fatigue performance of a wide range of factors, such as mix design, moisture condition and age, can be accounted for by corresponding variations in static strength. An indication is given of how design curves for fatigue can be derived, from which an assessment may be made of the likelihood of fatigue cracks occurring during the useful life of a structure.

3.1 INTRODUCTION

In recent years there has been an increasing trend towards the use of more rational design methods for civil engineering structures, in which the probabilities of occurrence of various loading and environmental conditions are taken into account, and in which structural strength and performance are related more directly to the properties of the constituent materials. The limit state design approach, which covers serviceability requirements as well as ultimate collapse conditions, is a logical development of this trend. In the UK, limit state design has been used for concrete buildings since 1972[1a] and more recently has been applied to design standards for road and railway bridges of all types.[1b]

With the introduction of limit state principles, fatigue has become increasingly important as a design condition for those types of structure which have to withstand significantly large numbers of live loads during their service life. This is particularly true for road and airfield pavements, bridges and offshore structures, all of which are subjected to large numbers of external loads of varying magnitude during their operational lifetimes. Fatigue is also relevant to the design of earthquake-resistant structures where, although the numbers of dynamic load fluctuations may be relatively small, the stresses induced in the structure may be quite high.

In general, the implication of possible fatigue failure of a structure may range from catastrophe to mere nuisance. Between these two extremes lies a large region of unserviceability, where the consequences of local fatigue failure may result in the need for expensive remedial action, even though there is no immediate danger to human life. Failure of concrete by fatigue comes almost entirely into this category. With unreinforced concrete pavements, for example, fatigue cracking can lead to local spalling of the surface and deterioration of the subgrade by penetration of rainwater. In the case of reinforced concrete bridge decks, cracking may lead not only to surface deterioration but also to corrosion of the reinforcement, with consequent loss of strength; a problem which has been particularly severe in the USA. Apart from the risk of corrosion, the fatigue performance of reinforcing bars themselves is likely to be affected by the extent to which the surrounding concrete may crack under service loading. For any concrete structure that is subject to cyclic loading it is important, therefore, for the designer to consider the likelihood and possible consequences of fatigue failure of the concrete.

The fact that concrete can suffer from fatigue under repeated stress was recognised as long ago as 1903, when van Ornum[2] produced the first fatigue curve for concrete cubes repeatedly loaded in compression at a frequency of four cycles per minute. Van Ornum also drew attention to the fact that repeated stressing changed the shape of the stress–strain curve, with a progressive decrease in stiffness under cyclic loading. Investigations into the flexural fatigue properties of concrete appear to have begun in the 1920s, particularly in Germany and in the

USA,[3] when fatigue started to be taken into account in the design of concrete highway pavement slabs. Since that time, a good deal of research effort has been expended throughout the world in studying the performance of various types of concrete, much of the effort in the post-war period being centred at the University of Illinois.[4]

It is intended in this chapter to review some of the more recent work in the UK on fatigue of plain concrete, with particular reference to time-dependent effects. Relevant research has taken place at the Transport and Road Research Laboratory (TRRL), the Building Research Establishment, the University of Leeds and elsewhere. As pointed out in an earlier review by the author in 1968,[5] although a great deal of research has been carried out on the behaviour of concrete under fatigue loading, over a period of more than 70 years, it has been difficult to draw specific conclusions. This is because of the variety of testing procedures used and, in many cases, the lack of detailed information on the form of the test specimen, the curing conditions, the age of the concrete at the time of testing and precise details of the method of loading.

In a properly coordinated test programme designed to investigate several different variables it is desirable to minimise the variability of the test specimens by closely controlling the mixing, compaction and curing of the concrete used. In the TRRL programme, for example, all the aggregate was graded and dried before mixing, and the cement for the whole programme (totalling several hundred small beams and associated test cubes and cylinders) was taken from a single production batch. The preparation of the test specimens and all the testing were performed by the same two operators. As a result of careful control it was possible to achieve an overall coefficient of variation of indirect tensile strength of 5·4 % over a total of 608 control cylinders, tested in batches of 4 over a period of about 2 years. The aim of the specimen production programme was to provide a stock of test beams which would be at the right age for testing when they were required, and from which a random selection of individual beams could be made. Batches of between 6 and 10 nominally identical specimens were then selected for replicate testing under predetermined conditions over a total period of about 3 years, with a more limited programme extending to 5 years.

3.2 PRESENTATION OF FATIGUE TEST DATA

Fatigue performance is usually expressed in terms of the familiar endurance curve (sometimes known as the S–N curve), in which the mean value of the number of cycles to failure under a particular condition of loading is plotted against the magnitude of the cyclic load applied or against a derived stress level. For practical reasons, particularly for very long endurances, it is sometimes necessary to terminate a test before fatigue failure has occurred. Such tests are usually termed 'runouts'.

In plotting endurance curves for concrete the cyclic loading may be expressed as a stress amplitude or more usually as a maximum stress, frequently given as a proportion of the stress at failure under static loading. Typical endurance curves are illustrated in Figs. 1, 2 and 3, for compression, flexure and direct tension respectively. The 'true' curves of Figs. 2 and 3 are explained later. Cycles to failure, N, are usually plotted on a logarithmic scale. Loads or stresses, S, are sometimes plotted on a linear scale, sometimes on a logarithmic scale.

Although the range of stress fluctuation is the parameter which has the biggest effect on the fatigue performance of any material, the absolute values of maximum and minimum stress in a load cycle may also be important. In the case of concrete the fatigue performance

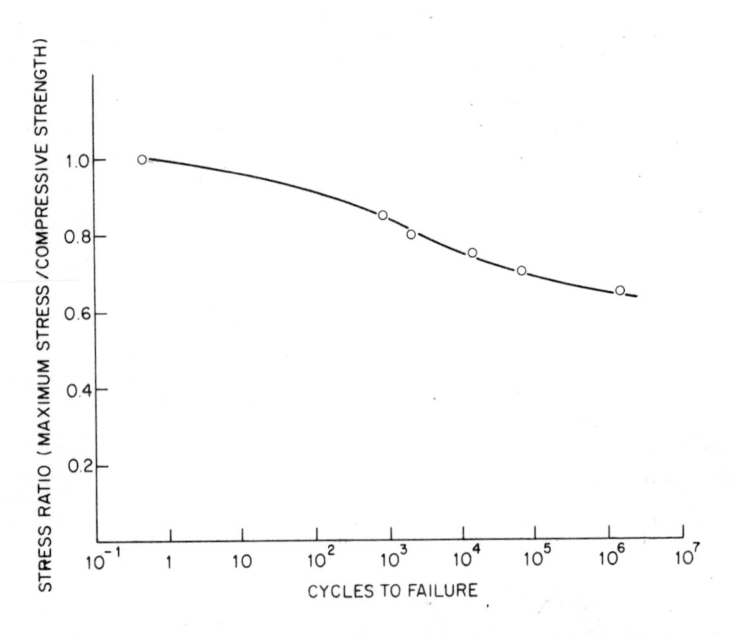

FIG. 1. Endurance curve for prisms in compression at 3·2 Hz. Each point is the log mean of 6 tests. After Bennett.[30]

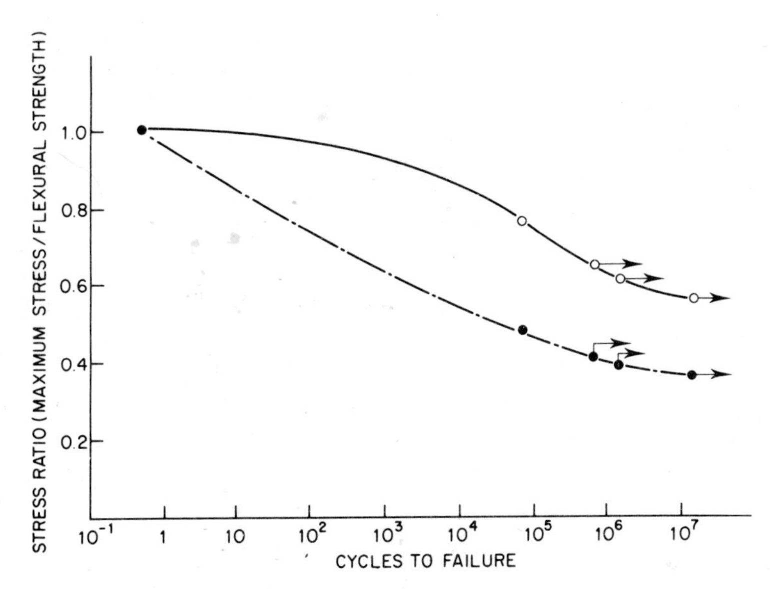

FIG. 2. Typical endurance curves for flexural loading at 20 Hz. Each point is the log mean of 5 or more tests. →, includes some runouts; ——, conventional endurance curve; — - —, true endurance curve. After Galloway and Raithby.[7]

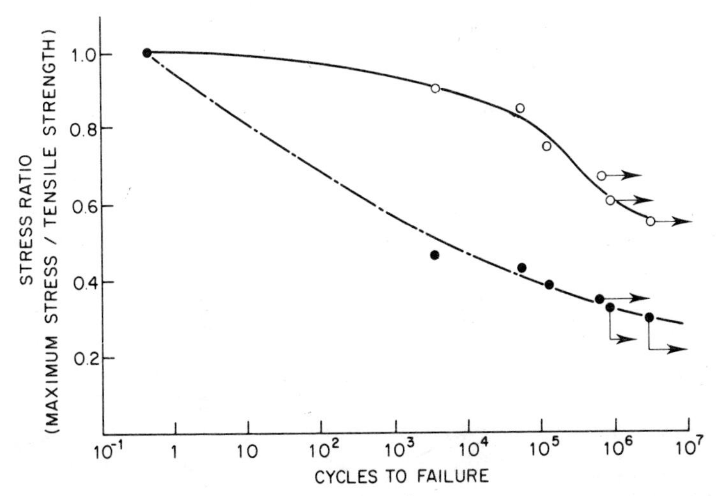

FIG. 3. Endurance curves for direct tension on lean concrete at 15 Hz. Each point is the log mean of 3–6 tests. →, includes some runouts; ——, conventional endurance curve; — - —, true endurance curve. After Kolias and Williams.[10]

under constant-amplitude cyclic loading is usually dependent on both the stress amplitude and, to a lesser extent, on the mean stress level. There are various ways in which these effects may be expressed. One is to show a series of endurance curves for different mean stress levels; another is to plot endurance curves for various values of the ratio (R) of minimum to maximum stress. In this form the following special cases may be noted:

$R=0$ represents repeated loading, from zero to some peak stress value; here the mean stress is equal to half the stress range.

$R=-1$ represents alternating load, where the minimum and maximum stresses are of equal magnitude but of opposite sign; here the mean stress is zero.

$R=+1$ represents a constant applied load with no stress fluctuation, since the minimum stress is equal to the maximum stress; this could be a creep condition where the component is subjected to sustained loading.

In practice, fatigue conditions for the majority of concrete structures will lie between $R=0$ and $R=+1$, when live loads are superimposed on dead loading arising from self weight and superimposed dead load. In a few cases, such as with lateral wave loading on offshore oil structures or seismic loading from earthquakes, alternating loading with approximately zero mean stress ($R=-1$) may be appropriate.

Very little information has been published on the effects of mean stress level on fatigue performance of concrete. Most research workers have done fatigue tests under loading conditions that approximate to repeated loading ($R=0$) in which the minimum stress in the loading cycle is close to zero. (It is rarely precisely zero because of practical difficulties in controlling fatigue test loading equipment at very small loads.)

In one of the few papers reporting the effects of mean stress on fatigue performance of concrete, Murdock and Kesler[6] show a series of endurance curves for various stress ratios and conclude that the results can be expressed in the form of a modified Goodman diagram from which the fatigue strength can be deduced for various values of mean stress. Figures 4 and 5 show this type of presentation for the results of fatigue tests on concrete in compression and bending respectively. These diagrams show combinations of stress range and maximum (or minimum) stress which will produce mean fatigue lives of particular values (2 million cycles in the case of Fig. 4; 10 million in the case of

Fig. 5). Such curves cannot be plotted directly from the fatigue test results. They are derived by cross plotting from a family of endurance curves representing different mean stresses or stress ratios.

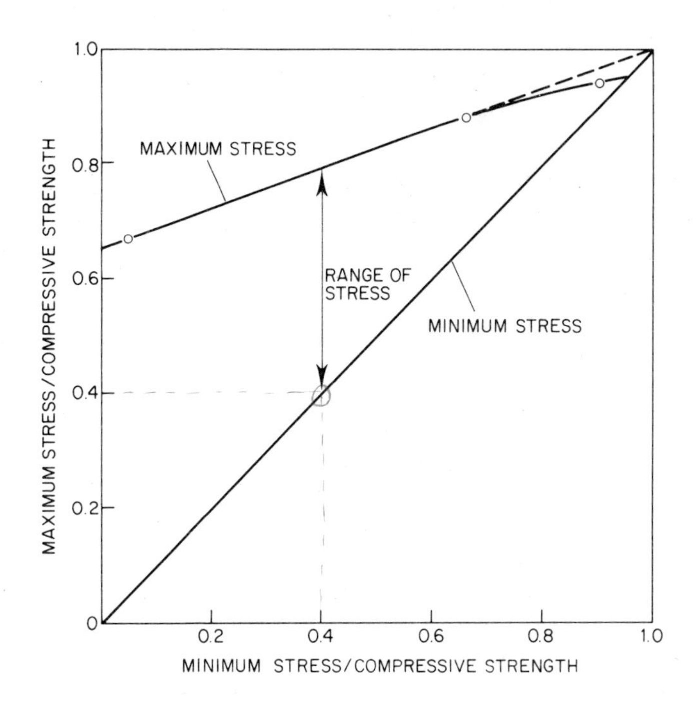

FIG. 4. Fatigue strength at 2×10^6 cycles of prisms in compression. Mean static strength, 17·6 N/mm^2.

The results of fatigue tests on a number of nominally identical test specimens when tested under the same loading conditions show considerable scatter. With concrete it is not unusual for the longest life to be at least a hundred times the shortest in a small batch of results. There is some experimental evidence to show that the scatter of individual results at any particular condition of loading conforms to a normal (or 'Gaussian') distribution of the logarithm of the number of cycles to failure. This is usually referred to as a log normal distribution. The results of groups of tests can then be expressed in terms of a geometric or log mean endurance (which is the antilogarithm of the mean value of the logarithms of the individual results) and a standard deviation of log endurance, i.e.

$$\text{geometric mean endurance, } \bar{N} = \text{antilog}\left(\frac{1}{n}\sum \log N\right)$$

$$\text{standard deviation, } s_{\log N} = \sqrt{\frac{\sum(\log N - \overline{\log N})^2}{(n-1)}}$$

where N represents the number of cycles to failure in a particular test, n is the number of tests, and $\log N = 1/n \sum \overline{\log N}$ is the mean value of $\log N$.

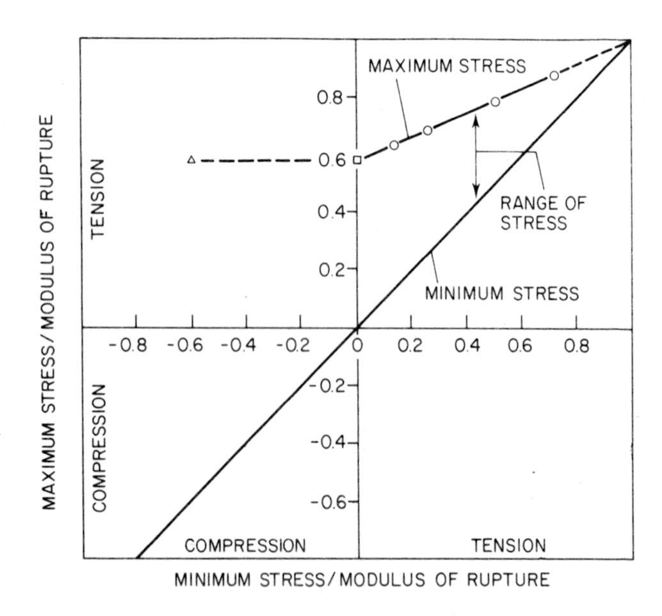

FIG. 5. Fatigue strength at 10^7 cycles of simply supported beams. Crushing strength of concrete 31 N/mm^2.

The scatter may also be expressed in terms of a coefficient of variation, which is the ratio $s_{\log N}/\overline{\log N}$, usually given as a percentage.

Typical values of coefficient of variation of log endurance for plain concrete in flexure lie between 10 and 20 %. This is roughly equivalent to a coefficient of variation of fatigue strength of 7–15 %, where fatigue strength is defined as the cyclic stress level required to give a particular value of mean life to failure (e.g. 10^4 or 10^6 cycles). Comparable values of coefficient of variation of static flexural strength are likely to lie between 5 and 10 %.

From a knowledge of the standard deviation of log N it is possible to derive additional endurance curves representing various probabilities of survival and these can be used as a basis for design.

Various forms of fatigue test have been used to investigate the effects of repetitive loading on concrete—for example, direct compression, direct tension, indirect tension (splitting) and flexure. Fatigue failures can be produced in all these types of loading and the appropriate endurance curves can be drawn. Figures 1–3 are typical curves for compression, flexure and direct tension respectively. Some explanation of the two curves in Figs. 2 and 3 is necessary. Conventional fatigue endurance curves of the type shown are incorrect in one important respect. The cyclic stresses are expressed as a proportion of the static failure stress, but the rate of loading applied in conventional static tests is very much slower than the rate of loading during fatigue tests. The difference may in fact be several orders of magnitude. It is well known that the ultimate strength of concrete is highly dependent on the rate of loading.[7,8] At the equivalent of a test frequency of 20 Hz, for example, with the ultimate load being reached in 0·025 s, the strength may be some 50 % greater than under the standard static test conditions defined in BS 1881[9] in which failure occurs in about 2 min. If the fatigue results are expressed in terms of true strength values achieved at rates of loading comparable to those applied during the fatigue tests, a rather different form of endurance curve results. Few examples of 'true' fatigue curves for concrete are to be found in the literature. Two such curves are shown, in Fig. 2 for flexural loading on a pavement quality concrete[7] and in Fig. 3 for direct tension loading on lean concrete.[10] However, in view of the fact that conventional slow-speed strength tests conforming to a British Standard (BS 1881) are well known and easily reproducible, there is some advantage in continuing to relate fatigue performance figures to the slow-speed strength values. This procedure has been adopted for the rest of this chapter.

'Conventional' endurance curves, such as those shown in Figs. 1–3, show marked similarities in form for a variety of types of loading if the loading is expressed in terms of the mean nominal static strength under the same conditions. This is not altogether surprising since, under any type of loading, concrete tends to fail basically by cracking in directions perpendicular to the maximum principal tensile strain. Thus, in the case of direct compressive loading, tensile strains are set up at right angles to the direction of loading, which will eventually lead to failure by cracking parallel to the direction of loading. Somewhat similar conditions arise in the indirect tensile (or cylinder-splitting) test whilst under direct tension loading, failure occurs by cracking in a direction perpendicular to the applied load.

3.3 TIME-DEPENDENT AND STRESS-HISTORY EFFECTS ON FATIGUE PERFORMANCE

In practice, service loading of a structure, such as a road pavement or bridge, is not applied in a regular cyclic pattern over a period of a few hours or days as in the conventional constant amplitude fatigue tests. Live loads vary in intensity and occur intermittently, with rest periods between individual load pulses over the whole life of the structure. (This may be as long as 120 years in the case of a highway bridge.) The rate of loading varies with the speed of the traffic; ageing of the material may occur and the structure itself will be subject to wetting and drying and to variations in temperature. These are all factors which may affect the behaviour of concrete in the structure. Recent research has provided test data which will help to determine some of these effects so that the long-term fatigue behaviour of concrete in real structures may be assessed more realistically.

3.4. VARIABLE AMPLITUDE LOADING AND CUMULATIVE FATIGUE DAMAGE

There is remarkably little published information on the behaviour of concrete under variable amplitude loading in spite of the fact that service loading histories on any concrete structure liable to fatigue are certainly not represented by constant amplitude loading at a constant frequency. Service loads are more likely to be intermittent in nature, with randomly varying amplitudes and often with randomly varying time intervals between loading pulses.

As far as the author knows no random loading fatigue tests have been performed on concrete and the only variable amplitude loading tests reported have been for block programme loading with only two stress levels.[11,12]

The wide scatter in results makes interpretation difficult, but these tests suggest that for design and assessment purposes the Palmgren–Miner linear cumulative damage hypothesis,[13] works fairly well for concrete. It is doubtful whether any more sophisticated damage law is justified in the present state of knowledge. The Miner hypothesis relates the damage caused by n cycles of loading of a particular amplitude to the endurance under that loading as

$$d_f = \frac{n_f}{N_f}$$

where

> d_f = fatigue damage resulting from applied loading at stress level f, n_f = number of cycles applied at stress level f and N_f = mean number of cycles to failure at stress level f.

The accumulated fatigue damage caused by cycling at several different stress levels is therefore

$$D = \sum_{f=1}^{j} \frac{n_f}{N_f}$$

where j = number of stress levels. Failure occurs when $D = 1$.

Although on the face of it there would be attractions in using a random loading sequence for fatigue tests relating to the performance of concrete structures the interpretation of such tests presents great difficulty in view of the time-dependent nature of the properties of concrete. Thus in a real structure under a given loading, the stress distribution may change with time as the modulus of elasticity varies with age and with moisture movements, while the strength of the concrete is also varying. The only true form of representative loading would therefore be to reproduce service loading histories in real time and with realistic moisture movements and temperature changes. This would be quite impracticable and by definition would not constitute an accelerated test procedure. In view of these difficulties it is probably better to stick to the simple constant amplitude type of loading, covering a sufficiently wide range of mean and alternating load conditions, and to use a simple cumulative fatigue damage relationship to estimate the probable fatigue behaviour of the concrete during the lifetime of the structure being considered.

3.5 RATE OF LOADING AND REST PERIODS

Although it has been known for a long time that the compressive and flexural strength of concrete increases with an increase in rate of straining, the effects on fatigue performance were thought to be less marked.[5] Recent work at the Building Research Establishment,[8] however, indicates a substantial effect of rate of loading on the fatigue performance of concrete prisms in compression, if a sufficiently wide range of loading rates is considered. It was found that a reduction in loading rate by a factor of 100, from 50 N/mm^{2}s to 0·5 N/mm^{2}s, resulted in a tenfold reduction in the number of cycles to failure. Similar conclusions have been reported by Awad and Hilsdorf[14] for 'low cycle' (i.e. high stress) fatigue tests under compressive loading. These results suggest that conventional fatigue tests may overestimate the endurance of concrete under slow cyclic loading, particularly for high cyclic stresses.

In flexure tests on small beams at the Transport and Road Research Laboratory[7] only small and statistically insignificant differences in fatigue performance were found, when the frequency of loading was varied between 4 and 20 Hz, even though the flexural strength at equivalent rates of loading might be expected to differ by about 15 %. These results confirmed an earlier conclusion of Kesler and his coworkers,[4] based on tests at 1·2 and 7·3 Hz, that the loading frequency has no appreciable effect on fatigue life. Kesler has pointed out that under flexural loading, as long as the applied stresses lie within the linear portion of the stress–strain curve (usually up to 40–60 % of the ultimate failure stress), variations in rate of loading have little effect on strain response but once microcracks start to spread, the strain response is affected markedly by the rate of loading. At the load levels applied in the TRRL tests, which were roughly 35–50 % of the mean ultimate strength at the appropriate rate of loading (see the 'true' endurance curve of Fig. 2), strains would tend to remain fairly elastic until near the end of the test. The rate of loading might, then, be expected to influence only the later stages of the test and the overall life to failure might not be greatly affected within the range 10^5–10^7 cycles. It is worth noting that the rate of loading effects observed in Refs 8 and

14 for concrete prisms in compression apply strictly to endurances of less than 10^5 cycles; at higher endurances[8] the effects tend to disappear.

At very high rates of loading results may become affected by temperatures induced by internal friction. At 150 Hz, for example, internal temperatures as high as 38 °C have been measured.[15]

For most laboratory fatigue tests, test frequencies of up to 20 Hz are likely to be acceptable but care must be taken in applying such results to the performance of structural components where the rate of loading in service is much lower. The laboratory results could seriously overestimate the life under service conditions.

The introduction of rest periods between loading cycles seems unlikely to have a very significant effect on fatigue performance of concrete. The results of a few tests carried out on a limestone concrete[16] are given in Table 1. These show an apparent reduction in

TABLE 1

EFFECTS OF REST PERIODS IN FLEXURAL FATIGUE TESTS[16]

Loading condition	*Number of tests*	*Mean endurance cycles*	*Standard deviation of log endurance*
Continuous cycling at 20 Hz	10	8590	0·99
0·5 s rest after each cycle	10	3380	0·60
2·0 s rest after each cycle	10	2550	0·56

fatigue life when rests of up to 2 s were introduced between each successive loading cycle, but the differences are not statistically significant. All beams were tested wet at a nominal maximum cyclic stress of 4·59 N/mm^2 after 13 weeks curing under water.

Earlier tests at the University of Illinois[11] had indicated that periods of rest of up to 5 min after each block of 4500 loading cycles prolonged the fatigue life under flexural loading, particularly at the lower stress levels. Rest periods longer than 5 min, up to 27 min, had no further effect. The frequency of the rest periods appeared to be more important than their duration.

3.6. AGE AT TIME OF TESTING

Few research workers appear to have looked closely at the effects of age on fatigue performance of concrete, although it is well known that compressive and flexural strengths increase markedly with age. For long-life structures subject to fatigue loading it is important to know how the fatigue resistance of concrete is likely to change. Tests have recently been made at TRRL to determine the flexural fatigue performance of three types of concrete, having widely differing flexural strengths, after curing under water for periods of up to 5 y.[17] Results for one concrete (made with a flint gravel aggregate) are given in Fig. 6, which shows mean endurance plotted against age at the time of testing for two cyclic loading conditions, giving peak stresses of 3·2 and 3·53 N/mm^2 respectively.

For a given stress level there is a marked and rapid increase in cycles to failure as the age of the concrete increases. At the same time there is

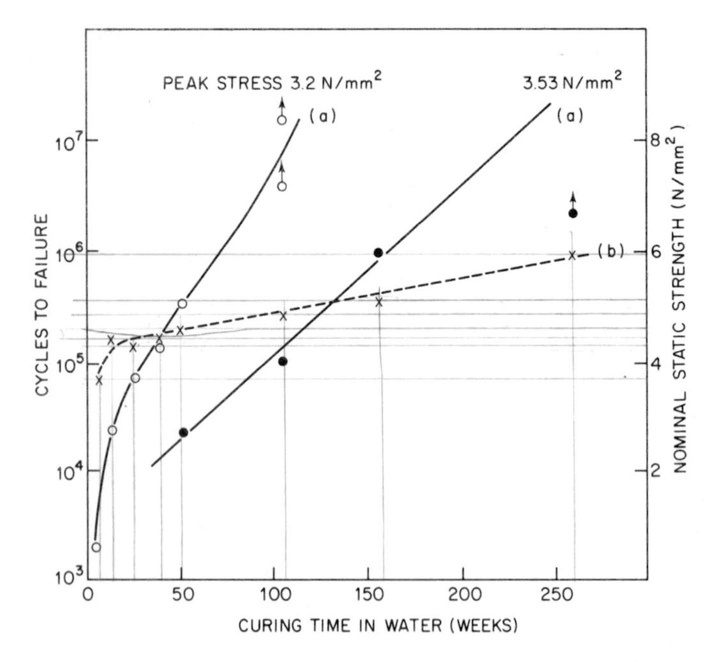

FIG. 6. Effect of age on flexural static strength and fatigue endurance. Each point is the mean of 10 tests. →, includes some runouts; (a) fatigue tests; (b) static strength tests. After Galloway *et al.*[17]

a corresponding increase in the static strength, so that the applied fatigue loading represents a diminishing proportion of the static strength at the time of testing. Qualitatively similar results have been obtained with a much weaker lean concrete and with a stronger structural concrete containing limestone aggregate. If the results of all these tests are expressed in terms of the flexural strength at the appropriate ages, they tend to lie close to a single endurance curve. Figure 7 shows such a curve which covers a total of about 200 test results on the 3 different concretes at ages ranging from 4 weeks to 5 years.

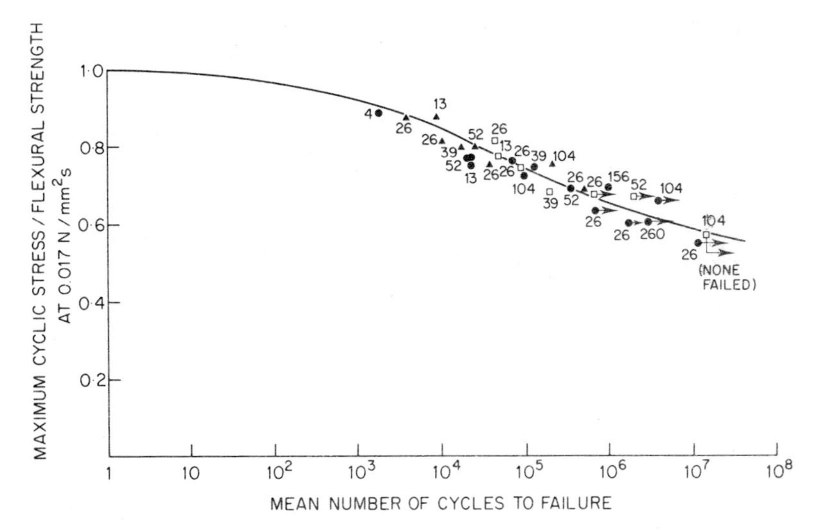

FIG. 7. Flexural fatigue performance at 20 Hz for three concretes at various ages: all cured under water; ●, concrete PQ1 (flint gravel), 4 weeks to 5 years; ▲, concrete PQ2 (limestone), 13 weeks to 2 years; □, lean concrete LC1 (flint gravel), 13 weeks to 2 years; →, includes some runouts. Numbers beside points denote age in weeks. After Galloway *et al.*[17]

3.7 MOISTURE CONDITION AND METHOD OF CURING

The achieved strength of concrete depends on the moisture balance in the cement matrix. If there are no significant moisture gradients or external restraints against shrinkage the strength tends to rise as moisture is lost from the bulk of the concrete. In practice, drying from free surfaces inevitably leads to variations in moisture intensity through the concrete, resulting in differential shrinkage which may induce local stresses high enough to cause microcracking at, or near, the surface. If such conditions occur, then it may be expected that the static strength and fatigue performance will be reduced. The effect is illustrated in Fig. 8, which gives fatigue endurance curves derived from flexural tests on small concrete beams which had all been cured under water for 26 weeks.[18]

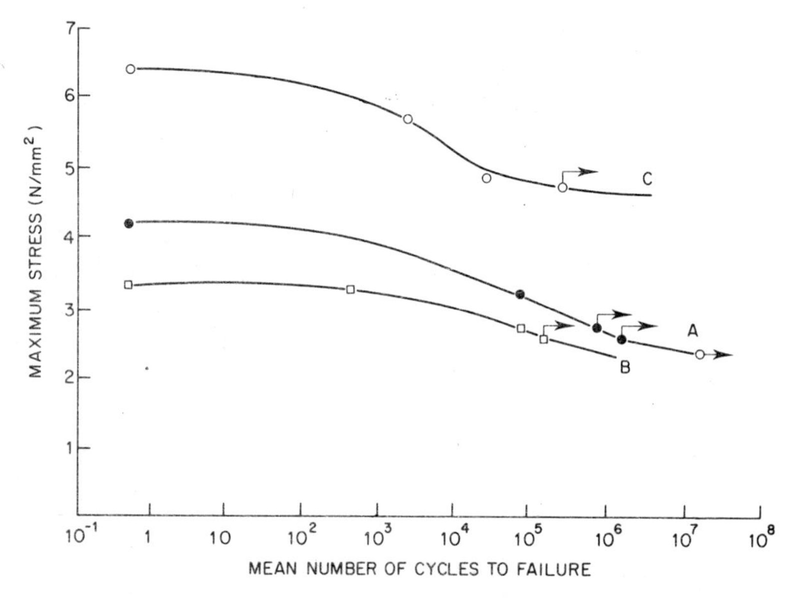

FIG. 8. Effect of moisture condition on flexural fatigue performance. All specimens cured under water for 26 weeks before being tested. →, includes some runouts. A, saturated; B, surface-dry; C, oven-dried. After Galloway *et al.*[18]

In test series A the beams were kept wet throughout the period of testing by first sealing them in polyethylene bags containing some free water. For series B the beams were left to dry in the laboratory atmosphere for a further week before being tested. For series C the beams were dried in an oven for a week at a temperature of 105 °C before being tested. There are evident differences in performance due to moisture variation. Because of this, and the difficulty of defining the conditions in a partially-dried test specimen, most of the basic fatigue tests carried out in the TRRL research programme on plain concrete have been performed on concrete beams cured under water and kept wet throughout the test period.

If moisture condition at the time of test is so important then it might be expected that static and fatigue strength would also depend on the method of curing. Tests have been made[18] on specimens which were all aged for 26 weeks at a temperature of 20 °C before being tested, but the methods of curing used varied from air-drying to immersion in water, with various intermediate combinations of air and water curing and with various methods of sealing the free surfaces of the concrete. Both modulus of rupture and fatigue performance showed considerable variation according to the method of curing that had been used. In general, the best results were obtained from specimens that had been cured for between 1 and 4 weeks in water, with the remaining period of cure being in air. The worst results were with complete air curing, which resulted in modulus of rupture and fatigue strength values some 25 % lower. Sealing the free surfaces of the concrete before curing in air gave intermediate values of strength and fatigue performance. If the cyclic loads applied in the groups of tests applicable to the various methods of curing are expressed in terms of the mean static flexural strength of beams subjected to the same conditions of curing, then the fatigue results fall quite close to the curve shown in Fig. 7.

3.8 EFFECTS OF PREVIOUS STRESS HISTORY

If concrete is subjected to a period of cyclic loading of an intensity less than the fatigue strength corresponding to 10^7 cycles and then loaded to failure under static conditions, the residual strength after cycling may be greater than that of similar specimens which have not been cycled. This phenomenon of 'understressing' has been observed both in compression[19] and in flexural tests,[18,20] and may result in increases in strength of as much as 15 %. The effect is in some ways analogous to strain hardening in metals, and is probably due to the blunting of microcracks in the cement matrix due to local plastic strain, thus leading to a slightly increased strength resulting from a reduction in effective stress concentration. The effects observed are too small to have any practical significance in design or in the assessment of structural performance.

3.9 STRESS–STRAIN RELATIONSHIP IN FATIGUE

Stress–strain curves for concrete loaded steadily to failure usually have the form shown in Fig. 9, which also shows the influence of drying on stress–strain relationships in flexure, the drier beams being less stiff

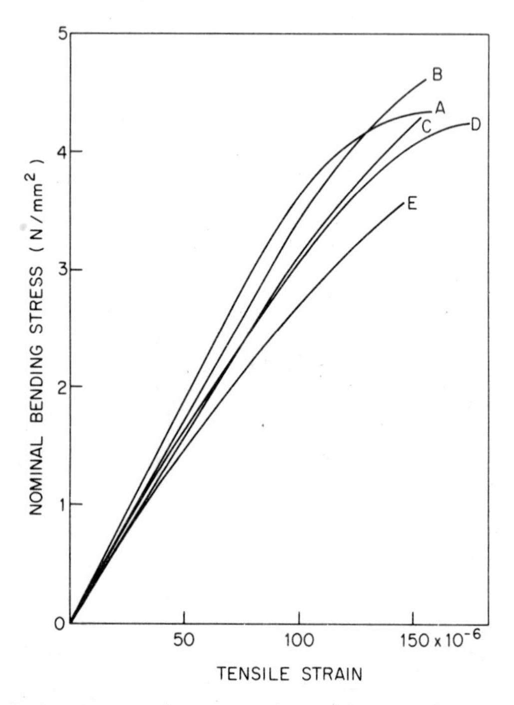

FIG. 9. Typical stress–strain curves for various curing conditions. 'Static' flexural strength tests at 0·017 N/mm^{2}s. Time in water and air (weeks): 26 and 0 (curve A), 13 and 13 (curve B), 4 and 22 (curve C), 1 and 25 (curve D), 0 and 26 (curve E) respectively. After Galloway *et al.*[18]

than the wetter ones. The beams were all 6 months old at the time of test. One had been cured entirely in air; the others had been removed from the curing tank after 1, 4, 13 and 26 weeks, the remainder of the time being spent in air at a relative humidity of 65 % and a temperature of 20 °C.

After an initial linear portion, the typical stress–strain curve becomes increasingly non-linear until, finally, failure takes place when certain cracks reach critical dimensions. The onset of non-linearity is due to the formation of microcracks within the cement matrix in regions of high local stress. Some of these develop and join up to form a network of macrocracks, which may be visible under a low-powered microscope. Initially these cracks are stable, i.e. they do not extend catastrophically with a small increase in load. With further increase in load one or more of these macrocracks will eventually extend rapidly, leading to unstable fracture when the rate of release of strain energy becomes sufficient to create new fracture surfaces. The cracks then become self-propagating. This tends to occur when the tensile strain reaches about 150 microstrain. The effect of the formation of stable cracks is to relieve local stress concentrations and to give the concrete a pseudo-ductility. In this mechanism the aggregate plays an important part in providing crack barriers, which limit the extent to which small cracks can grow. A similar sort of mechanism of failure is thought to take place under fatigue loading, in which slow extension of zones of macrocracking takes place as a result of repeated stressing until one or more cracks reach a critical dimension and failure ensues.

Several investigators have remarked on changes that have taken place in the shape of the stress–strain curve as a result of compressive fatigue loading. The early work of van Ornum[2] indicated that under repeated compression, the curve changed from initially convex to a linear relationship and then to a concave or even an S-shaped curve as the number of loading cycles was increased. This sort of behaviour has been confirmed by other research workers. Under constant amplitude flexural loading, changes in shape are less marked[20] but there is a distinct change in slope of the dynamic stress–strain curve,[16] indicating a reduction in effective modulus of elasticity as fatigue progresses. This is shown in Fig. 10 for a concrete containing limestone aggregate and tested after 13 weeks and after 52 weeks immersion in water. The stress–strain curves have been derived from continuous trace records of load and strain during fatigue tests at 20 Hz. The older concrete is slightly stiffer initially, but by the time fatigue failure is imminent both test beams have nearly the same stiffness. The strain at failure is roughly the same in each case.

If the applied loading is controlled so as to maintain a constant stress amplitude throughout the test, the resulting peak strain tends to increase as the number of load cycles applied is increased. This increase in strain is made up of two parts, one an increase in the strain amplitude ('elastic' strain) the other a residual non-recoverable strain resulting from cyclic creep. Typical changes in the elastic strain range are shown in Fig. 11 for specimens tested in flexure after different conditions of curing. The tests were all performed under the same nominal cyclic stress but the fatigue lives varied from about 5000 to nearly 4 million cycles. (This was a reflection of the fact that the 'static' flexural strength varied considerably with the different curing conditions.) Although the initial strain range is much the same for each specimen the rate of change of strain with increase in loading cycles

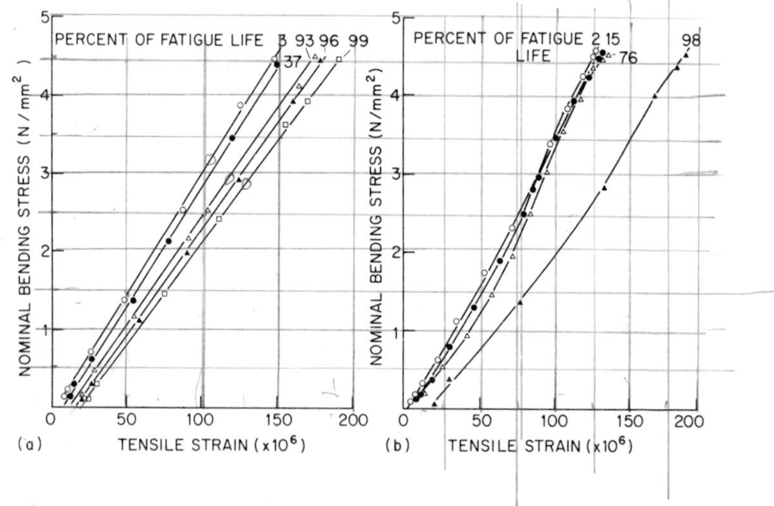

FIG. 10. Dynamic stress–strain curves from flexural fatigue tests at 20 Hz. After 13 weeks (a) and 52 weeks (b) immersion in water. After Raithby and Galloway.[16]

applied varies considerably. Neither these nor measurements taken during other parts of the TRRL's research programme showed any consistent pattern of change, other than a general tendency for strains to increase more rapidly during the later stages of fatigue testing. It was not possible to determine the precise value of strain at failure, but with few exceptions strains reached 150–200 microstrain shortly before fatigue fracture occurred.

These results show that, as a general rule, fatigue failure in flexure tends to occur very soon after the elastic tensile strain reaches 150–200 microstrain, irrespective of the intensity of the applied loading. This compares with a figure of about 150 microstrain usually achieved under static loading. The similarity of tensile strains at failure under static and fatigue loading suggests a similar mechanism of failure, in which some critical level of cracking has to be reached before fracture can occur.

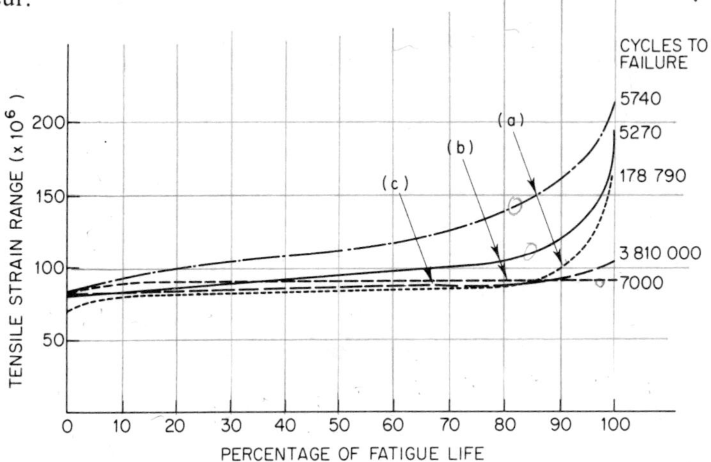

FIG. 11. Variation of elastic strain with load cycles applied during flexural fatigue tests for five individual test specimens. (a) 26 weeks fog cured; (b) 26 weeks in polyethylene cover; (c) 13 weeks in water, 13 weeks in air. After Galloway *et al.*[18]

3.10 FIBRE-REINFORCED CONCRETE

If plain concrete fails by cracking when the local tensile strain reaches a certain critical value, it ought to be possible to improve its fatigue performance by altering the stress–strain characteristics so as to increase the critical failure strain. One way of doing this is by incorporating short fibres into the mix, for example by using chopped steel wires[21] or alkali-resistant glass fibres.[22] Although the introduction of fibres does not significantly increase the cracking strain of the matrix the overall strain at rupture is increased considerably, being a function of the elasto-plastic properties of the fibres and the effectiveness of the bond between the fibres and the surrounding concrete. The incorporation of fibres into the mix imparts increased strain capacity to the concrete.

It has been stated by the American Concrete Institute[23] that the fatigue strength of concrete containing randomly dispersed steel fibres ranges from 90 % of the first crack strength at 2 million cycles to 50 % at 10 million cycles. However, the tests on which these conclusions were based were in fact performed on cement mortar beams containing no coarse aggregate. In these tests the first crack strength was defined as the stress at which the load–deflection curve became non-linear, and the fatigue life was the number of cycles at which the deflection started to increase rapidly. Neither of these conditions represents complete rupture of the test beam, which continued to carry load by transfer of stress from the cracked mortar to fibres bridging the cracks. Somewhat similar conclusions have been reached about the fatigue strength of glass-reinforced cement[24] where the flexural fatigue strength at 1 million cycles was found to be approximately 90 % of the limit of proportionality determined under static loading.

Very little information has been published on the fatigue performance of fibre-reinforced concrete containing normal sized aggregate. A limited investigation of flexural fatigue, carried out as part of a full-scale trial of a fibre-reinforced concrete overlay for resurfacing a cracked, reinforced concrete highway pavement,[25] indicated the possibility of a substantial improvement in mean fatigue life when compared with ordinary plain concrete. The inclusion of 1·3 % by weight of randomly dispersed chopped steel wires in a paving-quality concrete gave a fatigue life that was some five times greater than that of plain concrete having the same flexural strength, when loaded to about 80 % of the corresponding modulus of rupture. This suggests that fibres are effective in restricting the growth of fatigue cracks and result in a more ductile type of fracture, similar to that which occurs in fibre-reinforced concrete under static loading.[21,26]

The potential advantages of the use of fibres to extend the fatigue life of concrete should be tempered with caution until more evidence is available on the fatigue behaviour of the fibres themselves when embedded in concrete. The effects of local corrosion of the fibres, together with possible changes in bond strength with time, are particularly important here. There is very little evidence so far on what the magnitude of such effects might be, although it is well known that the strength properties of glass-reinforced cement do show some reduction with time of exposure to air and water.[22]

In structural concrete, it may be possible to delay or eliminate fatigue cracking more effectively by placing fibres preferentially in zones subject to high tensile strains. Fibrous cement-based composite materials, such as asbestos cement, glass-reinforced cement or wire-reinforced cement show an effective 'ductility' much greater than that of cement mortar or concrete, so that if they are used to form a surface skin (as proposed by Dave,[27] for example) there is a prospect of increasing the apparent strain at failure and hence of prolonging the fatigue life. In fact, microcracks in the cement paste will still tend to form at about the same strain as with unmodified cement paste, but the fibrous layer will not fracture until much higher strains are reached.

3.11 LIGHTWEIGHT AGGREGATE CONCRETE

There is an increasing interest in the use of structural lightweight concrete based on expanded clays and shales and on pelletised fuel ash, but little information is available on the fatigue performance of such concretes using currently available lightweight aggregates. In 1961 Antrim and his coworkers[28] concluded that the fatigue behaviour in compression of lightweight concrete made from expanded shale aggregates was not significantly different from the performance of normal weight concrete, when expressed in terms of the corresponding compressive strength.

Reference 8 includes some results of compression fatigue tests on lightweight concrete made from sintered pulverised fuel ash (pfa) which show slightly inferior fatigue performance when compared with a gravel concrete and a limestone concrete.

During full-scale structural tests at TRRL on prestressed lightweight-concrete bridge beams made from sintered pfa,[29] cracks have been propagated under repeated loading. The numbers of cycles to initiate the fatigue cracks were not significantly less than values estimated for conventional concrete from Fig. 7. However the type of failure produced was much more brittle than with natural stone aggregates. Crack propagation was more rapid and there was little sign that the aggregate formed any effective barrier to cracking. Failure of aggregate particles in the path of the cracks was general. By the time fatigue failure of the prestressing tendons occurred many pieces of concrete had fallen away from the tension flange of the beam, exposing the reinforcement. It might be inferred that lightweight concretes based on sintered pfa have a poorer overall fatigue performance than concretes containing natural aggregates, but more extensive testing is needed to justify such a conclusion.

3.12 ESTIMATION OF FATIGUE PERFORMANCE FOR STRUCTURAL DESIGN

If fatigue cracking is going to occur in concrete structures it is more likely to be due to repeated flexure than to direct tension or compression, although in certain cases the latter cannot be ruled out entirely. The simple flexural test is probably the most useful type of test for the study of fatigue behaviour of concrete.

The recent studies of fatigue of concrete referred to in the preceding pages have confirmed that under constant amplitude repeated loading, the effects of a wide range of factors, such as mix proportions, aggregate type, moisture condition, age, etc., reflect quite closely changes that occur in the static strength of similar test specimens which have had the same treatment. If the cyclic stress levels applied in particular fatigue tests are expressed as a proportion of the static failure stress under the same conditions (i.e. the same type of concrete, the same age and the same curing treatment) then the mean fatigue results for each level of loading tend to fall close to a single fatigue curve of the form indicated in Fig. 7.

The results of Fig. 7 suggest that the fatigue performance of a particular type of concrete may be estimated for various ages (taking account of the appropriate curing conditions) from a knowledge of the quasi static flexural strength under the same conditions, using well established standardised testing techniques. Such a curve could form the basis of a design curve to be used in fatigue assessment.

In line with the limit state approach to structural design a 'characteristic' fatigue design curve may be constructed to represent a 95 % probability of survival. Such a design curve passes through endurance values that are 1·64 standard deviations less than the mean endurances on a logarithmic scale. A curve of this form is illustrated in Fig. 12 (curve A). An alternative approach is to base the design curve on 1·64 standard deviations below the mean fatigue strength at any particular endurance (curve B). Where appropriate, the effects of mean stress may be allowed for by using a modified Goodman diagram of the form indicated in Fig. 5.

Bearing in mind the complex nature of stress fluctuations that occur in practice and the fact that the strength of concrete varies with time and with change in moisture state throughout the life of the structure, it is reasonable to assume for design purposes that the simple Miner Hypothesis of cumulative damage applies. Thus the accumulated fatigue damage at time T is given by

$$D=\sum\frac{n_f}{N_f}, \text{ and the mean life to failure is } T/D$$

where n_f = number of cycles of stress level f occurring in time T and N_f = mean life to failure at stress level f.

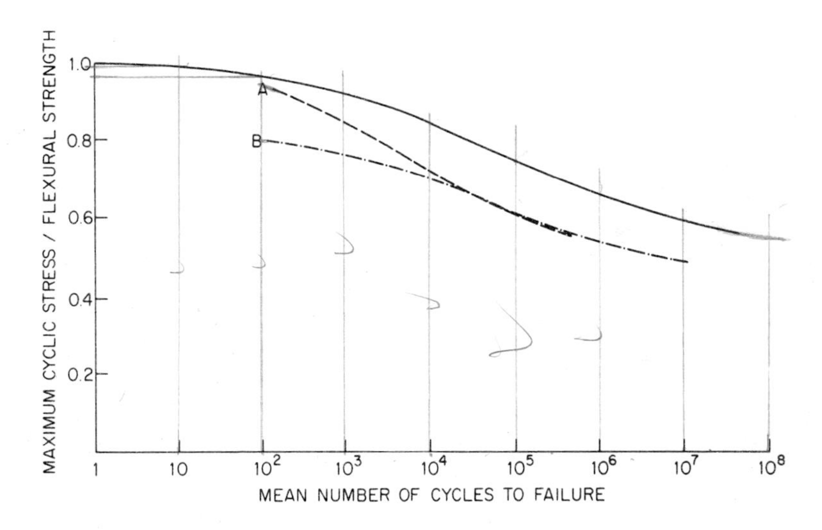

FIG. 12. Design curves for fatigue in flexure (95 % probability of survival). ——, mean endurance curve (from Fig. 7); — — —, characteristic curve based on coefficient of variation of 15 % on log endurance at all stresses; —·—·—, characteristic curve based on coefficient of variation of 11 % on fatigue strength at all endurances.

Allowance will have to be made for the fact that N_f will vary according to the age of the concrete.

From a knowledge of how the static strength of the particular concrete mix varies with time, curing conditions, etc. it is possible to estimate what the fatigue characteristics of the concrete are likely to be throughout the life of the structure and hence to estimate the probability of fatigue cracks developing during its service life.

ACKNOWLEDGEMENTS

Thanks are due to my colleagues Mr J. W. Galloway and Mr H. M. Harding for carrying out much of the experimental work to which reference has been made. Any opinions expressed in this chapter are those of the author personally and do not necessarily represent the official views of the Transport and Road Research Laboratory.

REFERENCES

1(*a*) BRITISH STANDARDS INSTITUTION (1972). *The Structural Use of Concrete.* CP 110, B.S.I., London.

1(*b*) BRITISH STANDARDS INSTITUTION (1978), *Steel, Concrete and Composite Bridges.* BS 5400. B.S.I., London.

2. VAN ORNUM, J. L. (1903). Fatigue of Cement Products. *Trans. ASCE,* **51,** 443.
3. CLEMMER, H. F. (1922). Fatigue of Concrete. *Proc. ASTM.,* **22** II, 409.
4. LLOYD, J. P., LOTT, J. L. and KESLER, C. E. (1968). Fatigue of Concrete. *Univ. of Illinois Eng. Expt. Stn. Bulletin 499.*
5. RAITHBY, K. D. and WHIFFIN, A. C. (1968). Failure of Plain Concrete under Fatigue Loading—a Review of Current Knowledge. *Road Research Laboratory Report LR 231.* TRRL, Crowthorne, Berks.
6. MURDOCK, J. W. and KESLER, C. E. (1958). Effect of Range of Stress on Fatigue Strength of Plain Concrete Beams. *J. Am. Concrete Inst.,* **30,** No. 2, 221.
7. GALLOWAY, J. W. and RAITHBY, K. D. (1973). Effects of Rate of Loading on Flexural Strength and Fatigue Performance of Concrete. *Transport and Road Research Laboratory Report LR 547.* TRRL, Crowthorne, Berks.
8. SPARKS, P. R. and MENZIES, J. B. (1973). The Effect of Rate of Loading upon the Static and Fatigue Strengths of Plain Concrete in Compression. *Mag. Concr. Res.,* **25,** No. 83, 73.
9. BRITISH STANDARDS INSTITUTION (1970). *Methods of Testing Concrete* BS 1881. Part 4. B.S.I., London.
10. KOLIAS, S. and WILLIAMS, R. I. T. (1978). Cement-bound Road Materials: Strength and Elastic Properties Measured in the Laboratory. *Transport and Road Research Laboratory Report SR 344.* TRRL, Crowthorne, Berks.
11. HILSDORF, H. K. and KESLER, C. E. (1966). Fatigue Strength of Concrete under Varying Flexural Stresses. *J. Am. Concrete Inst.,* **63,** No. 10, 1059.
12. BALLINGER, C. A. (1971). Cumulative Fatigue Damage Characteristics of Plain Concrete. *Highway Research Record No. 370,* p. 48. Highway Research Board, Washington DC.
13. MINER, M. A. (1945). Cumulative Damage in Fatigue. *Trans. ASME,* **67,** A159.

14. AWAD, M. E. and HILSDORF, H. K. (1974). Strength and Deformation Characteristics of Plain Concrete Subjected to High Repeated and Sustained Loads. *Abeles Symposium on Fatigue of Concrete, ACI Publication SP-41*, p. 1.
15. ASSIMACOPOULOS, B. A., WARNER, R. F. and EKBERG, C. E. (1959). High Speed Fatigue Tests on Small Specimens of Plain Concrete. *J. Prestressed Concr. Inst.*, **4** (2), 53.
16. RAITHBY, K. D. and GALLOWAY, J. W. (1974). Effects of Moisture Condition, Age and Rate of Loading on Fatigue of Plain Concrete. *Abeles Symposium on Fatigue of Concrete, ACI Publication SP-41*, p. 15.
17. GALLOWAY, J. W., HARDING, H. M. and RAITHBY, K. D. (1979). Effects of Age on Flexural, Fatigue and Compressive Strength of Concrete. *Transport and Road Research Laboratory Report LR 865*. TRRL, Crowthorne, Berks.
18. GALLOWAY, J. W., HARDING, H. M. and RAITHBY, K. D. (1979). Effects of Moisture Changes on Flexural and Fatigue Strength of Concrete. *Transport and Road Research Laboratory Report LR 864*. TRRL, Crowthorne, Berks.
19. BENNETT, E. W. and RAJU, N. K. (1971). Effect of Understressing on the Deformation and Strength of Plain Concrete in Compression. *Proc. Int. Conf. on Mechanical Behaviour of Materials, Kyoto*, Vol. IV, p. 278. The Society of Materials Science, Japan.
20. YOSHIMOTO, A., OGIRO, S and KAWAKANI, M. (1972). Microcracking Effect on Flexural Strength of Concrete after Repeated Loading. *J. Am. Concrete Inst.*, **69,** No. 4, 233.
21. SWAMY, R. N. and MANGAT, P. S. (1975). The Onset of Cracking and Ductility of Steel Fibre Concrete. *Cement and Concrete Research, Vol. 5*, p. 37, Pergamon Press Inc, New York.
22. (1976). A Study of the Properties of Cem–FIL/OPC Composites. *Building Research Establishment Current Paper CP 38/76*. BRE, Garston.
23. ACI COMMITTEE 544. (1973). State-of-the-Art Report on Fibre-Reinforced Concrete. ACI *J. Am. Concrete Inst.*, **70,** No. 11, 729.
24. HIBBERT, A. P. and GRIMER, F. J. (1975). Flexural Fatigue of Glass-Fibre-Reinforced Cement. *J. Mat. Science*, **10,** No. 12, 2124.
25. GALLOWAY, J. W. and GREGORY, J. M. (1977). Trial of a Wire-Fibre-Reinforced Concrete Overlay on a Motorway. *Transport and Road Research Laboratory Report LR 764*. TRRL, Crowthorne, Berks.
26. EDGINGTON, J., HANNANT, D. J. and WILLIAMS, R. I. T. (1974). Steel Fibre Reinforced Concrete. *Building Research Establishment Current Paper No. 69/74*. BRE, Garston.
27. DAVE, N. J. (1967). Fibre Reinforced Cement (FRC) Composite Concrete Construction—a New Approach. *RILEM Symposium on Fibre Reinforced Cement and Concrete 1975, Vol. 2*, p. 615. The Construction Press Ltd, Lancaster.
28. GRAY, W. H., MCLAUGHLIN, J. F. and ANTRIM, J. D. (1961). Fatigue Properties of Lightweight Aggregate Concrete. *J. Am. Concrete Inst.*, **58,** No. 2, 149.
29. HOWELLS, H. and RAITHBY, K. D. (1977). Static and Repeated Loading Tests on Lightweight Prestressed Concrete Bridge Beams. *Transport and Road Research Laboratory Report LR 804*. TRRL, Crowthorne, Berks.

30. Bennett, E. W. and Raju, N. K. (1969). Cumulative Fatigue Damage of Plain Concrete in Compression. *Int. Conference on Structure, Solid Mechanics and Engineering Design in Civil Engineering Materials*, Southampton University. Wiley, New York.

Chapter 4

THE PERFORMANCE OF STRUCTURAL CONCRETE IN A MARINE ENVIRONMENT

R. D. BROWNE and A. F. BAKER

Research and Development Department, Taylor Woodrow Construction Ltd, Southall, UK

SUMMARY

Concrete has been used for many years in marine construction and has proved to be a durable material, requiring little maintenance. However, in order to ensure durability, attention must be given to the specification of concrete for marine works with due regard to those factors which may mar its performance in practice. Factors which dictate the quality, and hence durability, of concrete are described, together with the techniques available for the inspection and repair of concrete structures.

4.1 INTRODUCTION

Structural concrete in the marine environment differs from concrete in other environments in that, underwater, the material is continuously exposed to the action of seawater, and above water, to spray or airborne salt. In addition, in cold climates freezing conditions may exist. For offshore structures, repeated wave loading requires dynamic fatigue assessments and under high water flow conditions, abrasion from sand or hydraulic cavitation needs consideration.

The material has been used successfully for over 80 years for marine application in the construction of sea defences, ships, barges, floating harbours, defence forts, lighthouses and, more recently, for offshore drilling and storage platforms by the oil industry.

On the other hand, for jetties and wharves, piling and soffits to decks have deteriorated requiring repetitive maintenance, due to corrosion of reinforcement in the splash zone.

In its application to the marine environment, whether estuarial, coastal or offshore, care is needed to ensure long-term durability particularly where its application is extended to more onerous conditions. This is best achieved by understanding the possible factors which can mar the performance of concrete and not just accepting codes of practice rules.

Not only is it necessary to appreciate the factors which cause deterioration of the concrete, but also those which affect corrosion of steel present, whether internally as reinforcement and prestress, or externally, as attachments or fixings.

4.2 THE MARINE ENVIRONMENT

The marine environment can be divided into three main zones:

(1) Above high tide level (splash zone), where build-up of salt spray, wetting–drying and freeze–thaw cycles can occur.
(2) The inter-tidal zone, where the concrete is kept in a mainly wet state, but intermittently exposed to the air.
(3) The totally immersed zone, where oxygen availability for steel corrosion is limited, but where hydrostatic pressure increases with depth resulting in rapid penetration of seawater into concrete.

In the limited number of cases where deterioration has occurred, the cause may be attributable to the presence of sea salts in the concrete or to freeze–thaw cycling damage. Marine growth[1] generally has not caused damage, and in fact may further protect the structure. Higher temperatures can enhance the steel corrosion in tropical and Middle East climates, not only due to the direct influence of temperature on the corrosion process, but also to the faster drying out of the concrete, and salt accumulation on surfaces from evaporation in the splash zone.

Seawater generally contains about 3.5 % of inorganic salts, the principal compounds being sodium chloride and magnesium sulphate.

Table 1[2] summarises the main ionic concentrations of the salts present for the North Sea, Atlantic Ocean, Baltic Sea and Persian Gulf; they can vary substantially with location.

TABLE 1

CONCENTRATION OF SOLUBLE SALTS FOR VARIOUS SEAS[2]

Ion	*Concentration* (g/100 cm^3)			
	North Sea	*Atlantic Ocean*	*Baltic Sea*	*Persian Gulf*
Sodium	1·220	1·110	0·219	1·310
Potassium	0·055	0·040	0·007	0·067
Calcium	0·043	0·048	0·005	0·050
Magnesium	0·111	0·121	0·026	0·148
Chloride	1·655	2·000	0·396	2·300
Sulphate	0·222	0·218	0·058	0·400
Total	3·306	3·537	0·711	4·275

The dissolved oxygen content in seawater varies with temperature and depth. It can decrease from 7 ppm at the surface to about 3 ppm at 100 m depth.[3]

4.2.1 Deterioration Process

In order to ensure that concrete structures give satisfactory performance, in addition to minimum maintenance, it is necessary to understand the principal factors controlling concrete deterioration in seawater.

Empirical relationships which are defined in the laboratory are only useful if the overall process of deterioration is understood. The use of small samples and concentrated salt solutions under accelerated exposure conditions not only exaggerates the rate of concrete deterioration, but may also indicate problems of concrete deterioration not normally found in practice. A number of exposure trial tests are available[4] stretching back over many years of observation. The findings of tests such as these are perhaps more valid when considering marine concrete durability.

In order to examine durability it is necessary to define the concept of a durable material for marine applications. Concrete is a durable material if its quality and performance remain acceptable for the design life of the structure, whether this is five or fifty years. The durability of concrete in seawater is only questionable if it deteriorates to a significant extent within the design life of a concrete structure.

Deterioration can be viewed as a two part process where t_1 is the time taken for the marine environment to penetrate to a point of interest within the concrete and t_2 is the time taken for a material change of significance to the structure to take place at that point. Therefore, to ensure durability

$$\text{design life} < t_{1(x)} + t_{2(x)}$$

where x is the point of significance from the surface of the concrete, and where

$$t_{1(x)} = \text{function}\left(\frac{(\text{cover})^2}{\text{permeability/diffusion coefficients}}\right)$$

$$t_{2(x)} = x - \Delta x > \text{design value of a property at } x \text{ (strength, loss in steel section, spalling pressure, etc.).}$$

Therefore, measures to prolong t_1 and t_2 will increase the durability of concrete and prolong its service life. The values of both t_1 and t_2 change for different concretes. Similarly, for any one concrete structure both t_1 and t_2 are influenced by whether the concrete is in the splash, tidal or underwater zones. In the underwater zone, for example, t_1 is small as the concrete may rapidly become saturated by seawater under hydrostatic pressure. However, t_2 is large because oxygen availability for corrosion is severely limited. Measures to increase the time t_1 are primarily concerned with maintaining the quality and integrity of the cover concrete. Measures to increase the time t_2 include coatings and cathodic protection, air entraining agents and the use of durable aggregate and low C_3A, pozzolanic or slag cements.

4.3 CONCRETE

4.3.1 Chemical Attack on Concrete

The sulphate salts present in seawater (Table 1) can attack the constituents of Portland cement (Table 2). Sulphate attack on land-

TABLE 2

TYPICAL PHASE COMPOSITION OF UK CEMENTS

Compound	*Phase composition (% by wt)*	
	Ordinary Portland cement	*Sulphate resisting Portland cement*
Tricalcium silicate (C_3S)	48	58
Dicalcium silicate (C_2S)	24	16
Tricalcium aluminate (C_3A)	12	3
Tetracalcium alumino-ferrite (C_4AF)	9	17
Gypsum (SO_3)	2·5	2

based structures disrupts and eventually cracks and spalls concrete due to the expansive growth of the mineral ettringite ($C_3A.3CaSO_4.31\,H_2O$). However, structures in the marine environment normally show no evidence of expansion and cracking due to sulphate attack, as the mineral ettringite is significantly soluble in seawater and can therefore be leached from the concrete. In addition, unlike sulphate attack on land, the presence of sulphate as magnesium sulphate in seawater[5] breaks down the basic cementitious structure of concrete by attacking the calcium silicate component of the cement. Therefore, marine sulphate attack leaches and softens the concrete rather than cracking it.[5]

Tests on small-scale laboratory mortar bars immersed in sulphate solution (Fig. 1) show an extremely good correlation between expansion and the C_3A content of the cement used.[6] Exposure trials worldwide, however, on specimens produced using cements of different types (OPC, SRC, high alumina and slag cements) show that in addition to C_3A content, the quality of the cover concrete is a vital factor in ensuring durability.

Leaching and softening of the cement binder, due to marine sulphate attack is enhanced by the action of chloride salts which further soften the concrete[7] and by leaching of lime within the concrete in the splash, tidal and underwater zones. Tests carried out by the US navy[8] on 67-year-old concrete have also shown that seawater attacks both the hydrated aluminates and also the hydrated calcium silicate components of set cement, although no significant loss of compressive strength resulted.

Aggregates used in concrete are generally not attacked chemically by seawater, although internal aggregate reaction may occasionally occur where certain siliceous aggregates have been used with high alkali content cements. The alkalies present in the concrete originally may

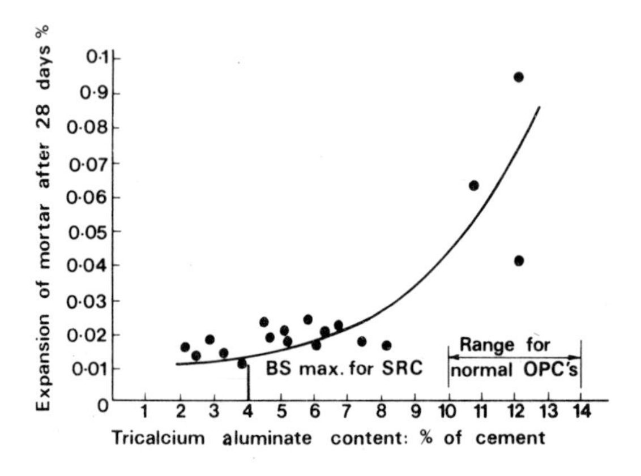

FIG. 1. Effect of tricalcium aluminate on sulphate resistance of Portland cement.[6]

attack the aggregate producing expansion, particularly in damp concrete, and the cracking associated with this reaction may eventually damage the cover to the reinforcement. In Denmark, Germany and the USA, this reaction has produced significant deterioration although in the UK at present this problem is not significant. However, any minor damage to the concrete cover by cracks will open up the structure to further chemical attack and the effects of freeze–thawing, wave erosion, and other physical processes.

4.3.2 Physical Deterioration of Concrete

Concrete may be physically damaged by exposure to the marine environment. Physical damage can occur due to freeze–thaw cycling of the concrete as water present in pores within the concrete expands on freezing (9 % volume increase)[9] causing disruption of the cement. The damage caused by cyclic freeze–thawing is influenced considerably by original water–cement ratio. At water–cement ratios above approximately 0·45, voids within the concrete constitute a continuous capillary system. Below 0·45, minute isolated voids are formed within which the freezing point of water is depressed.[10] Concrete's susceptibility to freeze–thaw damage is therefore influenced by the original water–cement ratio (Fig. 2)[11] and also by the type of aggregate used. Concrete may be protected from freeze–thaw damage by using an air-entraining agent which forms discrete voids (0·05 mm)[12] within the cement to accommodate ice expansion. The effect of freezing and

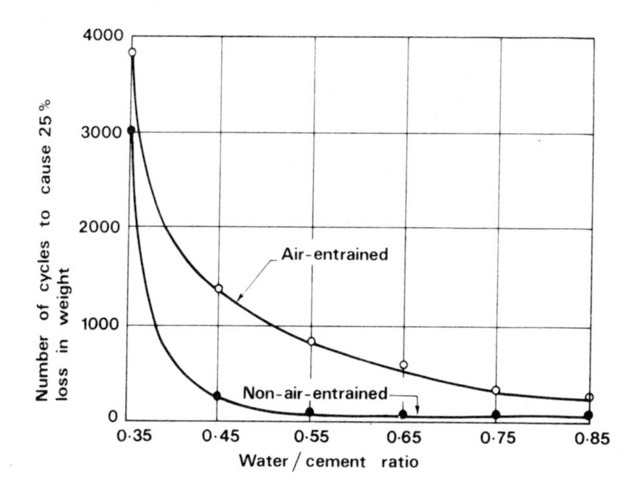

FIG. 2. Influence of water–cement ratio on the frost resistance of concrete moistcured for 28 days.[11]

thawing of concrete in seawater is much more severe than in fresh water.[13]

The action of physical agents tends to accelerate deterioration resulting from other factors, i.e. chemical attack and reinforcement corrosion. Microcracking and softening of the concrete is enhanced by freeze–thaw action, alternate wetting and drying, and erosion. Concrete kept continuously immersed is maintained in a relatively uniform environment when compared to splash and tidal zone concrete. However, with time (and depending on cover quality and hydrostatic pressure) seawater will penetrate the cover to varying degrees. The presence of water within the concrete lowers its strength by 10 % over dry concrete,[14] although this effect does not appear to be enhanced by hydrostatic pressure at depths of up to 150 m.[15] The fatigue strength of concrete under dynamic wave loading conditions can be reduced by hydrostatic water saturation.[16]

4.3.3 Penetration of the Environment

In the atmospheric zone the rate of penetration of salts and oxygen is a function of the pore structure built up within the concrete and the degree of saturation of porespace with water. The movement of salts and oxygen is primarily a diffusion process influenced by the concentration of the diffusing elements, the diffusion constant of the concrete, and time. In the submerged zone, hydrostatic pressure becomes the dominant motive force and thus the rate of penetration is influenced by the actual external water pressure and the concrete's permeability.

Under hydrostatic pressure, the rate of penetration of water into concrete can be defined by a modification of Valenta's equation[17] which takes into account the concrete's void content.

$$K = \frac{d^2 v}{2ht}$$

where K = permeability coefficient (m/s), d = depth of saturation (m), v = void content of concrete, h = hydrostatic head (m), and t = time (s).

The protection offered by the cover concrete to the reinforcing steel is dependent on its ability to prevent chloride salts reaching the reinforcement.

Once the chlorides reach the reinforcement the salt concentration may increase to a level where activation of the corrosion process begins, i.e. the steel loses the alkaline passivity provided by the cement and a corrosion cell is initiated.

The concentration of chlorides to be expected within concrete varies, depending on the height of the concrete above or below mean sea level. Above water, salt build-up may increase with height of the concrete above sea level due to salt spray accumulation on the concrete surface above the average wave height. The depth of chloride penetration with time can be determined using Fick's second law of diffusion.[18]

$$\frac{\mathrm{d}c}{\mathrm{d}t} = D\frac{\mathrm{d}^2 c}{\mathrm{d}x^2}$$

where x = distance from concrete surface, t = time and D = diffusion constant. The distribution of chlorides with time can be derived from the above equation as follows:

$$C_{xt} = C_e \left[1 - \text{erf} \frac{(x)}{\sqrt[2]{Dt}} \right]$$

where C_e = equilibrium chloride level on concrete surface (equal for all t), C_{xt} = chloride level at position x and time t, and erf = error function (from tables).

This derivation from Fick's second law can be used, together with experimental data to predict the penetration depth of a chloride concentration front of 0·40 % by weight of cement through poor quality concrete (Fig. 3) with time. In this particular case, using

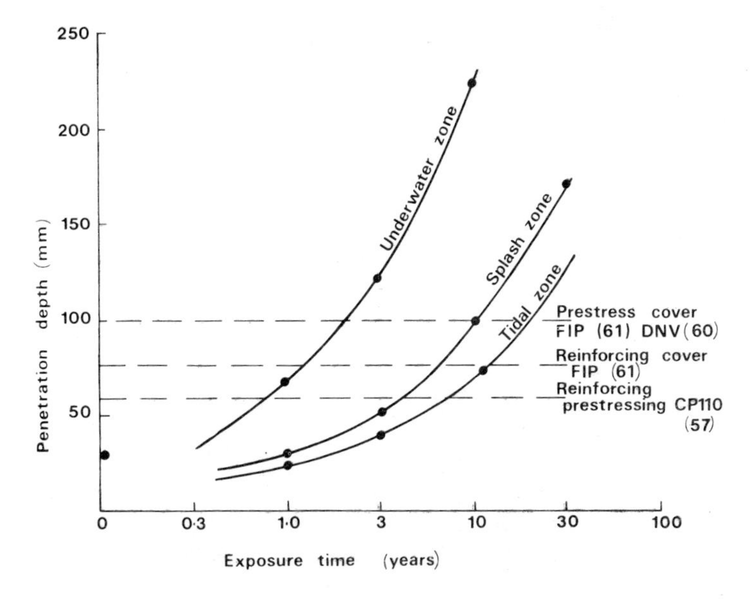

FIG. 3. Penetration of the activation front (activation front taken as equivalent to Cl^- content of 0.40 % by weight of cement). Assuming: (i) Diffusion coefficient of $5{\cdot}5 \times 10^{-8}\ cm^2/s$ in the atmospheric zone. (ii) A surface salt concentration of 4 % by weight of cement in splash and 2 % by weight of cement in tidal concrete. (iii) An underwater concrete permeability coefficient of 6×10^{-13} cm/s.

assumed values, splash and tidal concrete may become active after approximately 7–12 y with 75 mm poor quality cover. Concrete underwater rapidly becomes saturated with seawater to the depth of reinforcement, although subsequent corrosion may be negligible, as explained later, due to the lack of oxygen in seawater necessary for electrochemical corrosion.

4.3.4 Permeability

The permeability of concrete is dependent primarily on the permeability of the aggregate (Table 3)[19] and the cement binder. In order to increase

TABLE 3

AGGREGATE PERMEABILITY VALUES AND EQUIVALENT WATER–CEMENT RATIOS[19]

Type of rock	*Coefficient of permeability (m/s)*	*Water–cement ratio of mature paste of the same permeability*
Dense trap	$2{\cdot}47 \times 10^{-14}$	0·38
Quartz dolerite	$8{\cdot}24 \times 10^{-14}$	0·42
Marble I	$2{\cdot}39 \times 10^{-13}$	0·48
Marble II	$5{\cdot}77 \times 10^{-12}$	0·66
Granite I	$5{\cdot}35 \times 10^{-11}$	0·70
Sandstone	$1{\cdot}23 \times 10^{-10}$	0·71
Granite II	$1{\cdot}56 \times 10^{-10}$	0·71

concrete quality in the cover zone, low permeability aggregate should be used, together with a low water cement ratio mix which produces a discontinuous cement pore structure (Fig. 4).[20] Data from a number of tests on steady state concrete permeability have shown, however,[21–24] that permeability measurements vary by 10 000–1 for concrete cast using a low water/cement ratio mix (Fig. 5). It is necessary, therefore, to examine this critical property of concrete further.

In order to produce a dense, low permeability, concrete cover, two further factors are of critical importance in controlling the permeability of the *in situ* material:

(1) Efficiency of compaction.
(2) Efficiency of curing.

It is essential that a mix design for offshore concrete should take into account the ability of the mix to be efficiently compacted so that there is a minimum voidage. Sufficient cement should be available within the wet concrete to over-fill the voids left when the aggregate is completely

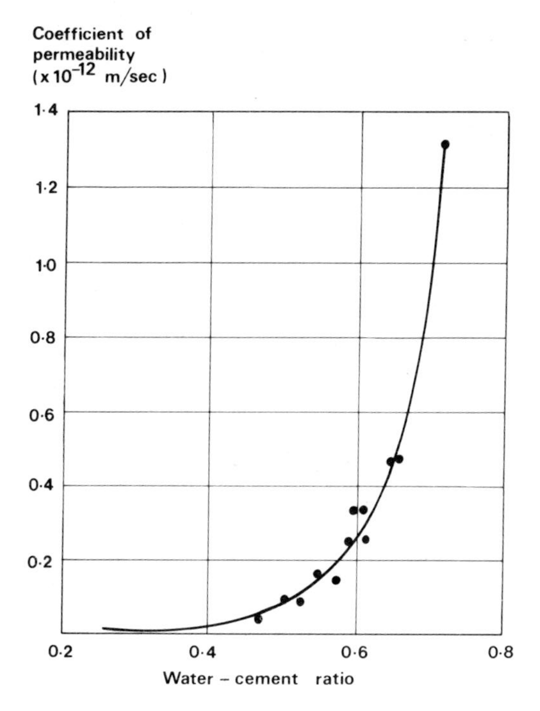

FIG. 4. The effect of water–cement ratio on the permeability of mature cement paste.[20]

compacted. Compaction by vibration is more efficient than manual compact in laboratory specimens.[25] With site concrete, incomplete compaction can lead to much greater variability, and differences in permeability of several orders of magnitude are easily obtained. The benefits achieved by using high cement contents in offshore construction are, to a large extent, due to better compactability of the mix leading to less voidage in the concrete cover to steel.[4]

After placing, concrete should be well cured, if possible underwater. Concrete cured in air (site practice), does not undergo complete hydration and cannot, therefore, develop an optimum discontinuous pore structure. Experience has shown that the permeability of concrete cured in a high humidity environment for 28 days reduces dramatically on subsequent storage underwater due to the blocking up of the pore network within the hardened concrete by further hydration of residual

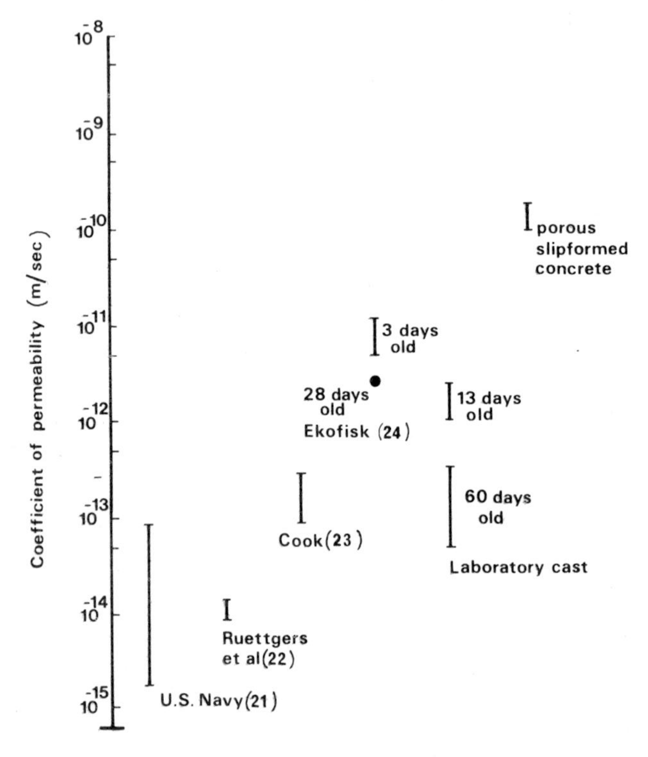

FIG. 5. Comparison of concrete permeability data. Water–cement ratio = 0·4–0·5 in all cases.

unhydrated cement. Thus the permeability of submerged concrete, from boat hulls to deep water offshore structures, may be less than anticipated by simple permeability measurements on air cured materials.

Figure 6 shows the depth of penetration of seawater into concrete after 1 year, and its dependence on hydrostatic pressure and permeability. With low quality (10^{-8} m/s) concrete, water will have penetrated 8 m after 1 year at 50 m depth, whereas with lower permeability, good quality concrete, penetration will occur to only

30 mm with a 50 m head of water. When the previous statements on pore blocking are taken into account it can be seen that good quality cover concrete which continues to hydrate underwater, should afford the reinforcement a high level of protection from the hostile marine environment.

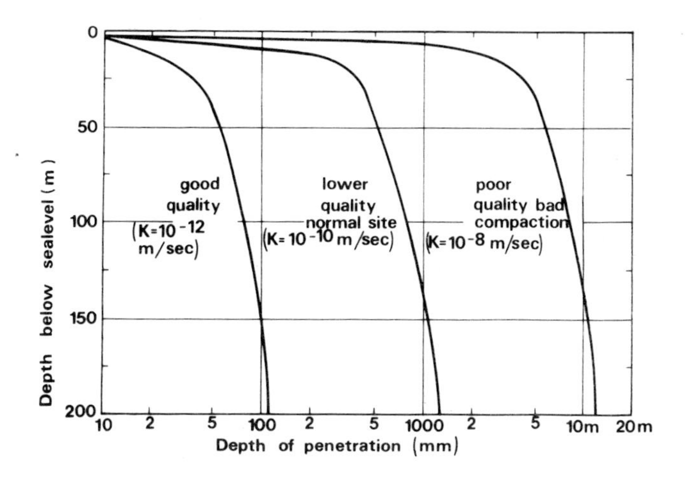

FIG. 6. The effect of concrete variation on the penetration rate of seawater after immersion for 1 year.[26]

A final important factor in cover permeability is the treatment given to the concrete when it is placed. Cover concrete is not generally very thick so that damage to the surface using slipform construction[26] or bleeding can increase the permeability of the outer concrete layers, thus reducing the effective cover thickness. On the other hand, concrete which is finished by hand trowelling can lead to even greater compaction density. It is believed[26] that the high durability of ferrocement boat hulls (25 mm thick) which have a low cover is, in part, a result of hand trowelling producing good concrete compaction,[4] and also possibly, in part, of pore blocking due to continual hydration of the cement underwater.

4.3.5 Erosion

The resistance of concrete to erosion is to a large extent determined by the type of aggregate used, although changes in cement can also influence resistance[14] due to differences in their susceptibility to chemical deterioration by seawater. Hard durable aggregates such as flints and granite appear to produce durable concrete. If the basic integrity of the concrete cover fails by cracking, delamination and spalling, then these features act as centres for preferential rapid erosion. In addition to the normal action of wind, wave and sand scour, concrete can also be damaged by cavitation effects where extremely high flow rates are encountered.[27] Specific instances of this type of damage have occurred in turbine chutes and dam spillways associated with hydroelectric power construction.

4.4 STEEL IN CONCRETE

4.4.1 Corrosion Mechanism

The corrosion rate for bare steel in seawater is shown in Fig. 7.[28] An increase in temperature or an increase in velocity of water against the steel will increase the rate of corrosion still further.

Mild steel reinforcement is normally produced from low carbon steel. Higher carbon steels are used for high tensile reinforcement and high carbon steel containing manganese and silicon for prestressing wire and bar. All steel is vulnerable to corrosion, especially in environments

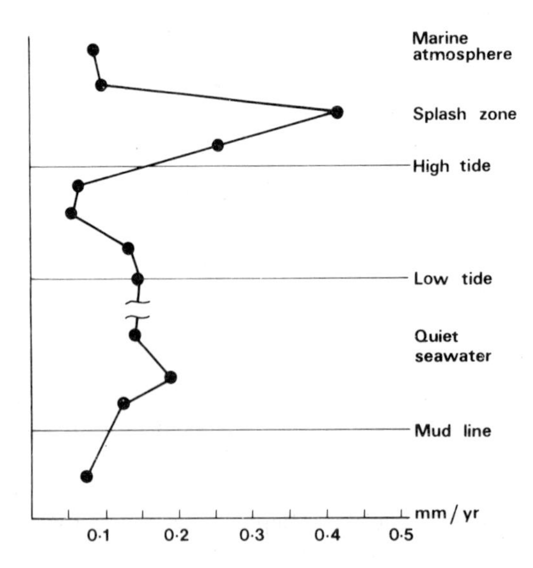

FIG. 7. Typical corrosion rate of steel piling in seawater.[28]

where chloride salts are present. However, when embedded in concrete, steel is passivated or protected by the high alkalinity of the surrounding cement (at a pH of 12·5 or greater). This high pH can be maintained for a considerable period of time, even with leaching and carbonation of lime, as there is an enormous reservoir of lime within the concrete. The presence of chlorides can destroy this protection if the concentration of chloride ion by weight of cement reaches approximately 0·40 %.[29] Once passivation is lost, by leaching of lime or carbonation or chloride build-up, then corrosion of the steel can commence. The process is shown diagrammatically in Fig. 8.

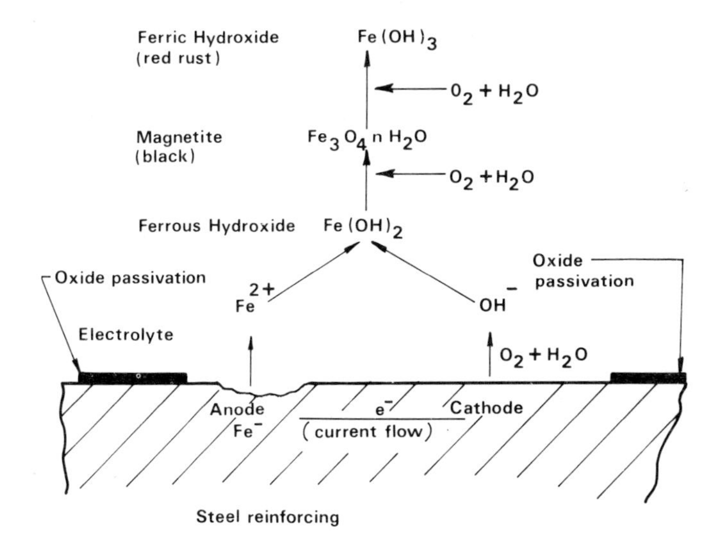

FIG. 8. Corrosion process of the steel surface.

'Activation' of the reinforcement occurs when the passivation effect of concrete on steel is lost. The steel is passified by the development of a barrier layer of Fe_2O_3 (iron oxide)[25] on the steel or chemically bound hydroxyl groups (OH^{-2})[31] which prevents current flowing from anodic to cathodic areas using the moist concrete as an electrolyte. Carbonation of the lime in cement reduces its pH, allowing corrosion currents to form. Chlorides depassivate steel by destroying the necessary integrity of this passivating layer,[32] although the presence of chlorides also lowers the pH of a saturated lime solution.[33] Once a corrosion current has formed, Fe^{2+} is now available at anodic areas where it combines with hydroxyls to produce iron hydroxide ($Fe(OH)_2$) which is precipitated in the electrolyte.[30] In the concrete this then oxidises to haematite (red iron oxide) or magnetite (black iron oxide) where oxygen availability is less.

The volume of the corrosion products is 2–3 times greater[30] than that of the original steel, so that tensile forces are set up around the steel leading to the eventual cracking and spalling of the cover. Once cracking and spalling are initiated then corrosion can proceed rapidly, especially in the marine environment, as cover protection is not now available to the steel. The extent of corrosion necessary to cause spalling is relatively small. For a 6 mm bar with 30 mm of cover, spalling can occur when the steel has corroded to a depth of 0·1 mm.[35]

The primary factors believed to control steel corrosion in concrete are:

(1) Depassivation of the steel by chlorides derived from seawater.

(2) The development of electrolytic cells in areas of differing potential, caused by changes in the concentration of chlorides and oxygen around the bar.

(3) The reduction in resistivity of the concrete due to the presence of chlorides and water which allow a substantial corrosion current to be maintained.

(4) The presence of oxygen in cathodic/anodic areas (see Fig. 8) to form the corrosion products.

As was explained in the introduction, reinforcement corrosion or any other deterioration process in concrete can be subdivided into the time it takes to initiate corrosion (deterioration, t_1) and the subsequent time (t_2) for this deterioration to become significant. Once activation of the steel has occurred (t_1) then the rate of corrosion (t_2) is dictated by points (2)–(4). Electrolytic cell development may be enhanced if there are variations in the quality of the concrete cover along a steel bar, as poorer areas will allow more oxygen and chlorides to reach the bar than areas of dense low permeability cover. The amount of oxygen necessary to maintain a high corrosion rate is substantial. It can be calculated from the equation for the combination of iron and oxygen to form Fe_2O_3 and is approximately 1·5 litres O_2 per gram of iron oxidised to Fe_2O_3.

4.4.2 Corrosion Rate (t_2)

The magnitude of the corrosion current which is set up through the concrete (the electrolyte) from anodic to cathodic areas of steel, controls the corrosion rate. The magnitude of this corrosion current depends on the potential developed between anodic and cathodic areas and the electrolytic resistance (resistivity) of the concrete. Resistivity is dependent primarily on the moisture content of the concrete but is influenced significantly by chlorides within the concrete at higher moisture levels (Fig. 9) An empirical scale[34] of measured resistivity against steel corrosion is given in Table 4.

The corrosion rate which can be supported varies through the structure. In the atmospheric zone (splash and above) the rate of corrosion (t_2) appears to be controlled primarily by the moisture state of the concrete cover, as the corrosion rate is most rapid when the concrete is partially saturated, has a low resistivity and is yet dry enough to allow oxygen to diffuse rapidly as a gas through the cover.

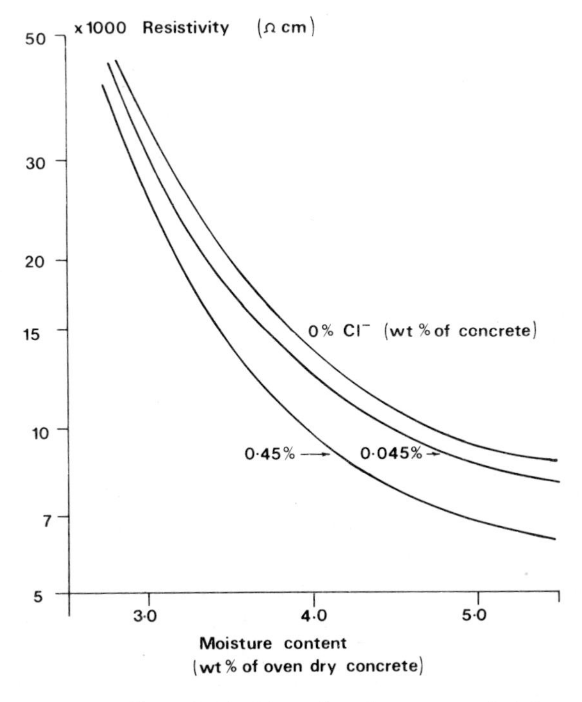

FIG. 9. Effect of chloride and moisture on resistivity.

TABLE 4

CONCRETE RESISTIVITY LEVELS RELATED TO EMBEDDED STEEL VULNERABILITY[34]

Resistivity (ohm cm)	*Rate of corrosion supported*
20 000	Negligible
10 000–20 000	Low
5 000–10 000	High
5 000	Very high

Underwater, concrete loses its passivation rapidly if the cover permeability is low, and the steel becomes active (i.e. depassivated with a potential of at least 200 mV above the surrounding concrete). The speed at which corrosion can proceed in saturated concrete is rate-controlled by oxygen availability. The average concentration of oxygen is low (8 ppm)[2] in seawater and is reduced even further by an increase in water temperature, and depth below mean sea level.[35] This appears to result in a subsequently negligible corrosion rate,[36] due to oxygen stifling of cathodic and anodic areas on the reinforcement.

Concrete in the tidal zone normally remains saturated with water so that the rate of gas diffusion through the concrete to the steel is reduced.[36] Gas diffusion through saturated concrete appears to be much slower than through dry concrete.[37]

4.4.3 Spalling in Reinforced Concrete

Once corrosion is initiated, by failure of the cover to hold back the environment, then expansive stresses will be built up in the oxide layers formed on the reinforcement. Whether the cover will subsequently spall depends not only on the rate of corrosion of the reinforcement, but also the thickness of cover, the strength of the concrete and the shape of the concrete member itself.[4] Failure may occur initially at the corners of concrete members where oxygen can penetrate from two faces simultaneously and where salt may have become concentrated by run-off down the vertical faces of beams (Fig. 10). Additionally, the quality of cover at corner sections is often less than on flat surfaces due to compaction difficulties. Slabs may take longer to fail by delamination parallelling the reinforcement. Damage to marine concrete is often restricted to splash and tidally exposed sections, and is more frequent on the underside of deck beams and slabs and rectangular piles.[38] In these splash zone areas salt can accumulate to very high concentrations by evaporation, especially in hot climates. Coastal structures in the UK may show damage after 10–20 years[38] in vulnerable sections on corners in splash zone. In tidal and underwater zones up to 70 years' exposure may produce little deterioration due to factors previously described.

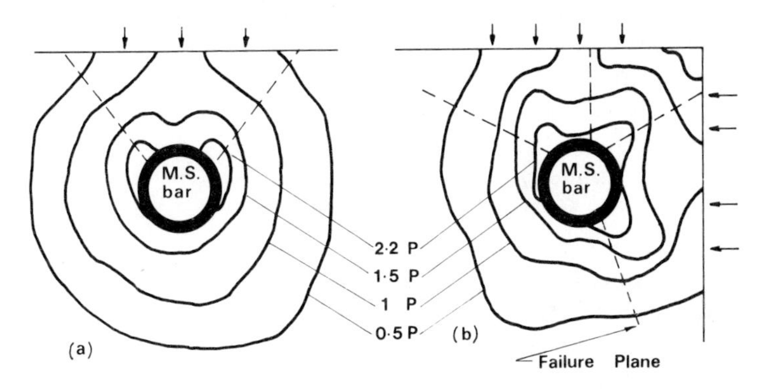

FIG. 10. Possible stress contours around reinforcing bars due to salt and oxygen penetration uniaxially on slab (a) and biaxially at corners (b). P = pressure at reinforcement surface.

The amount of corrosion necessary to cause spalling can be determined experimentally. It has been estimated that 0·15 g of steel loss per cm of bar causes a bursting pressure of 4·5 N/mm^2 which will crack concrete. The pressure developed may also be dependent on the bar diameter and the time the cover is stressed, as stress relief by creep may occur.

4.4.4 Prestress Damage

Damage due to reinforcement corrosion can be seen visually as cracks and spalls resulting in a loss of section in the affected concrete member. Damage is normally easily observed and a decision on remedial measures taken long before the loss in section can become structurally significant.

However, in prestressing systems, corrosion of the highly stressed steel resulting in a reduction of section can lead to failure. Although great care is taken to ensure that the ducts containing prestressing cables or bars are completely filled with protective grout, air voids have been observed in bridges on land and it is believed that similar situations could exist in tendons in offshore structures. In addition ducts are, in certain cases, made of helical steel and may be penetrated by seawater under high hydrostatic pressure,[36] and it is doubtful whether grout reaches all interstices within a prestressing tendon composed of many wires. Voidage in prestressing ducts which can be penetrated by seawater could lead to failure of prestressing systems once corrosion has been initiated. There is, as yet, no evidence that offshore prestressed structures are at risk, although no examination has been undertaken of embedded prestress as non-destructive testing techniques are not at present available to do so.

There is a possibility of stress corrosion cracking in prestressing steel leading to sudden failure only when using unsuitable steels.[39]

4.4.5 Corrosion Potentials

The relationship between splash and tidal zones and the effects of salt build-up, oxygen availability, and the influence of section dimensions can all be clearly seen, if a surface electrical potential map is drawn up of the structure under consideration. Such a map is shown in Fig. 11[34] where potentials between the steel and concrete have been recorded between a copper sulphate electrode on the surface of concrete and embedded steel reinforcement. Contours join areas of equivalent potential and therefore delineate areas of 'active', i.e. corroding, and passive (passivated) steel. It can be seen from Fig. 11 that areas of high potential are associated with areas of concrete containing chlorides above 0·40 % by weight of cement. Potential measurements will indicate areas of 'active' reinforcement and are therefore used as an NDT (non-destructive test) survey technique. Table 5 shows the potential measurements measured against a copper cell and the confidence limits for corrosion.[29]

4.4.6 Macrocell Corrosion in Concrete

Corrosion initiation occurs once steel passivation is removed and a corrosion current can then flow through the steel from anodic to cathodic areas. In concrete offshore structures, the size of these corrosion cells may be considerable since, if the reinforcement is linked to external steel or to seawater by an area of poor concrete, then these points may become the anode of a cell with the cathode being the remainder of the reinforcement. Even if oxygen availability for the cathodic reaction (Fig. 8) is limited, the large cathode–small anode area ratio could lead to significant corrosion rates.[40] This idea of macrocell corrosion systems could be significant where external steelwork, for example riser pipes in offshore platforms, is inadvertently connected to the reinforcement resulting possibly in corrosion of the external steelwork.[36] Little evidence of this phenomenon has so far been

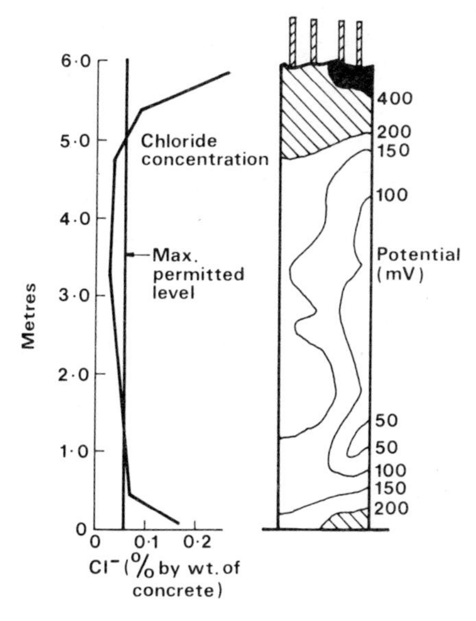

FIG. 11. Potential and chloride measurements on partially completed columns in the Middle East adjacent to shoreline.[34]

TABLE 5

STEEL POTENTIALS VERSUS STATISTICAL CHANCE OF CORROSION

Range of potential (vs CSE[a]) *(mV)*	*Chance of corrosion* *(%)*
more positive than −200	5
−200–300	50
more negative than −350	95

[a]CSE—Copper sulphate electrode.
NB: 100 mV = 0·1 V.

reported from existing structures. Similarly, large corrosion currents may also be set up where steel protection systems are accidentally connected to reinforcement, or where welding is employed using any available steel as the earth in the arc welding process.

4.4.7 Fatigue Properties of Reinforced Concrete

From an analysis of wave loading on offshore concrete structures it can be seen that a large number of loading cycles will occur (2×10^8 in 25 years).[41] This cyclic stressing may produce significant levels of fatigue loading for certain parts of offshore structures such as the base of towers. When considering reinforced or prestressed concrete sections, steel stress levels are a greater percentage of the ultimate stress than the stresses imposed on the concrete. It is therefore generally sufficient to consider steel fatigue properties as the controlling factor in the performance of the structural element.

The analysis of stress corrosion is based on the summation of the partial lives expected at each cyclic stress level (Miner's Rule),

$$1 > \sum \frac{n}{N} = \frac{n_1}{N_1} + \frac{n_2}{N_2} + \frac{n_3}{N_3}$$

where n = number of cycles over a particular stress range and N = number of cycles to failure at that range.

In normal design practice the limit for n/N is usually taken as 0·2.

In order to carry out the analysis it is therefore necessary to know the fatigue limit N at each stress level in the stress loading spectrum. Stress range versus number of cycle plots (S–N curves) have been determined experimentally for a number of steels protected and unprotected by concrete cover and immersed in fresh and salt water (Figs. 12 and 13). It can be seen from Fig. 12[42] that the number of cycles which can be maintained at high stress ranges (300 N/mm^2) without failure falls dramatically as steel is immersed in fresh and salt water, e.g. 10^7 cycles of bare steel immersed in seawater reduce the tolerable stress range to 100 N/m^2. It would appear from initial work (Fig. 13)[25] that the provision of sufficient concrete cover to the steel (75 mm) even when this is seawater saturated and the concrete purposely cracked, increases both the number of cycles and the stress range over which cycling can be accommodated without failure.

4.5 SPECIFYING MATERIALS AND WORKMANSHIP

The concept underlining the approach to concrete deterioration, which was outlined in the introduction, must be borne in mind when specifying materials and workmanship, as it highlights the importance of concrete cover in the overall durability of a structure to be placed in the marine environment.

In view of the susceptibility of the hydrated cement to attack from sea salts, specifically magnesium sulphate on the hydrated calcium aluminate, a cement with a low tricalcium aluminate (C_3A) content should be specified. A considerable number of exposure trials[43–55] have

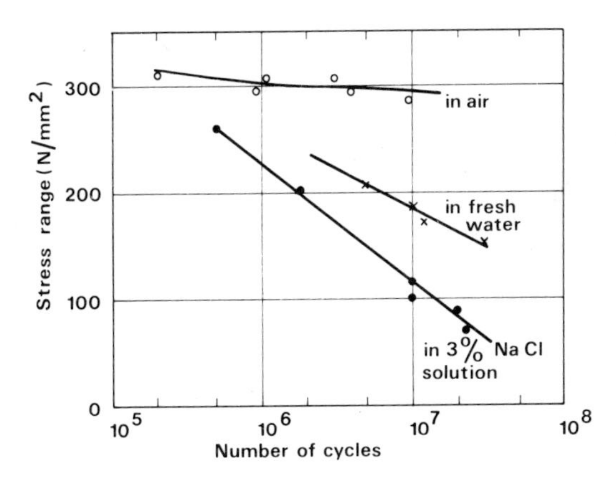

FIG. 12. *S–N* curves for 0·35 % carbon steel in air, fresh water and sodium chloride solution.[42]

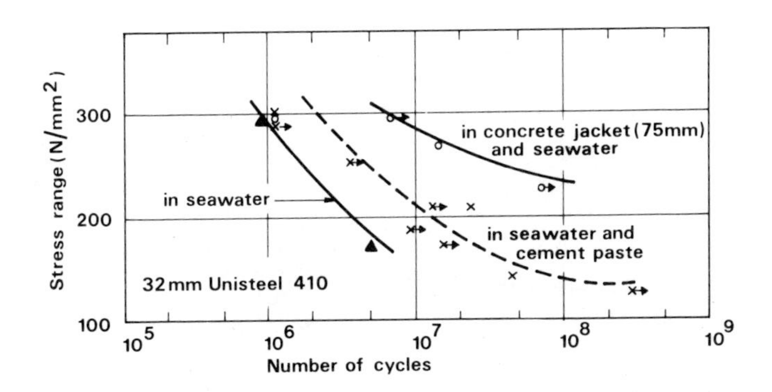

FIG. 13. *S–N* curves for steel with and without concrete protection.[26]

shown a good correlation between C_3A content and durability, providing other factors influencing durability are also taken into account. The figure of not more than 8% C_3A (Fig. 1) is reasonable[51,52] providing the water/cement ratio is kept low[4] as this is also a dominant factor in concrete permeability. Similarly it is also worth noting that certain concrete aggregates have high permeabilities compared to hydrated cement paste (Table 3).

In critical areas such as the splash zone, or under certain circumstances, additional measures may be specified. The use of dense durable aggregates, low water/cement ratios and an air entraining agent are protection against freeze–thaw damage. Cement replacement material such as PFA (pulverised fuel ash) and granulated blastfurnace slag can both be used to enhance concrete workability and may also improve concrete's resistance to chemical attack by seawater.[5] Admixtures to increase the workability of low water/cement mixes may also be valuable, especially in difficult or heavily reinforced concrete sections where compaction is a problem.

The contamination of aggregates and mixing water by salts can occur in certain areas of the world. A number of authorities[56,57] specify maximum chloride and sulphate contents for aggregates in order to limit the total salts by weight of cement, which can be accepted in hardened concrete. Measures to produce high quality cover are pointless if contaminated aggregates or water are used. For sodium chloride the critical level of salt depends on whether the salt was added at the mixing stage, i.e. by contaminated aggregates or water, or whether the salt had penetrated from the surface at a later date. This distinction is drawn because salt present in the mix initially may combine with hydrating cement and therefore be removed in part from the concrete pore solutions. A five-fold decrease in soluble chloride concentration has been found in chloride contaminated cement as the C_3A content of the cement was increased from 0 to 12·6%.[58,59] The suitability of the final mix design and the ease with which it can be placed and compacted are at least as important to the quality of the cover as the materials used. Specifying authorities normally base their recommendations on a maximum water/cement ratio, minimum cement content and minimum characteristic strength. It is believed[4] that the benefit derived by using a high cement content is largely the result of achieving a high compacted concrete density. In order to ensure good concrete durability the quantity of cement paste should exceed the void volume of the compacted aggregate to be used. For example, concrete made with a rounded gravel aggregate (contains 22 % voids) requires approximately 30 % of cement by volume. This mix results in an over filling of voids of 8 % and this ensures ease of compaction on site and a dense, low permeability mix, particularly in the cover zone. Water required to produce high workability can be minimised by the use of water reducing or superplasticising admixtures.

A number of authorities[57,60–62] have laid down specifications for concrete to be used in marine environment (see Table 6). It can be seen

TABLE 6

CONCRETE DESIGN SPECIFICATIONS FOR THE SPLASH ZONE OF MARINE STRUCTURES

	Cover (mm)	*Crack widths (mm)*	*Maximum water–cement ratio*	*Minimum cement content (kg/m³)*	*Permeability (m/s)*
DNV[60]	50	–	0·45	400	10^{-12}
FIP[61]	75	0·004 × cover (0·3)	0·45	400	–
CP110 Very severe[57]	60	0·004 × cover (0·3)	0·5	320–410	–
ACI[62]	65	–	0·40	360	–

that all specifications recognise the importance of cover thickness, cement content and cement type, although only the DNV specification recognises that the recommendations are designed in part to achieve a minimum permeability. However, it can be seen from Fig. 5 that this specification for a permeability of 10^{-12} m/s is $\frac{1}{100}$ of that which might be expected for normal air cured site concrete. It should, therefore, be apparent that specifications limited only to materials, depth of cover, etc. cannot be relied upon to produce a satisfactory level of cover concrete quality. Specifications should not be blindly applied without some knowledge of the processes at work in the deterioration of marine concrete.

The interrelationship between cement contents, water/cement ratio, cement types, and the importance of these factors in concrete durability are shown in Fig. 14 where minimum cement content is plotted against maximum water/cement ratio for a number of exposure tests carried out during the past 30 years, where no concrete damage was reported.[43–55] It can be seen that points for durable marine concrete fall into a broad band where high minimum cement contents have been used with low original water/cement ratios. More chemically resistant cements such as high alumina, sulphate resisting and Portland blastfurnace cement allow lower cement contents to be used together

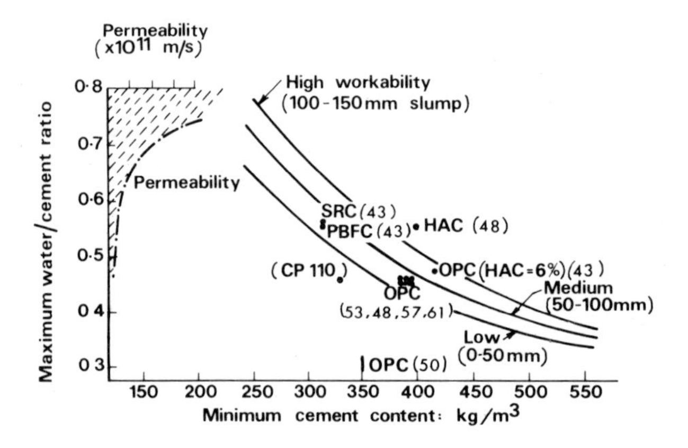

FIG. 14. Summary of recommendations of critical mix values to ensure marine durability. Reference numbers in brackets.[4]

with higher water/cement ratios. For most cements the limiting mix values fall within the workability lines, showing that if the water/cement ratio is controlled below a maximum value, and a medium workability mix produced, then the minimum cement content needed to satisfy the mix properties is automatically achieved. Further, where high strength concrete is used, which will demand a high cement content, then a lower durability cement may well perform satisfactorily.

Finally, it is well worthwhile specifying some form of quality control or inspection before the structure is actually exposed to marine conditions. By identifying and repairing areas of honeycombing, checking expansion joints and making sure the cover to steel is as specified using a cover meter, any last improvements to the cover protection can be made.

4.6 DESIGN SPECIFICATION

Attention to design is an important factor in marine concrete durability. The designer can include protective measures which will increase concrete durability, especially in critical areas such as the splash and tidal zones. Attention to details of design, notably density and layout of reinforcement, will also ensure that no problems are encountered on site with efficient compaction of concrete in susceptible areas such as the undersides of deck beams and on column corners. Similarly, in areas which are particularly vulnerable such as deck beams, soffits, etc. in the splash zone, provision of sloping concrete surfaces which allow spray to run off rather than collect on the concrete, can prevent high salt concentrations building up on the concrete by evaporation. In addition, corner sections allow biaxial penetration of salt and oxygen to occur at corner reinforcement.

4.6.1 Cover Thickness

The cover thickness decided on in design should be related to the quality and hence permeability of the concrete to be used, and the severity of exposure, as this determines the time (t_1) in the concept of deterioration outlined in the introduction. Figure 15 shows a possible relationship between recommended cover thicknesses given in CP 110 for an assumed concrete permeability of 10^{-10} m/s and durability.[36] It is interesting to note that the slope of the lines of constant durability fall as exposure becomes more severe. Therefore, slight changes in concrete permeability require considerable changes in cover thickness to maintain a constant degree of protection. Cover to steel in vulnerable areas such as corners should be increased or corner sections eliminated in design, for three reasons:

1. (1) Increasing cover allows compaction to be carried out more effectively in inaccessible areas.
2. (2) Expansive corrosion pressures can be contained more effectively (Fig. 10).
3. (3) Biaxial penetration of salts to the corner reinforcing bars can be minimised.

4.6.2 Cracking

In design it is generally accepted that all concrete structures will contain cracks, either of the macro or micro variety.

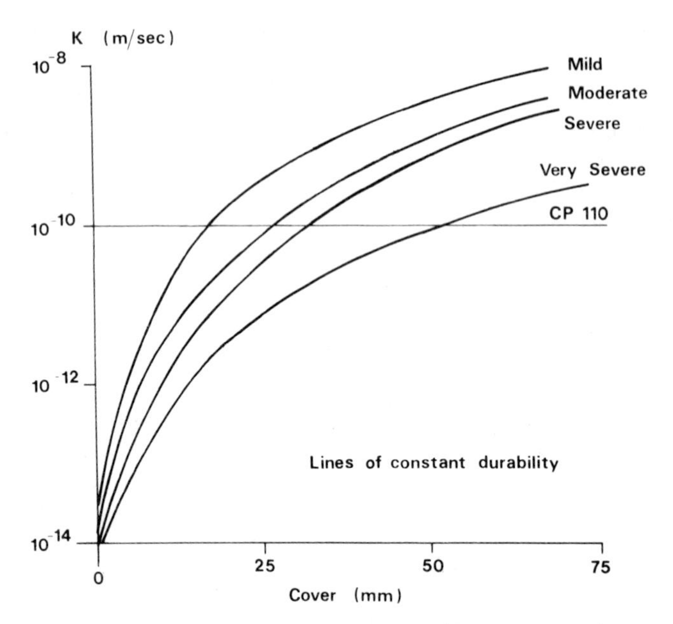

FIG. 15. Possible effect of concrete permeability cover and exposure on durability.

There are three principal causes:[63]

(1) Structural loading
(2) Drying shrinkage
(3) Thermal expansion (differential)

Unlike in steel structures the presence of cracks in concrete structures does not generally impair its performance, as the concrete is primarily carrying compressive loads, and in the tensile crack prone areas most of the stress is taken by the reinforcing steel. It is usually impossible to predict the exact location of cracks although careful reinforcement design, mix design and attention to casting and curing of the concrete will ensure that cracking is kept to a minimum prior to the exposure of the concrete to seawater.

Considerable research has been undertaken on the influence of crack widths on embedded steel corrosion.[30] Laboratory experimentation in salt solution[64] shows that there is a definite correlation between crack widths and corrosion for cracks above a certain size (0·2 mm) due perhaps to loss of lime protection to steel (Fig. 16). However, many of these exposure trials carried out[32] have shown little correlation between crack width (up to 1·5 mm) and corrosion for cracks propagated normal to the line of reinforcement. It appears that for short periods of exposure in the laboratory there is a correlation between crack widths (normal to reinforcement) and corrosion which is not maintained over

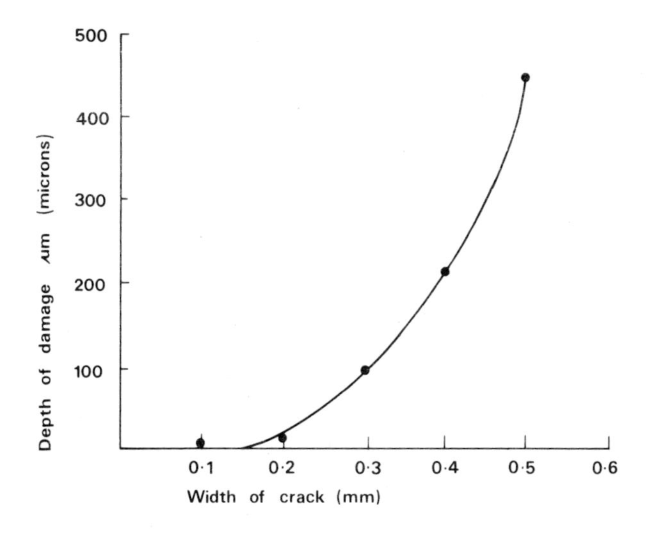

FIG. 16. Average corrosion of reinforcement in cracks after immersion in 6 % NaCl solution for 6 months.[64]

the design life of a concrete structure. Therefore, although codes of practice limit crack widths allowable in design (the CP 110 limit is 0·004 times cover) such recommendations to check widths may not be of great value.[30] At the present time little work has been carried out on steel corrosion due to cracks parallel to the reinforcement as they are not generally load induced. These cracks are likely to be more significant to corrosion of steel.

Although cracked concrete is more likely to deteriorate than uncracked concrete, the amount of corrosion appears to be structurally insignificant.[30] The explanation of this phenomenon is based on the fact that with a narrow crack, there is a negligible supply of oxygen through dense concrete to cathodic areas of the corrosion cell (see Fig. 8), the anode being the steel exposed in the crack.

4.6.3 Additional Protection During Design

In addition to design specification relating to cover, reinforcement layout, and allowable cracking, a number of other techniques are available and can be specified at the design stage in order to give increased protection to concrete, especially to the reinforcing system, i.e.,

(1) Coatings;
(2) Cathodic protection by impressed currents or sacrificial anodes.

Coatings can be used either to provide additional protection directly to the steel, e.g. by galvanising [65,66] or as sealants painted on to the concrete or steel[66,67] to prevent the ingress of salt, moisture, oxygen, etc. necessary for the corrosion reaction to continue at a significant rate. It has been shown that concrete, contaminated with chloride, can be protected from reinforcement corrosion by coatings (Fig. 17) which prevent sufficient oxygen supply to the steel to maintain a high corrosion rate. This effect has also been observed in membrane coatings for concrete.[68]

Coatings applied to reinforcement may be effective in preventing corrosion but care must be exercised to prevent damage prior to concreting, and loss of bond strength between steel and concrete may occur. Alternatively, galvanised reinforcement can be used, although in certain situations the galvanising may result in the evolution of hydrogen during hardening of the concrete. This can cause loss of bond of the concrete to the reinforcement. However, the addition of small quantities of cadmium or traces of chromates present in most UK cements may effectively inhibit this effect,[69] although other evidence[70] indicates that hot dip galvanised reinforcement can produce cracking and loss of bond to concrete due to corrosion of the zinc coating and that the effect is not prevented by the presence of chromates in the cement.

4.6.4 Cathodic Protection

Cathodic protection of steel in concrete was suggested for prestressed concrete pipes back in the early 'sixties. However, its use as a remedial measure has at present only been determined on a laboratory scale.[71] It was found that corroding steel in salt laden concrete could be repassivated by controlling its potential at a level of -710 mV vs the copper/copper sulphate electrode. Although usually applied underwater, cathodic protection of reinforcement has been tested above water on bridge decks in Canada with some success[72] by using a low resistivity (about 10 ohm cm) asphalt coke mix to spread the current.

For offshore platforms it is almost certain that the underwater sections of concrete will be under the influence of cathodic protection systems, due to accidental (or otherwise) connection of cathodically

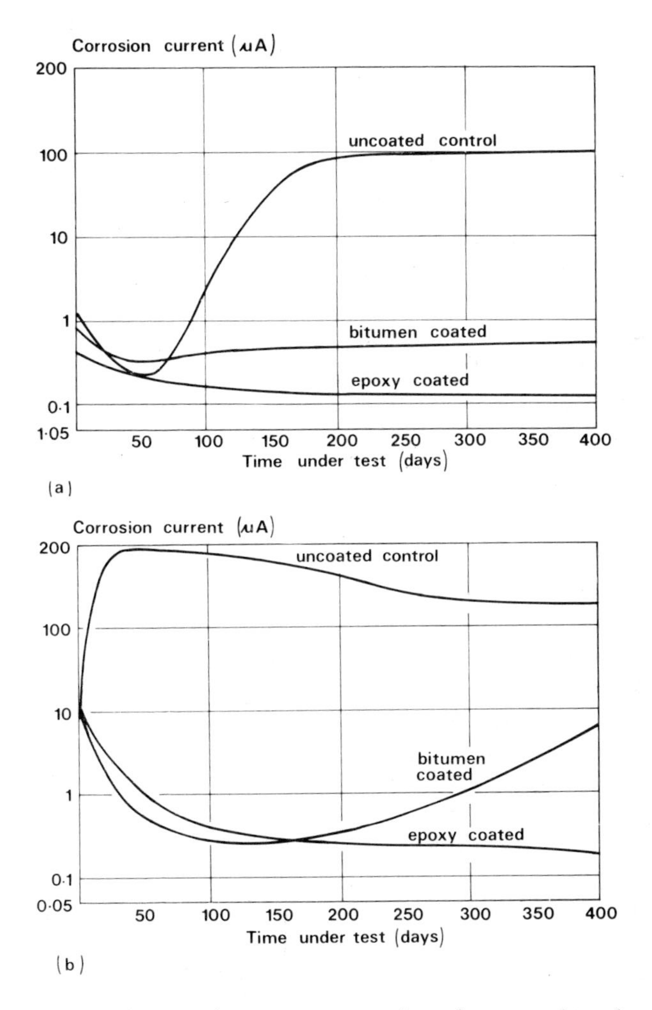

FIG. 17. Measured corrosion current vs time for coated and uncoated good/poor concrete containing steel. (a) Results of current vs time for good concrete. (b) Results of current vs time for poor concrete.

protected external steelwork to the reinforcing grid. The steel can also be protected with distributed sacrificial anodes.[73] There will possibly be quite a substantial loss of current to the reinforcement (the DNV regulations[60] suggest 1–1·5 mA/m^2 of concrete surface) amounting to up to $\frac{1}{3}$ of the total current required to protect a steel structure. However, the soundness of such a system still has to be fully evaluated as there are data[74] to suggest that concrete covered steel could still be corroding even at the normal cathodic protection level of −850 mV vs the silver–silver chloride electrode.

4.7 INSPECTION TECHNIQUES

The need to detect early deterioration in reinforced or prestressed concrete has increased considerably, as the material is being used under much more severe exposure conditions such as in the North Sea.

4.7.1. Concrete Inspection

Various non-destructive testing techniques are available to assess concrete quality, especially the cover concrete:

(1) Impact devices, such as the Schmidt hammer;[75] for measuring concrete elasticity.
(2) Ultrasonic pulse velocity; the velocity of an ultrasonic pulse passing through concrete can be measured using a proprietary 'Pundit' apparatus, and is related to *in situ* strength and density.
(3) X-ray photography; detection of voided or poorly compacted concrete.[76]
(4) Gamma radiography; reading can be related to density of concrete.[75]
(5) Cover depth of steel; using the proprietary cover meter.

These techniques used in conjunction with each other can provide an overall assessment of the state of the concrete cover and identify areas of potential deterioration which can then be examined more closely by:

(*a*) *Chloride drillings.* A small rotary percussion drill can be used to obtain powder samples of concrete at various depths. Chemical analysis for chloride can then be carried out and the extent of chloride penetration with depth to the reinforcement calculated. This simple technique provides valuable information on the chloride levels to be expected at reinforcement level, whether this is significant, and the extent to which the concrete may need to be cut back to remove the chloride contamination.

(*b*) *Cores.* One valuable but somewhat destructive method of assessing deterioration is to drill cores from the concrete. These can then be carefully analysed in the laboratory and the cause of deterioration of either the concrete or steel firmly determined. The extent of chemical deterioration of concrete can only be determined from core analysis, by using techniques such as X-ray diffraction and thin section microscopy. The depth of concrete carbonation can be measured using phenolphthalein and the permeability of the concrete determined.

In certain circumstances coring may be the only way to establish the reasons for concrete deterioration and it is vital that the reason for deterioration is known, i.e. chemical deterioration, expansive aggregates, and carbonation before remedial measures are undertaken.

4.7.2 Steel Inspection

(*a*) *Potentials* (*see previous section*). Above high tide level, corroding areas of reinforcement can be detected by measuring the potential of the steel using a copper/copper sulphate reference electrode as discussed earlier (Fig. 11).

(*b*) *Resistivity* (*see previous section*). Concrete's resistivity can be measured using a four probe system,[77] which indicates the susceptibility of the concrete to encourage corrosion of embedded steel.

(*c*) *Linear polarisation.* The measure of the electrolytic potential of steel does not give an idea of its corrosion rate. This rate can be estimated by polarisation techniques where changes in potential are measured against an applied current using a counter and reference electrode system.[36] The corrosion rate measured is, however, also a fraction of the area of steel polarised during the test and this may be difficult to quantify.

Special NDT techniques are also necessary for the underwater inspection of concrete structures. At the present time, underwater electrochemical testing is difficult as seawater short circuits the areas of concrete under examination unless these are isolated.

4.8 REPAIR TECHNIQUES

Repair techniques which can be applied to concrete structures that have deteriorated, due to chemical attack from seawater, or reinforcement corrosion, can be sub-divided into two types:

(1) *Repair methods which cut back concrete for eventual replacement.* Replacement of deteriorating concrete can only succeed if the cause of deterioration is known and the remedial measures are then applied to a sound concrete surface. The main conclusion of a study on repairs to concrete structure carried out in the UK[38] was that the major cause of concrete deterioration was reinforcement corrosion. Deck beams were normally the most seriously affected parts of structures, and repair techniques only appeared to be successful where the underlying cause of deterioration, i.e. the reinforcement, was substantially renewed or cleaned up and passivated if possible. Guniting and epoxy repairs were often unsuccessful if applied purely to 'patch up' a structure, as the underlying cause of deterioration, e.g. chloride contaminated concrete and steel, still remained.

Where other causes of concrete deterioration are concerned, such as sulphate attack and alkali reactivity, it is essential to establish first the cause of deterioration and secondly the extent. Once the cause of the deterioration has been identified, concrete can be replaced using properly tested and approved materials.

(2) *Novel repair methods.* A number of novel repair methods have been suggested to protect reinforced concrete from corrosion. Cathodic protection has been discussed in a previous section as have coatings which are applied to concrete or steel to prevent chloride moisture or oxygen access necessary for electrochemical corrosion. In additon, a repair technique has also been described[78] in which active steel is repassivated using ion exchange resins. The concrete is initially covered in a mixture of ion exchange resin and calcium hydroxide solution (the electrolyte). This layer is then covered by metal anodes across which a potential of 100 V is applied using the reinforcement as the cathode. The applied potential produces a migration of chlorides into the resin

towards the metal anodes. Potentials taken before and after treatment showed that most of the reinforcement had returned to a passive state.

4.9 CONCLUSION

Concrete has been used successfully for marine construction for many years and, in most cases, has proved to be an ideal construction material requiring little in the way of maintenance. The durability of good quality concrete designed and constructed with regard to all the points mentioned previously is not in doubt, as can be seen from exposure trials. However, the term 'concrete' includes an enormous range of good, average and extremely poor materials with a wide variation in properties, and a corresponding variation in durability for different exposure situations.

The idea that good quality concrete is more durable is clear, although the methods by which quality and therefore durability are achieved is less obvious. Achieving high quality concrete, especially in the cover to the reinforcement begins with materials' selection and mix design. The objective at this stage is to produce a dense, impermeable, chemically durable concrete with a minimum voidage, which on site can be easily and efficiently compacted into formwork and yet has a low water/cement ratio. This may involve the use of normal workability admixtures. Attention to the layout and density of the reinforcement and the shape of concrete members must also be considered. When placed, the concrete must be cured thoroughly by ponding, spraying, etc. with fresh water, or effectively sealed with a proprietary curing membrane in order to get complete hydration of the cement and as low a concrete permeability and diffusivity as possible.

The specifications available (DNV, FIP, CP 110, ACI) lay down minimum cement contents, maximum water/cement ratios and minimum cover to reinforcing and prestressing steels, but fail to stress what these recommendations are actually aimed at, i.e. producing a high quality concrete with as low a permeability as possible which is sufficiently thick over the reinforcing to protect it against marine salts and freeze–thaw conditions.

At present, specifications merely state that a concrete, made with such a cement content and water content, achieving a minimum strength of $x\,N/mm^2$ at 28 days should be used. Recently, however, as can be seen from Table 6, the required permeability of concrete has been specified in one case, as this property is a more reliable indicator of the quality of concrete necessary for marine works. However, there are no guidelines as to how this low permeability is to be achieved.

The inspection of offshore and coastal concrete structures can be more thoroughly carried out by using a number of NDT techniques. In addition to a simple visual survey an NDT examination can undoubtedly highlight those areas of a structure which need to be repaired, as well as those areas which, although not visually deteriorated, require additional protective measures before damage occurs, such as coatings or cathodic protection.

Repair of concrete structures will always be an extremely expensive business and it is therefore well worthwhile carrying out a thorough inspection and repair programme to avoid high repetitive maintenance costs during the life of a structure.

REFERENCES

1. Biczoc, I. (1967). *Concrete Corrosion and Concrete Protection.* Akademiai Kiado, Budapest. pp. 353–366.
2. Sverdrup, H. V., Johnson, M. W. and Fleming, R. H. (1946). *The Oceans, their Physics, Chemistry and General Biology.* Prentice-Hall, Englewood Cliffs, NJ, USA.
3. Myers, J. J., Holm, C. H. and McAllister, R. F. (Eds.). (1969). *The Handbook of Ocean and Underwater Engineering.* McGraw-Hill, London. 1094 pp.
4. Browne, R. D. and Domone, P. L. (1974). *Conference on Offshore Structures.* Institute of Civil Engineers, London.
5. Mather, B. (1966). *Highway Research Record No. 113.*
6. Wolochow, D. (1952). *Proc. ASTM*, **52,** 250–66.
7. Neville, A. M. (1969). *J. Materials*, **4** (4), 781–816.
8. Haynes, H. H. and Zubiate, P. C. (1973). *Tech. Note No. 1308.* Naval Civil Engineering Laboratory, Port Hueneme, USA.
9. Powers, T. C. (1975). *SP. 47-1 ACI Durability of Concrete Publication.* Detroit, USA.
10. Powers, T. C. (1949). *Bulletin ASTM*, No. 158, 68–76.
11. US Bureau of Reclamation (1955). *Concrete Laboratory Report C-810.* Denver, Colorado, USA.
12. Neville, A. M. (1977). *Properties of Concrete.* Pitman Press, London.
13. Lyse, I. (1935). *J. Am. Concrete Inst.*, **32** (12), 1575–84.
14. National Building Studies (1953). *HMSO Research Paper No. 30.* HMSO, London.
15. Haynes, H. H., Highberg, R. S. and Nordby, B. A. (1976) *Note TN-1436.* Naval Civil Engineering Laboratory, California, USA

16. WAAGAARD, K. (1977). *Offshore Technology Conference, Vol. 4*, American Inst. of Mining Metallurgy and Petroleum Engineers Inc., Houston USA. pp. 341–50.
17. VALENTA, O. (1969). *Int. Symp. on the Durability of Concrete, Prague.*
18. BARRER, R. M. (1951). *Diffusion in and through Solids.* Cambridge University Press, Cambridge.
19. POWERS, T. C. (1958). *J. Am. Ceram. Soc.*, **41,** (1), 1–6.
20. POWERS, T. C., COPELAND, L. E., HAYES, J. C. and MANN, H. M. (1954). *J. Am. Concrete Inst.*, **51** (11), 285–98
21. HAYNES, H. H. and KAHN, L. F. (1972). *Technical Report R774.* Naval Civil Engineering Laboratory, California, USA.
22. RUETTGERS, A., VIDAL, E. and WING, S. (1935). *Proc. Am. Concrete Inst.*, **31,** 382–416.
23. COOK, H. K. (1951). *Proc. ASTM*, **51,** 1156–65.
24. MARION, H. and MAHFOUG, G. (1974). *Proc. Inst. Civil Eng.*, **56** (1), 497–511.
25. TROXELL, G., DAVIS, H. and KELLY, J. (1968). *Composition and Properties of Concrete*, 2nd Edition. McGraw-Hill, London. p. 267.
26. BROWNE, R. D. and DOMONE, P. L. (1975). *Int. Conf. on Underwater Construction Technology*, U.C. Cardiff, April. University College.
27. HOUGHTON, D. L., BORGE, O. E. and PAXTON, J. A. (1978). *Proc. Am. Concrete Inst.*, **75** (12), 664–7.
28. INTERNATIONAL NICKEL CO. (1967). A Guide to the Selection of Marine Materials. *Imco Handbook 4050.*
29. VAN DAVEER, J. R. (1975). *J. Am. Concrete Inst.*, **72** (12), 697–704.
30. BEEBY, A. W. (1978). Concrete in the Oceans, *Tech. Report No. 1.* Cement and Concrete Association, London.
31. TUTTII, K. (1977). CBI Research Translation 4:77 (1977). *Concrete in the Oceans Publication SG1/S6*, CIRIA.
32. BEEBY, A. W. (1978). *J. Inst. Structural Engrs.*, **S6A** (3), 77–81.
33. BROWNE, R. D. (1973). *Corrosion in the Marine Environment Symposium.* Inst. of Marine Engineers, London.
34. BROWNE, R. D. (1978). *Colloquium on Inspection and Maintenance of Structures.* IABSE, Pembroke College, Cambridge. p. 164.
35. FAIRBRIDGE, R. W. (Ed.). (1966). *Encyclopedia of Oceanography.* Van Nostrand Reinhold, London.
36. BROWNE, R. D., DOMONE, P. L. and GEOGHEGAN, M. P. (1977). *Offshore Technology Conference, Houston, Texas, USA.* American Inst. of Mining Metallurgy and Petroleum Eng. Inc.
37. LILLIAN, LUNG-YU CHOU CHEN and KATZ, D. L. (1978). *J. Am. Concrete Inst.*, **75** (12), 673–9.
38. DYTON, F. J. (1977). *Conference on Maintenance of Maritime Structures* Institute of Civil Engineers, London.
39. GILCHRIST, J. P. (1965). *Prestressed Concrete Development Group Research Report No. 1.* London.
40. ANON. (1976). Huge Corrosion Cells, a Hidden Danger. *Offshore Engineer*, October.

41. Bury, M. R. C. and Domone, P. L. (1974). *Offshore Technology Conference, Houston, Texas, USA.* American Inst. of Mining Metallurgy and Petroleum Eng. Inc.
42. Glikman, L. A. (1972). *Corrosion, Mechanical Strength of Materials* (translation from Russian by J. A. Shapiro). London Butterworth, Oslo.
43. Gjorv, O. E. (1968). *Durability of Reinforced Concrete Wharves in Norwegian Harbours.* Norwegian Committee on Concrete in Sea Water, Oslo. p. 208.
44. Expanded Shale, Clay and Slate Inst. (1960). *Story of the Selma-expanded Shale Concrete Endures the Ravages of Time.* Expanded Shale, Clay and Slate Inst., Washington, USA.
45. Mishutin, V. A. (1972). *Proc. Federation Internationale de la Precontrainte Symp. Concrete Sea Structures, Tbilisi.* FIP, London, pp. 173–6.
46. Gewertz, M. W., Tremper, B., Beaton, J. L. and Stratfull, R. F. (1957). *Highway Research Board Bulletin No. 182.*
47. Sansoni, R. and Nicotra, L. (1970). *Galvanised Reinforcement for Concrete.* Int. Lead Zinc Research Organisation, New York. Appendix D.
48. Eckhardt, A. and Kronstein, W. (1950). *Deutsche Auschuss für Stahlbeton,* 102.
49. Wesche, K. (1966). *RILEM Bulletin No. 32,* 291–4.
50. Campus, F., Dantinne, R. and Dzulynski, M. (1965). *RILEM Symposium, Behaviour of Concrete in Sea Water, Palermo.* RILEM.
51. Brocard, J. and Cirrode, R. (1966). *RILEM Bulletin No. 32,* 323–9.
52. Gjorv, O. E., Gukild, I. and Sundh, H. P. (1966). *RILEM Bulletin No. 32,* 305–22.
53. Cook, H. K. (1952). *Proc. ASTM,* **52,** 1169–81.
54. Kennedy, T. B. and Mather, K. (1953). *Proc. Am. Concrete Inst.,* **49,** 141–72.
55. Thomas, G. H. *Group Research Report.* British Steel Group.
56. Fookes, P. and Collis, L. (1976). *Concrete,* **10** (2), 14–19.
57. British Standards Institution (1972). *CP110, Code of Practice for the Structural Use of Concrete.* BSI, London.
58. Monfore, G. M. and Verbeck, G. J. (1960). *Proc. J. Am. Concrete Inst.,* **57,** 491.
59. Whiting, D. (1978). Paper No. 73, *Corrosion 78, Houston, Texas,* USA. National Association of Corrosion Engineers, Texas, USA.
60. Det Norske Verit (1977). *Rules for the Design, Construction and Inspection of Fixed Offshore Structures.* DNV, Oslo.
61. FIP (1977). *Recommendations for the Design and Construction of Concrete, 3rd Edition.* FIP, London.
62. ACI (1978). Guide for the Design and Construction of Fixed Offshore Concrete Structures. *ACI Journal Committee 357,* **75** (12), 684–707.
63. Browne, R. D. and Domone, P. L. (1963). *Materials for Underwater Technology Symposium.* Admiralty Materials Lab., Poole.
64. Novgorodski, V. I. Translation from *Journal of Practical Chemistry (Russian)* **40** (3), 555–60.
65. Baker, E. A., Money, K. L. and Sanborn, C. B. (1977). *ASTM STP 629.* ASTM, USA.

66. Bird, C. E. and Strauss, F. J. (1967). *Materials Protection*, July, 48–52.
67. Clifton, J. R., Beeghly, H. F. and Mathley, R. G. (1975). US *Report No. NBS BSS-6S*. National Bureau of Standards, Washington, USA.
68. Stratfull, R. F. (1959). *Corrosion*, **15** (6), 65–8.
69. Everett, L. H. and Treadaway, K. W. (1970). *BRE Current Paper 3/70*. Building Research Establishment, Garston, Watford, UK.
70. Corrosion Institute Denmark (1978). Galvanised Steel in Concrete, A Necessary Warning. *Electroplating and Corrosion Newsletter*. Corrosion Inst., Denmark.
71. Hausman, D. A. (1969). *Materials Protection*, October, 23–5.
72. Framm, H. J. (1976). Paper 19, *International Corrosion Forum, Houston, Texas*. National Association of Corrosion Engineers, Texas, USA.
73. Lowe, R. A. (1977). *Offshore Services*, December, 24.
74. King, R. A., Nabizadah, H. and Ross, T. K. (1977). *Corrosion Prevention and Control*, **24** (2), 11–13.
75. Jones, R. (1969). *Non Destructive Testing of Concrete and Timber Symposium, London*. Inst. of Civil Engineers, London. pp. 1–7.
76. Brown, B. R. and Kelly, R. T. (1969). *Non Destructive Testing of Concrete and Timber Symposium, London*. Inst. of Civil Engineers, London. pp. 67–75.
77. Wenner, F. (1915). *Bull. of the Bureau of Standards*, **12,** 469–78.
78. Slater, J. E. and Lankard, D. R. (1976). Paper 20, *Corrosion 1976 Int. Corrosion Forum*. National Association of Corrosion Engineers, Texas, USA.

Chapter 5

CONCRETE UNDER COMPLEX STRESS

J. B. NEWMAN

Department of Civil Engineering, Imperial College of Science and Technology, London, UK

SUMMARY

The subject is introduced by a brief history of the testing of concrete under multiaxial stress and a discussion of why data produced by such tests are needed for structural design. In the following sections of the chapter considerable emphasis is given to the problems of testing techniques, specimen loading device interaction, and the need to use suitable methods to induce uniform and definable states of stress within elements of concrete under test. The main part of the text describes the stress–strain behaviour and failure mechanisms of concrete specimens under various boundary conditions of stress and strain, with particular reference to biaxial and triaxial states of stress. Finally, methods of representing strength and deformational data are discussed and comment made on the requirement to produce mathematical relationships which can be incorporated in computer-based techniques for structural analysis.

5.1 INTRODUCTION

The influence of combined stresses on the strength response of materials in structural engineering was first considered in the 18th Century, and over the years many failure criteria have been proposed for such materials. These criteria have been reviewed by Marin,[1] Nadai[2] and Timoshenko.[3]

For concrete, however, the first test to be used as a measure of the quality of concrete was the cube test proposed by Grant in 1872. Due to its simplicity of manufacture and test, it was adopted as a method for quantifying the 'strength' of the material. At about the same time, on a more fundamental level, Foppl reported the results of tests on hydraulic cement-based materials in which their behaviour under multiaxial compression was studied both by applying loads to cylinders surrounded by thick steel jackets and loading cubes biaxially. It is interesting to note that even at this early stage he recognised in his work the problem of frictional restraint at the boundary between the testing machine and the specimen. Considere in 1905 expressed the results of tests on mortar cylinders under axial compression and lateral hydraulic pressure in terms of the following type of relationship.

$$\text{Axial stress at failure} = K_1 + K_2 \times \text{confining pressure}$$

where K_1 and K_2 are constants.

Apparently work of this type continued only at the University of Illinois, and in 1928 Richart, Brandtzaeg and Brown[4] published the results of some tests which have become classics in the field of multiaxial testing. By loading cylinders axially in compression through rigid platens and with lateral confining pressures applied hydraulically, they studied the failure envelopes for various concretes under biaxial and triaxial compressive stress states. The principal outcome of this work was the now familiar equation for the ultimate strength envelope:

$$\sigma_1/\sigma_c = 1 + 4{\cdot}1\,\sigma_3/\sigma_c$$

where σ_c = uniaxial (cylinder) strength, σ_1 = axial compressive stress at ultimate, and σ_3 = lateral confining pressure.

This work was followed in 1929 by a further report[5] which presented the results of tests on spirally reinforced concrete cylinders simulating columns.

In the period up to the beginning of the Second World War only a limited number of investigations were conducted to study the behaviour of concrete under biaxial compression using cubes and slabs, and of mortars under triaxial compression using cylinders. Little work was carried out although interest was reawakened at the USA Bureau of Reclamation immediately after the Second World War.

In spite of the above studies it became generally accepted by the civil and structural engineer that, for design purposes, the mechanical properties of concrete could be adequately expressed in terms of the behaviour under relatively simple conditions of applied load such as uniaxial compression (using cubes or cylinders), and flexure (by bending beams). The results of such tests were incorporated as design parameters in codes of practice and they have served the extremely important function of providing a measure of the potential quality of the material. Indeed, by adopting standardised manufacturing and testing techniques, they have proved eminently satisfactory in this latter role.

However, it is known that the state of stress in such test specimens (a) is not easily defined in spite of the apparent simplicity of the loading technique, and (b) may not accurately reflect the condition of concrete under load in actual structures. Thus their use as a source of *design* information may be limited, unless the information obtained is used to describe a material from which laboratory test models are made to simulate elements or units in a structure. In fact, most codes of practice are derived from the results of empirical beam/column/slab tests which have been expressed as generalised design rules with concrete strength values incorporated as descriptors of the type of material used in the tests, rather than as measures of the fundamental material properties.

It has long been recognised that the behaviour of concrete under multiaxial loads can be markedly different from that under uniaxial boundary conditions. This aroused the interest of a number of researchers who directed their attentions towards evaluating the change in properties with stress state, with particular reference to ultimate strength. Such investigations were mainly academic in nature but, with improvements in design philosophy, structural analytical techniques and the use of concrete in more sophisticated structures, the interest in obtaining both stress and strain data under complex boundary conditions was re-activated in the late 1950s, after which research work was reported from the UK, USA, Japan, Australia, Germany, Italy, India and the USSR.

Perhaps the most widely used of the advanced analytical techniques is the finite element method which has been developed such that it is now possible to analyse three-dimensional problems with complex geometrical arrangements and boundary conditions. For the most realistic solutions this method requires information regarding the fundamental deformational properties of concrete under all stress states likely to be encountered by each element of material in the structural unit under consideration, throughout the entire loading regime.

Two main problems exist when attempting to provide information on the response of concrete and any other material to any particular type of stress state. First, that of inducing definable conditions in test specimens and secondly, presenting any data obtained in a form suitable for practical use. In the following sections testing techniques, the behaviour of concrete under complex stress, and methods of presentation will be discussed.

5.2 TESTING TECHNIQUES

5.2.1 Introduction

Most problems in structural engineering require the analysis of the response of an assemblage of members to an imposed boundary condition which is mostly specified in terms of a force (or stress), and sometimes as a displacement (or strain). Unless there is some experimental evidence to indicate the likely response of the particular geometrical arrangement to the given forces or displacements, then recourse must be made to a more fundamental solution by using the characteristics of the individual materials comprising the structure.

For such a fundamental approach, the stress–strain relationships of the materials must be determined by subjecting representative elements to the whole range of boundary conditions which it is possible to encounter in the structure, unless the materials are such that it is possible to deduce properties in one stress or strain state from those in another.

Since boundary conditions are more commonly specified as applied loads, it is generally assumed that the properties of a material can be determined merely by applying a load in an arbitrary manner to the boundary of a representative element of the material, and measuring the resulting behaviour. Such an approach, however, neglects the possible influence of the *structural characteristics of the material* on the type of stress or strain field induced in an element by the particular *method* in which the boundary load is applied.

An example of this can be seen when considering the way in which boundary loads are applied in standard tests used to determine material properties. This is normally accomplished by loading a specimen through a rigid steel platen arrangement. Such a method will apply a near-uniform boundary displacement but will only induce a uniform state of stress in an *ideally elastic, homogeneous and isotropic* material, and then only provided no secondary *restraint stresses* are present. For real materials, such as concrete, this may be far from the case since, for such a multiphase type of material, the difference in mechanical properties between the various constituents will induce considerable stress, and hence strain, disturbances. Usually no attempt is made to reduce such restraint stresses.

Other possibilities exist, the most extreme being that of applying uniform boundary stress to an element. For a multiphase material, the wide variation in properties of the various phases present will cause a non-uniform strain, and hence stress, distribution to be induced. This distribution, however, is likely to be different from that induced by the uniform boundary deformation condition. Of these two principal methods of applying a load to a material in a test, the engineer must decide that which will provide the most relevant information for his purposes.

Having decided, therefore, on the type of boundary condition which it is intended to apply to the material, it is necessary to assess the efficiency of any loading technique in attaining the desired state of stress or strain within the element under load for the results to be of practical use. This problem is discussed below.

5.2.2 Definition of the Problem

In evaluating the stress state within an element of material under load, the magnitude and direction of principal stresses should be determined at a sufficiently large number of positions within the volume of material for a given set of conditions on the boundary of the volume. However, to the author's knowledge, no technique has yet been developed which is capable of directly determining stress. Thus, unless an approximate analytical method is used, the only possible approach to the problem is to assess the stress state from measurements of strain.

Such an approach, although widely used, suffers from one fundamental disadvantage: namely, that of requiring a complete and accurate knowledge of the constitutive relations for the material in order to determine the stress state at each point which has given rise to the corresponding measured strains. In other words, we require to know the complete stress–strain characteristics of the material before tackling the problem.

This presents a seemingly insurmountable dilemma for an analytical or experimental investigation if the objective is to check the validity of tests which are intended to provide such complete stress–strain characteristics. Thus, only by use of simplifying approximations or assumptions can one arrive at some idea of the state of stress in a material.

The results of analytical and experimental investigations which have been carried out in an attempt to assess the stress state within specimens will be outlined, but before considering these, it is necessary to define the problems involved in transmitting loads from one material system to another.

5.2.3 Factors involved in Transmitting Load through Different Materials

In order to assess the fundamental load-carrying characteristics of a material it is usually required to subject the material to ideal boundary conditions of either uniform stress or uniform deformation. In order to attain these conditions it is necessary to transmit the load via another material resting against, or attached to, the boundary of the specimen under test. The material adjacent to the specimen may form an intermediate layer between the specimen and the main loading device and, indeed, may form only part of a composite subsidiary loading unit.

It is the incompatibility between the various materials forming the loading device and the material of the specimen, together with the conditions at the various interfaces, which can give rise to either positive (tensile) or negative (compressive) restraint stresses on the boundary of the specimen. Obviously, these restraints are undesirable, and should be minimised, but they cannot be avoided completely with practical loading systems. Since they will, inevitably, be present it is necessary to assess their magnitude and sign before evaluating the results of tests to establish the fundamental properties of materials.

5.2.3.1 *Uniform Boundary Displacement*

Many problems are associated with devices developed for applying a state of uniform displacement to the boundary of a material, since the load must essentially be applied to a specimen through platens which are rigid in the direction of loading. Under stress the lateral deformation of such platens will tend to be different from that of the specimen with the actual extent of the difference being influenced by the frictional characteristics of the platen/specimen interface. As a demonstration of the difficulties which can arise, the case of a solid platen will be considered: Assuming elasticity, the lateral strain in the platen for zero frictional restraint will be

$$e_{3\,\mathrm{plat}} = \frac{\sigma_3}{E_{\mathrm{plat}}} - \frac{\nu_{\mathrm{plat}}}{E_{\mathrm{plat}}}(\sigma_1 + \sigma_2) \tag{1}$$

where E_{plat} and ν_{plat} = elastic modulus and Poisson's ratio for platen material, and σ_1, σ_2 and σ_3 = principal stresses (compression positive)

but, since $\sigma_2 = \sigma_3 = 0$ for uniaxial stress, eqn. (1) becomes

$$e_{3\,\mathrm{plat}} = -(\nu/E)_{\mathrm{plat}}\sigma_1 \tag{2}$$

Similarly, the lateral strain in the specimen becomes

$$e_{3\,\mathrm{spec}} = -(\nu/E)_{\mathrm{spec}}\sigma_1 \tag{3}$$

If no material is interposed between the solid platen and specimen the loading device can be placed in one of the following categories:

(1) The platen strains laterally less than the specimen, i.e.

$$(\nu/E)_{\mathrm{plat}} < (\nu/E)_{\mathrm{spec}}$$

(2) The platen strains laterally more than the specimen, i.e.

$$(\nu/E)_{\mathrm{plat}} > (\nu/E)_{\mathrm{spec}}$$

If full or partial frictional restraint is developed at the interface then the differential movement can cause restraint stresses to be transmitted across the boundary to influence the specimen. The likely form of distribution of lateral restraint at the platen/specimen interface is shown in Fig. 1 which indicates zero restraint on the axis increasing to a maximum towards the corners. This trend has been confirmed by the results of the analytical and experimental investigations which will be described later.

Figures 1(a) and (b) show the form of lateral deformation profiles under uniaxial compression for cases (1) and (2), both without friction and with full friction. For the sake of simplicity it has been assumed that plane sections remain plane under load.

If the restraint stresses increase above the level capable of being maintained by friction at the platen/specimen interface, then the form of restraint distribution suggests that slippage should start near the

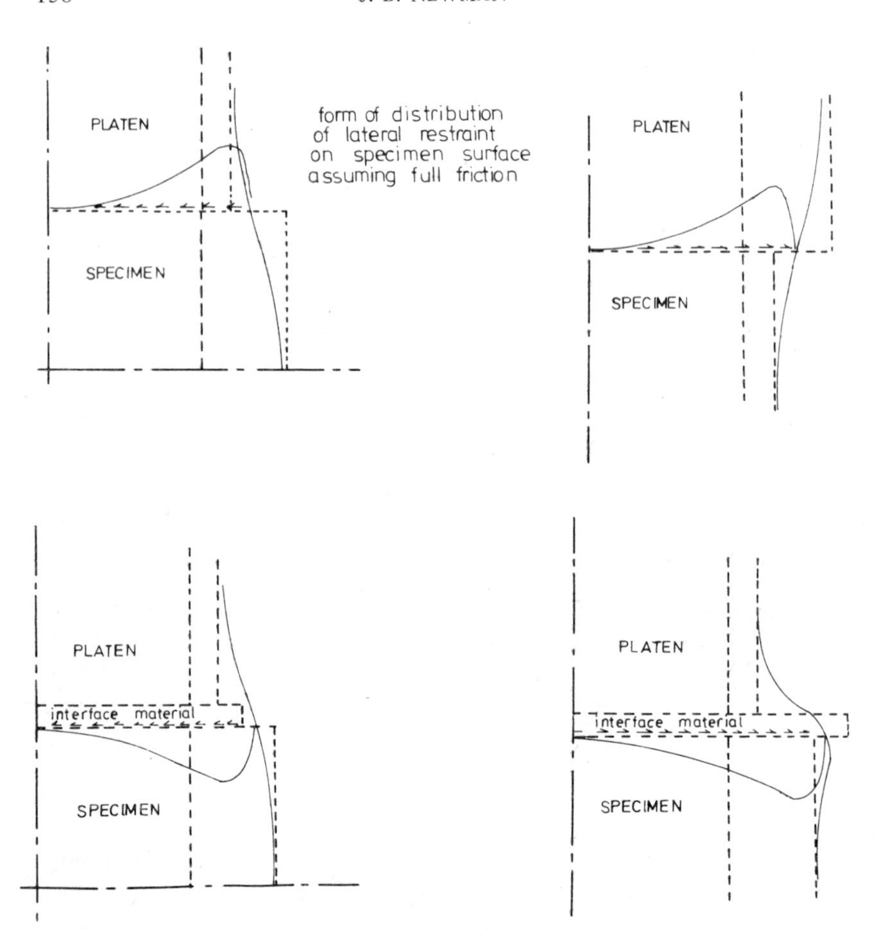

FIG. 1. Lateral deformation profiles for a rigid platen/specimen system under uniform compressive load with and without an intermediate material. (a) Case 1, $(\nu/E)_{plat} < (\nu/E)_{spec.}$ (b) Case 2, $(\nu/E)_{plat} > (\nu/E)_{spec.}$ (c) Case B, $(\nu/E)_{plat} < (\nu/E)_{mat} < (\nu/E)_{spec.}$ (d) Case C, $(\nu/E)_{mat} \geqq (\nu/E)_{spec.}$ ----, undeformed boundary and deformed boundary assuming no frictional restraint; ——, deformed boundary assuming full frictional restraint.

unloaded boundary and progress towards the axis. This process would undoubtedly cause considerable redistribution of stresses.

Now consider an intermediate layer of material interposed between the platen and specimen. The material can be classified in a similar manner to the previous cases namely:

(A) The material strains laterally to a degree equal to, or less than, the main loading platen, i.e.

$$(v/E)_{mat} \leqq (v/E)_{plat}$$

(B) The material strains laterally to a level intermediate between the platen and specimen.
(C) The material strains laterally to a degree equal to or greater than the specimen, i.e.

$$(v/E)_{mat} \geqq (v/E)_{spec}$$

Considering uniaxial compressive loads applied to concrete and similar materials the loads are normally transmitted through platens which are laterally stiffer than either intermediate material or the specimen. Case (A) is thus unlikely to be met in practice, and materials falling within Cases (B) and (C) only will be considered.

Figures 1 (c) and (d) show the forms of lateral deformation profiles under uniaxial compression for the above Cases (B) and (C), without friction and with full friction, again assuming plane sections remain plane. As previously discussed the restraint stress across the loaded face for the case of full friction developed at the interface is again assumed to be zero on the axis, reaching a maximum close to the unloaded boundary. As before, if the restraint stresses exceed those capable of being maintained by friction at any interface then slippage will occur, starting near the unloaded boundary and progressing towards the centre, resulting in a redistribution of stresses within the specimen.

The most important factors which may influence the restraint stresses transmitted to specimens under load through layered material systems are summarised in the following.

(*a*) *Incompatibility of* v/E *value between the various materials.* Typical values for materials which have been used by various researchers placed in order of increasing values of v/E, are given in Table 1. Unless the restraint stresses are relieved by some form of slippage at the specimen boundary *compressive* restraint will be induced on the loaded boundary of a material if loaded through a material in a higher position in the table, i.e. through one with a lower numerical v/E value. Conversely, if a material is loaded through a material below it in the table, *tensile* restraint stresses will be developed.

TABLE 1

VALUES OF ν/E FOR TYPICAL MATERIALS

Material	*Poisson's ratio*, ν	*Modulus of elasticity*, E		ν/E	
		(10^6 *lb/in*^{2})	(*kN/mm*^{2})	(10^{-10} *in*^{2}*/lb*)	(10^{-3} *mm*^{2}*/kN*)
Mild steel	−0·33	30·0	207	−1·1	−1·59
Zinc	−0·25	15·7	108	−1·6	−2·31
Limestone	−0·21	14·0	96·5	−1·8	−2·18
Copper	−0·343	17·5	120·5	−2·0	−2·84
Glass	−0·228	10·6	73	−2·15	−3·12
Brass	−0·35	14·59	100·5	−2·4	−3·48
Granite	−0·25	10·0	70	−2·5	−3·63
Concretes					
Natural aggregate	−0·15 to −0·20	6 to 2	41 to 14	−2·5 to −10·0	−3·62 to −14·5
Lightweight aggregate	−0·15 to −0·20	3 to 1	20 to 7	−5·0 to −20·0	−7·5 to −28·6
Aluminium	−0·345	10·2	70·5	−3·4	−4·9
Lead	−0·43	2·5	17	−17·0	−24·9
Bakelite	−0·32	1·0	7	−32·0	−46·4
Perspex	−0·33 to −0·35	0·89 to 0·4	6 to 3	−36·9 to −87·5	−53·8 to −126·9
ptfe (unfilled)	−0·42	0·58	4	−72·0	−105
Epoxy resin (Araldite)	−0·38 to −0·35	0·5 to 0·35	3·5 to 2·5	−76·0 to −100·0	−110 to −145
Celluloid (Xylonite)	−0·38 to −0·3	0·36 to 0·28	2·58 to 1·9	−105·6 to −107·1	−153 to −158
Rubber	−0·5	0·000 3	0·002	−166 000	−241 000

(*b*) *Frictional characteristics.* Restraint stresses will be reduced significantly if the frictional resistance across the specimen/intermediate material interface can be lowered. A decrease in the frictional resistance across the intermediate material/platen interface will also reduce the restraint, but this effect is likely to be less pronounced. In most practical cases, the distribution of restraint stresses at the various interfaces is likely to cause partial slippage which will considerably modify the stress distribution in specimens.

(*c*) *Thickness and modulus of rigidity of the intermediate material.* The two factors are closely linked since they control the magnitude of shear stress which can be transmitted across the intermediate material. A larger thickness and/or lower modulus or rigidity will reduce the restraint stresses. One important assumption has been made so far to simplify the problem, namely that constant values of ν and E be ascribed to the various materials forming the specimen and loading unit. However, for the case of a concrete or mortar specimen under load ν, and E change with stress level. This fact must be recognised when considering the restraint stresses imposed on the concrete, especially towards and beyond ultimate.

(*d*) *Surface irregularities.* Another assumption is that of the initial planeness of the various contacting surfaces. Obviously, conditions will be severely modified if the surfaces are initially diminsionally incompatible or if they become distorted during loading.

In the above, the case of solid platens with or without intermediate materials has been discussed and it has been shown that with such devices there is a high risk of inducing secondary restraints in specimens under load. Over the years, various devices have been suggested for overcoming this difficulty and one example which has been used by various researchers is the 'brush' platen originally devised by Hilsdorf (see Kupfer *et al.*[6]) at Munich University. These platens consist essentially of prismatic-shaped steel rods with each rod (bristle) aligned in the direction of loading. At the testing machine end each bristle is bonded to those adjacent over a portion of its length but is debonded over the remainder (see Fig. 4, p. 172). In this way the individual bristles can transmit axial load from the testing machine to the specimen but are free to flex laterally to accommodate specimen deformation.

5.2.3.2 *Uniform Boundary Stress*
The only practical way in which a state of uniform stress can be achieved on the boundary of a specimen of a heterogeneous material is by the use of some form of hydraulic loading. If a pure hydrostatic stress is required to be applied, then this can be relatively simply attained by immersing the specimen, surrounded by a suitable impermeable membrance if desired, in a fluid under pressure.

For a state of equal biaxial compression a similar technique can be used, provided the specimen can be isolated from the effects of the fluid pressure in one of the three principal directions.

For stress states other than the above, the problem becomes more complex since the hydraulic pressure must be applied in each direction independently. Thus, in any given direction it will be necessary to contain the fluid such that the pressure is applied orthogonal to the face of the specimen, while at the same time eliminating (a) any effects of extrusion between the surface of the specimen and the containment device, and (b) effects of intrusion into the specimen surface. Providing the above difficulties can be overcome then such a loading method should induce the desired boundary conditions close to uniform stress.

In fact several devices termed 'hydraulic' platens have been developed for the purpose of achieving independent and uniform states of boundary stress on specimens while minimising lateral restraints. They vary in type from those which apply load directly through fluid-filled flexible membranes[7] to those which apply load from flexible membranes through bundles of steel prisms[8] (see Fig. 4, p. 172) and some indication of their effectiveness is given in Section 5.4.

5.3 ANALYTICAL AND EXPERIMENTAL EVALUATION OF STRESS CONDITIONS WITHIN SPECIMENS

5.3.1 Analytical Investigations

The main conclusions to be drawn from analytical investigations[9] are as follows.

(*i*) *Uniaxial compressive loading through rigid platens* ($\nu/E_{\text{plat}} < \nu/E_{\text{spec}}$)

(a) Compressive lateral restraint stresses are induced at the platen/specimen interface. These stresses are zero on the axis and increase towards the periphery to values limited either by the difference in mechanical properties of the platen and specimen material or by the frictional properties at the interface.

(b) This restraint causes lateral compression stresses to develop near the loaded boundaries. This induces a state of triaxial compression within specimens of finite thickness.

(c) The lateral compressive restraint stresses reduce with distance from the loaded face and may reach tensile values near the centre of the specimen. For complete fixity at the loaded boundaries, these restraints may reduce to insignificance and result in a near-uniform state of near-uniaxial stress at a distance from the loaded face equal to the width or thickness of the specimen, whichever is the greater.

(*ii*) *Uniaxial compressive loading through rigid platens with soft intermediate materials* ($v/E_{\mathrm{mat}} > v/E_{\mathrm{spec}}$)

(a) Large tensile restraint stresses are induced near the loaded ends of the specimen which increase with increasing value of v/E for the intermediate material.

(b) Tensile restraint stresses decrease in magnitude with reduction in the thickness of the intermediate layer.

(*iii*) *Cylinders under multiaxial stress.* For confining pressure applied hydraulically and deviator load applied with rigid platens it has been shown that towards the ends of the specimen the stress disturbances are considerable but the central portion of the specimen is under a near-uniform stress state.

5.3.2 Experimental Investigations

All experimental methods for assessing the influence of the loading technique on the state of stress induced within a specimen of material are based essentially on estimating the stress from measurements of deformation. Such an estimation can only be considered valid (a) if the constitutive relations for the material under test are known with certainty, and (b) provided sufficient data are available regarding the distribution of strain throughout the entire material.

Since, for materials such as rock or concrete the first of the above requirements cannot be satisfied, and the second is extremely difficult to achieve, some workers have chosen to adopt other materials for evaluating the effectiveness of loading techniques. In fact, on the basis of the materials used, the investigations conducted to date can be divided into three main types as follows:

Type 1. The shape of specimen is modelled in a material exhibiting photoelastic properties to facilitate the measurement of internal strains.

Type 2. A material of known mechanical properties is used, such that the behaviour under any specified stress state can be calculated and compared with the behaviour observed under test conditions.

Type 3. A sample of material of largely unknown response to the range of induced stress states is used, and it is attempted to assess trends from measurements of surface strain or deformation.

Investigations using materials of Type 1 suffer from the main disadvantage that the behaviour of the model material, and the frictional characteristics at the loaded surfaces, may not resemble that of the material being modelled.

Type 2 materials can be chosen to approximate closely to brittle materials, such as rocks and concrete, within the 'elastic' range. Large differences are likely to be apparent both due to the disparity of frictional characteristics at the platen/specimen interface and of deformational behaviour at higher loads.

For investigations using materials within Type 3 it is impossible to deduce, with any degree of certainty, the actual state of stress within, or on, specimens but some evaluation of the *comparative* effectiveness of loading techniques can be assessed.

The investigations have consisted, basically, of measuring the deformational response of various shapes of specimen composed of the three main types of material under compression, with the following boundary conditions:

(a) Loading directly through rigid platens ($\nu/E_{\text{plat}} < \nu/E_{\text{spec}}$)

(b) Rigid platen loading through soft intermediate materials ($\nu/E_{\text{insert}} > \nu/E_{\text{spec}}$).

(c) Rigid platen loading through materials which 'bed-down' under load, i.e. those with a low E value and low value of ν (e.g. plywood, cardboard, etc.).

(d) Loading through platens of the same material as the specimen.

(e) Loading through non-rigid platens (i.e. 'brush' or 'hydraulic' platens).

From these tests the following main conclusions can be drawn regarding the effect of the above conditions.

(*i*) *Loading directly through rigid platens*

(a) Compressive lateral restraint stresses are induced at the platen/specimen interface. As predicted theoretically, these restraints increase from the axis towards the periphery and decrease towards the centre of the specimen.

(b) The restraint stresses, which increase with degree of roughness between the specimen and platen, tend to reduce the magnitude of the lateral deformation of the specimen.

(c) At low levels of load, the maximum lateral deformation occurs near the end zones of specimens giving characteristic 'bulged' lateral deformation profiles. At higher loads the position of maximum deformation changes to the central zone of the specimen.

(d) As the height/width ratio of a specimen is increased, the extent of the central zone under a near-uniform lateral deformation increases while the magnitude of the deformation within this zone decreases. At a H/W ratio of about 2·5, the central third of the specimen is under near-uniform strain which is in close agreement with theory.

(e) Failure initiates within the central zone of specimens and progresses towards the loaded ends where the restraint stresses inhibit fracture propagation.

(f) The failure load decreases with H/W ratio up to $H/W = 2{\cdot}5$ (approximately) after which it remains essentially constant.

(g) The effect of the degree of boundary fixity on the behaviour of a specimen of a given shape depends both on the type and degree of heterogeneity of the material of which the specimen is composed, together with the characteristics at the interface (i.e. whether ground or unground, etc.).

(h) Considerable stress disturbances occur if the contact surface between the specimen and platen is not accurately plane.

(*ii*) *Loading through 'soft' materials*

(a) As predicted theoretically, tensile lateral restraint stresses are induced at the interface between the intermediate material and the specimen due to extrusion of the insert material. These restraints decrease towards the centre of the specimen.

(b) The restraint stresses increase with value of ν/E insert (as shown theoretically) and tend to increase the lateral deformation of the specimen, the maximum of which occurs near the loaded ends.

(c) The restraint stresses increase with thickness of the insert material, which agrees with the results of theoretical analyses.

(d) The restraints apparently reduce with increasing H/W ratio, as evidenced by the increase in measured failure load. This increase occurs up to $H/W = 2{\cdot}5$ (approximately), after which the failure load remains essentially constant at a value below that for rigid-platen loading. It may be, however, that this behaviour is due to slippage at high loads since evidence of this has been observed in the loading surfaces of soft materials after test, and is predicted theoretically.

(e) Failure initiates within the end zones of the specimen and progresses towards the centre.

(*iii*) *Loading through materials which 'bed-down' under load.* These materials considerably reduce the strain concentrations within specimens, particularly for cases where the loading surfaces of either the platen or the specimen are not accurately plane.

The particular merit of such materials is that although their E-value is very low they do not induce significant tensile or compressive restraint stresses in the specimen since their lateral movement under load is exceptionally small. In fact the E-value is generally so small that, under the range of loads normally applied, these materials soon compress to a limiting deformation.

(*iv*) *Platens of the same material as the specimen.* These platens only reduce the stress concentrations if the contact surfaces are accurately plane and parallel. If this can be achieved then the complete platen/specimen assembly appears to act as a unit, with the characteristics of a monolithic specimen of the same total H/W ratio.

(*v*) *Non-rigid platens.* The following are the main conclusions which can be drawn from the reviewed investigations:

(a) For brush platens the lateral movement of the specimen may cause uneven flexing of the bristles, and thus tend to concentrate the load near the axis but insufficient data is available to quantify this effect.

(b) The stress is applied as a series of concentrated loads which may affect the behaviour of heterogeneous materials.

(c) Higher lateral deformations are induced in specimens by brush and hydraulic platens than by rigid steel platens.

This is confirmed by the originators of the brush platen who show that at an early stage of loading the strains are higher near the ends of the specimen than towards the centre position.[10]

5.3.3 Choice of Loading Technique

As we have seen, two fundamental possibilities exist for the application of load to specimens of a given material:

(a) A condition of uniform boundary stress
(b) A condition of uniform boundary displacement

The difference in response to a given volume of material to these states may be dependent on the heterogeneity of the material, in that a homogeneous material is likely to respond in a similar manner to both types of boundary condition, whereas a heterogeneous system will undoubtedly exhibit large differences.

These differences will be caused by the manner in which the load is transmitted through the structure of the material. For a material such as concrete, consisting of relatively stiff inclusions embedded in a softer matrix, a uniform boundary displacement will tend to concentrate the stress at the inclusions, whereas a uniform boundary stress will tend to induce higher strains in the matrix.

The differences in response of such a material will, however, probably reduce for material which is remote from the loaded boundaries. For a fine-grained material, where the degree of heterogeneity is low, the distance from the loaded boundaries required to ensure nearly identical behaviour under the different systems of load will be less than that required for a coarse-grained composite. In general, therefore, the response of a material to a given boundary condition of load will depend on the degree of heterogeneity of the material in relation to the size of the representative volume element under test. If the size of specimen is relatively small compared with the grain size, which may be inevitable in practice, some differences in behaviour are to be expected and a choice must be made regarding which type of loading condition more nearly represents the loading situation for which the data is to be used.

However, one further point must be considered before a choice of fundamental boundary condition is made, namely, the ability of any given loading device to attain the desired conditions without imposing secondary restraint effects.

With the above in mind we can discuss the results of tests carried out to investigate the behaviour of concretes under complex states of stress and appreciate the significance of the information obtained.

5.4 BEHAVIOUR OF CONCRETE UNDER BIAXIAL STRESS

5.4.1 Tests under Uniform Boundary Stress

Although such a boundary condition has been attempted using hollow cylinders under hydraulic pressure[11] it is generally considered that such a test induces considerable stress and strain gradients which would have a significant effect on the results. In the main, therefore, investigators have adopted solid cylinders tested in a hydraulic cell arrangement, with the component of pressure in the axial direction eliminated and with the specimen sealed against ingress of fluid.

Under such a test, Richart *et al.*[4] found that specimens of mortar and concrete exhibited equal biaxial strengths of approximately 1·3 × uniaxial, this proportion being greater for higher quality mixes. It is likely that the test imposed relatively high restraint stresses in the axial direction due to the friction between various components of the loading device.

Using a similar arrangement Chinn and Zimmerman[13] found various grades of concrete to give an equal biaxial compressive strength of approximately 1·25 × uniaxial. In this case the effect of axial restraint was estimated, and the concrete quality did not appear to influence the results to a significant extent.

Newman[9] investigating the effect of the type of boundary conditions on the behaviour of concrete under equal biaxial compression found that the strength of concretes under equal biaxial compression varied from 1·10 to 1·17 × uniaxial. It was also demonstrated that there were marked differences between the deformational and cracking behaviour of specimens under uniaxial stress and uniform strain conditions (see Section 5.4.3).

Recently, however, with the development of 'hydraulic' platens it has been possible to vary the applied boundary loads independently. For convenience the behaviour of concrete under such systems is included in the following section which is primarily concerned with uniform boundary displacement techniques.

5.4.2 Tests under Uniform Boundary Displacement

It is on such condition that much attention has been devoted since it enables the entire stress field to be investigated with relative ease. Two main loading methods have been adopted namely (a) rigid-platen devices with, or without, an intermediate layer interposed between them and the specimen, or (b) non-rigid platens.

5.4.2.1 *Strength Envelopes in Compression–Compression*

Figures 2 and 3 show typical strength results obtained in a number of investigations[14–22] from which the following main conclusions can be drawn:

(a) For rigid-platen or non-rigid platen loading the maximum ratio between biaxial strength and uniaxial strength is not attained under equal biaxial compression. In fact, in most cases, the maximum value is given under a stress ratio of approximately 2:1.
(b) Cement pastes under either rigid or non-rigid platen loading exhibit maximum biaxial strength ratios of between 1·07 and 1·12, with equal biaxial ratios between 0·95 and 1·02.
(c) Mortars and concretes give considerably higher maximum biaxial and equal biaxial strength ratios than pastes for all types of rigid and non-rigid platen loading, the ratio increasing slightly with maximum size of aggregate.
(d) Non-rigid platens, and rigid platens with a friction-reducing intermediate layer, give biaxial strength ratios lower than those obtained using rigid platens without lubricant.

To investigate further the differences between the behaviour of concrete under different boundary conditions, a cooperative programme of testing was carried out on an international basis with the author's laboratory taking part. The data from the tests have not been completely analysed to date but preliminary results were presented in 1976.[23] Basically, a single type of concrete cast in different moulds by one laboratory was tested on given dates under specified states of applied stress by the participants' own testing devices.

The basic differences between testing techniques were

(1) boundary conditions (displacement or stress)
(2) frictional restraints, and
(3) size and shape of specimen

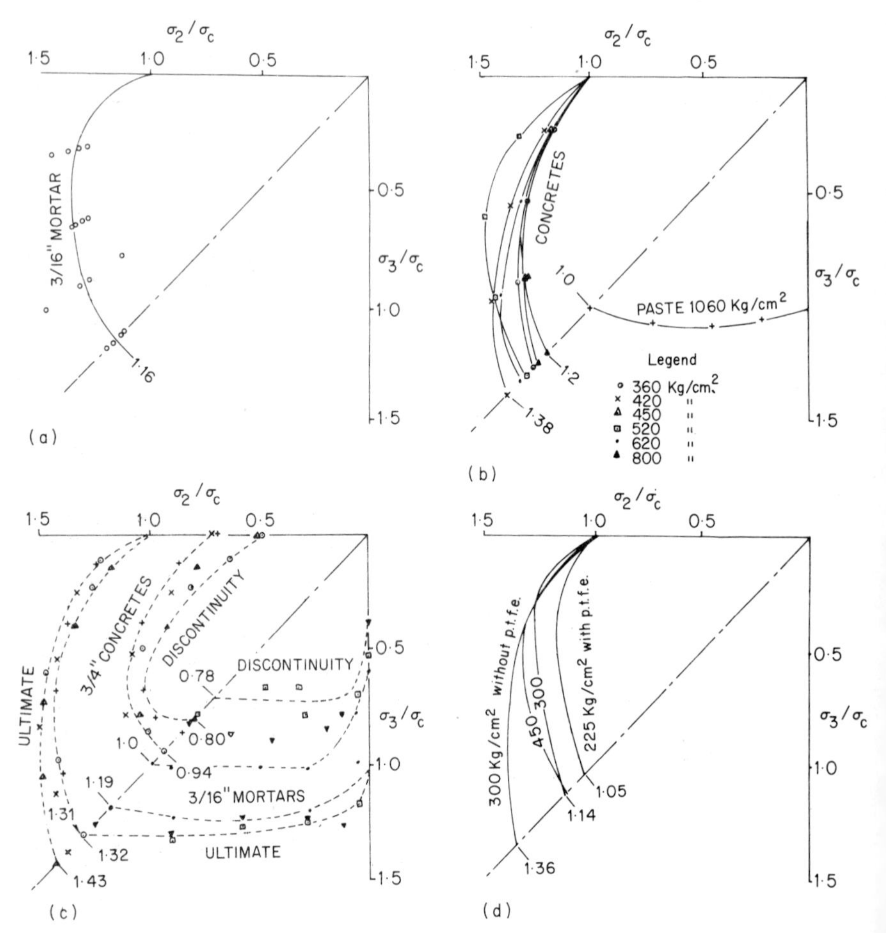

FIG. 2. Ultimate strength and discontinuity envelopes normalised with respect to the uniaxial compressive strength for concretes tested under biaxial compression. (a) Rigid platens with rubber sheets. After Wastlund.[15] (b) Rigid platens. After Weigler and Becker.[16,17] (c) Rigid platens with lubricant. After Vile.[14] (d) Rigid platens with and without ptfe/brass intermediate layer.[18]

Figure 4 shows the range of devices used in the investigation and Fig. 5 presents the normalised strength envelopes obtained for a mortar and a concrete. It is clearly demonstrated that larger increases in biaxial strength are given by testing devices which produce high restraint at the loaded surfaces of the specimen, and that mortar gives the most critical envelopes.

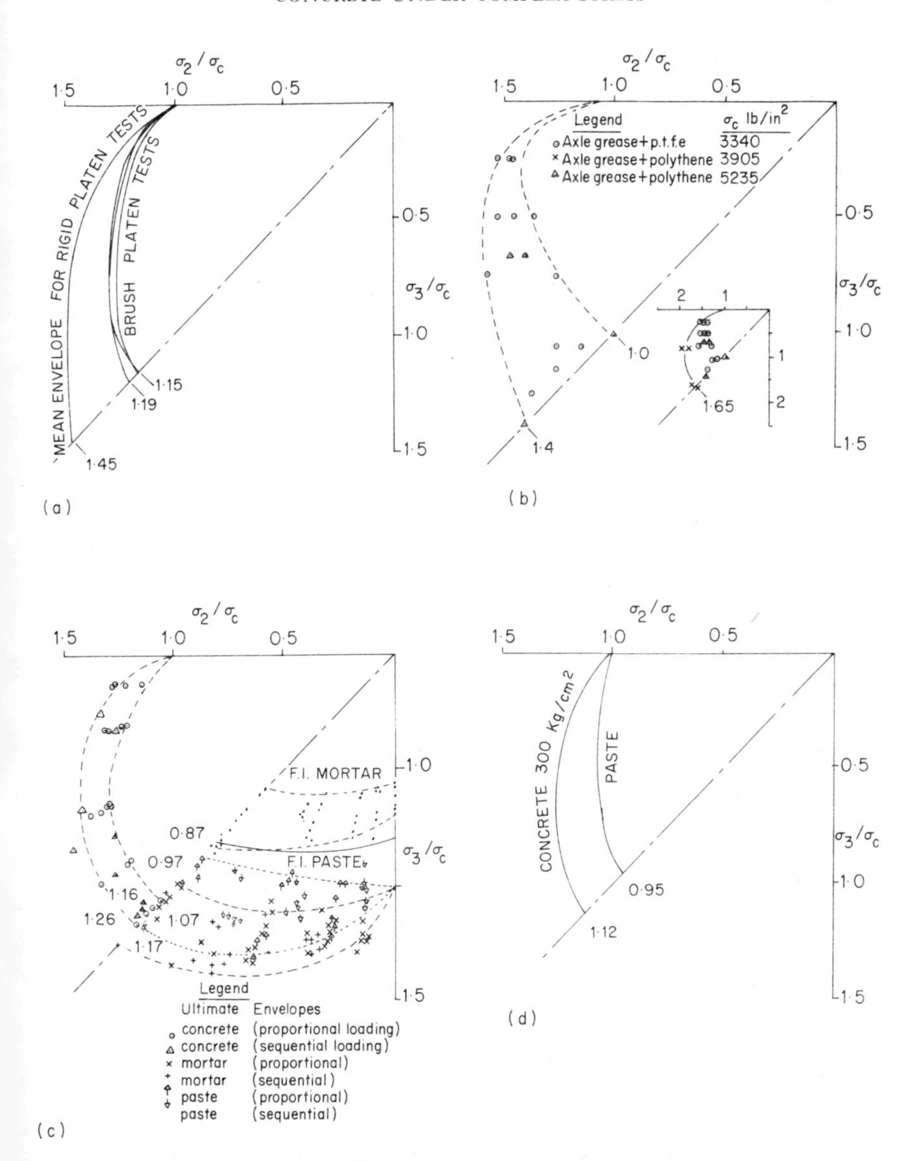

FIG. 3. Ultimate strength and fracture initiation envelopes normalised with respect to the uniaxial compressive strength for concretes tested under biaxial compression. (a) Rigid and brush platens. After Kupfer *et al.*[19] (b) Rigid platens with intermediate layers. After Mills and Zimmerman.[20] (c) Rigid platens with ptfe/grease/rubber intermediate layer. After Kobayashi and Koyanagi.[21] (d) Brush platens. After Stegbauer and Linse.[22]

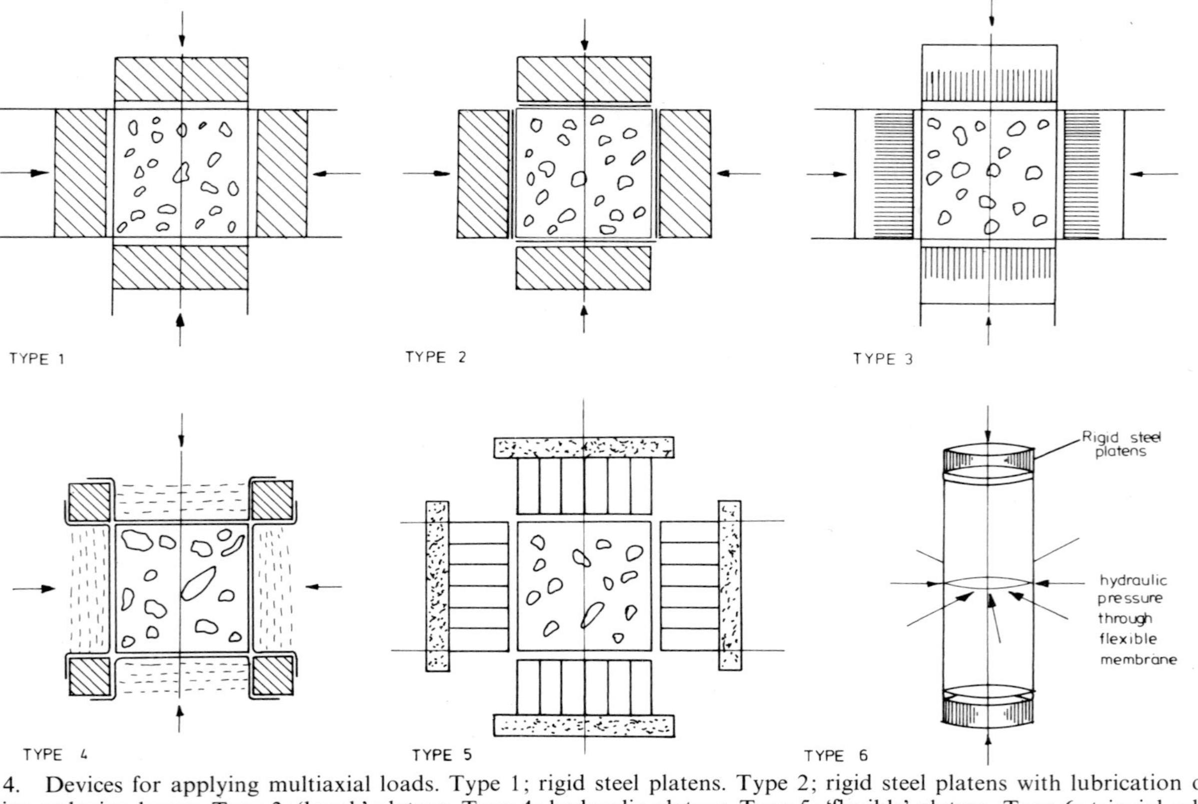

FIG. 4. Devices for applying multiaxial loads. Type 1; rigid steel platens. Type 2; rigid steel platens with lubrication or friction-reducing layers. Type 3; 'brush' platens. Type 4; hydraulic platens. Type 5; 'flexible' platens. Type 6; triaxial cell.

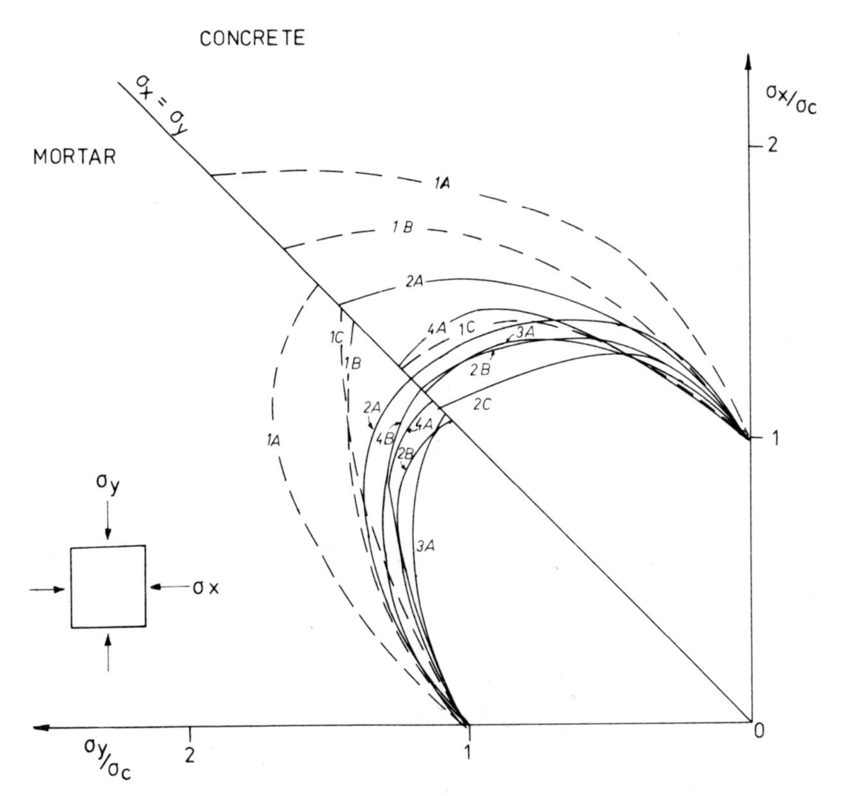

FIG. 5. Biaxial compressive strength envelopes for a mortar and a concrete tested using various types of loading device. Figures thus 3B refer to machine B of Type 3 as shown in Fig. 4. σ_c = uniaxial compressive strength measured using the same loading device.

5.4.2.2 *Strength Envelopes in Compression–Tension*

The behaviour of concrete within this portion of biaxial stress space is difficult to study since any slight restraint to movement at the compressively loaded boundaries will be a relatively large proportion of the applied tensile load. Nevertheless, a number of investigations have been reported using various loading techniques and a selection of these is given below.

Rigid platens with friction-reducing pads interposed at the compressively loaded boundaries.[14]
Hollow cylinders under compression/torsion.[24,25]
Brush platens.[6]

The results of these investigations are summarised in Fig. 6 which shows normalised strength envelopes. This figure demonstrates the wide disparity of envelopes obtained, although it must be emphasised that strict comparison of mixes is not possible. A number of tentative conclusions can be drawn from the data.

(1) For mixes with similar aggregate content the envelopes appear to be independent of concrete strength.
(2) Mixes with the lowest aggregate contents show the most critical envelopes.
(3) The shape of the envelopes is dependent on the testing technique adopted.

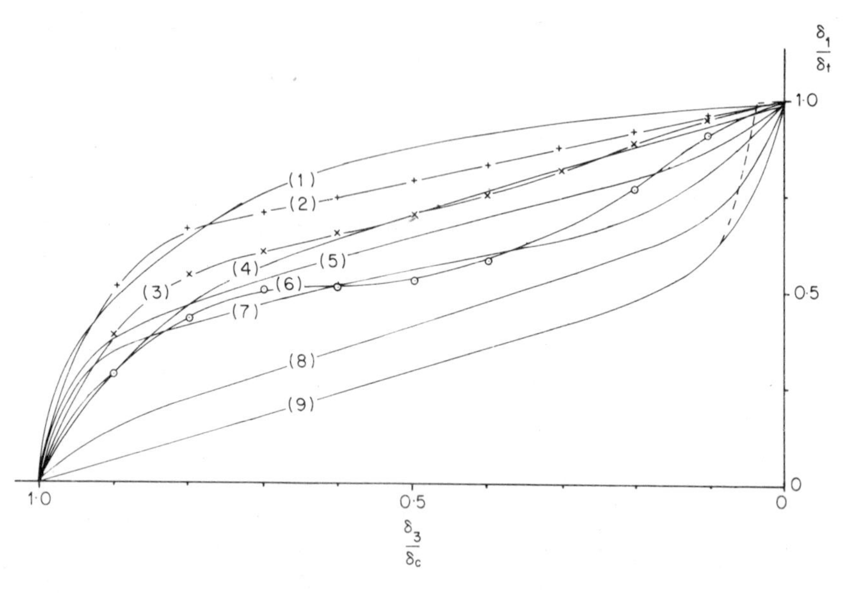

FIG. 6. Normalised ultimate strength envelopes for various concretes and mortars under biaxial compression/tension. (1)$V_f \simeq 0{\cdot}50$ (estimated tensile strength). After Bresler and Pister.[25] (2) $\sigma_c = 58\ \mathrm{N/mm^2}$. After Kupfer *et al.*[6] (3) After Kupfer *et al.*[19] (4) $V_f = 0{\cdot}51$. After Vile.[14] (5) Mix c ($V_f = 0{\cdot}53$, $\sigma_c = 16\ \mathrm{N/mm^2}$). After McHenry and Karni.[24] (6) After Kupfer *et al.*[19] (17) Mix a ($V_f = 0{\cdot}44$, $\sigma_c = 37\ \mathrm{N/mm^2}$). After McHenry and Karni.[24] (8) Concretes ($V_f = 0{\cdot}32$, $\sigma_c = 29$ and $42\ \mathrm{N/mm^2}$). After Vile.[14] (a) Mortars ($V_f = 0{\cdot}00$, $\sigma_c = 31$ and $49\ \mathrm{N/mm^2}$). After Vile.[14] σ_c, σ_t—uniaxial compressive and tensile strengths.

5.4.3 Deformational Behaviour

As for the case of strength envelopes the measured stress–strain relationships will depend (a) on the loading method used to achieve a given stress path and (b) the deviations from the required path. In addition, however, the relationships may also be affected by the deformation measuring technique employed. As an illustration of the trends observed under biaxial loading Figs. 7(a), (b) and (c) show relationships obtained by some laboratories participating in the international cooperative research project[23,26] for a concrete under ratios of biaxial to compressive stress, $\sigma_2/\sigma_1 = 0$, 0·33 and 1·0, respectively. These relationships illustrate that the deformational behaviour under biaxial compression is only slightly affected by changes in testing and measuring technique, as expected, but depends on the stress path (σ_2/σ_1) adopted.

As σ_2/σ_1 increases from uniaxial to equal biaxial loading the maximum compressive strain ε_1 becomes less compressive, the intermediate strain ε_2 becomes more compressive and the maximum tensile strain ε_3 becomes more tensile.

5.4.4 Failure modes

Fundamental experimental studies of the fracture processes in concretes[9] and other brittle materials at the interparticle or structural level have indicated that the materials of this class propagate cracks in the direction of maximum principal compressive stress or, in other words, orthogonal to the maximum principal tension. This can easily be seen for simple tension where the failure occurs on a plane at 90° to the direction of testing, but for compression the modes of failure are often difficult to reconcile with the postulated mechanism since testing effects can complicate the process. For example, the failure of a concrete cube loaded through rigid steel platens exhibits the well-known ‘hour-glass’ pattern which has led some workers to conclude that shear is the controlling factor. However, if the lateral restraint to boundary movement is reduced by use of an interfacial layer or non-rigid platen as discussed in Section 5.3 or if the length of the specimen is increased then the failure mode changes to that of cracking aligned mainly in the loading direction.

Figure 8 shows the modes of failure exhibited by concrete under biaxial states of stress where the secondary restraints are minimised and actual examples are shown in Figs. 9 and 10. It is evident that cracking does, in general, propagate in a direction orthogonal to the maximum

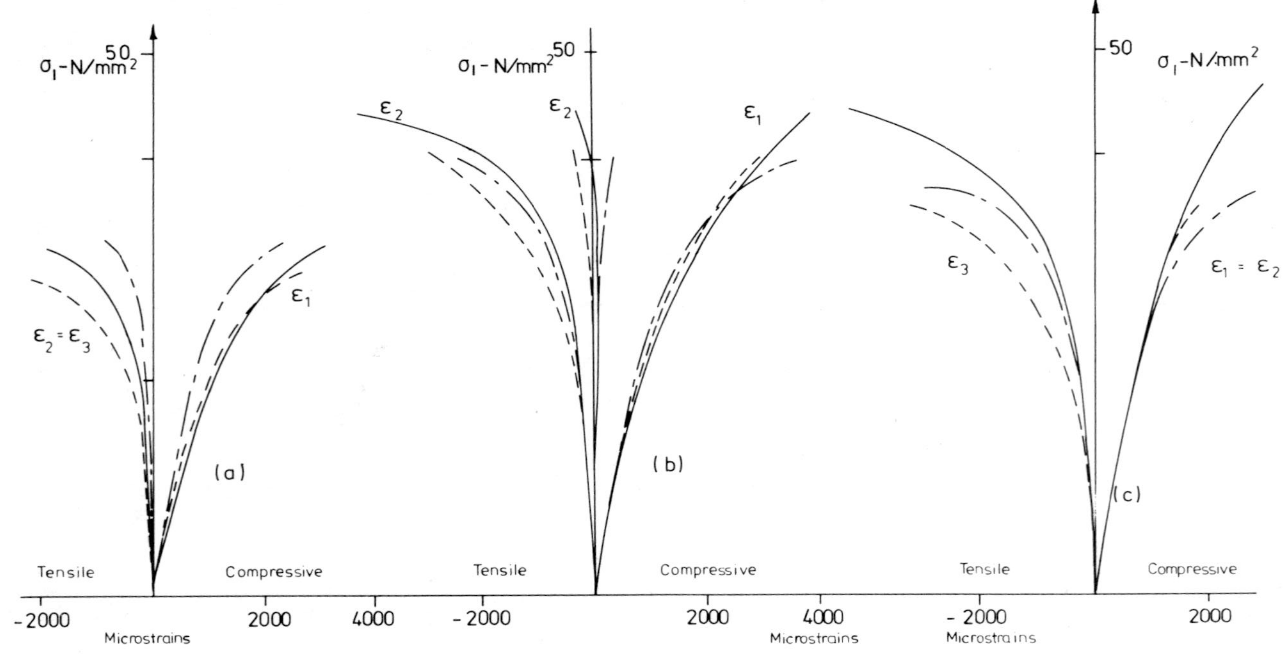

FIG. 7. Stress–strain. Relationships for a concrete under various biaxial compression–stress ratios. Loading device Type 2, ——; Type 3, ——— and Type 4, --- (see Fig. 4). (a) $\sigma_2/\sigma_1 = 0.0$ (b) $\sigma_2\sigma_1 = 0{\cdot}33$. (c) $\sigma_2/\sigma_1 = 1.0$.

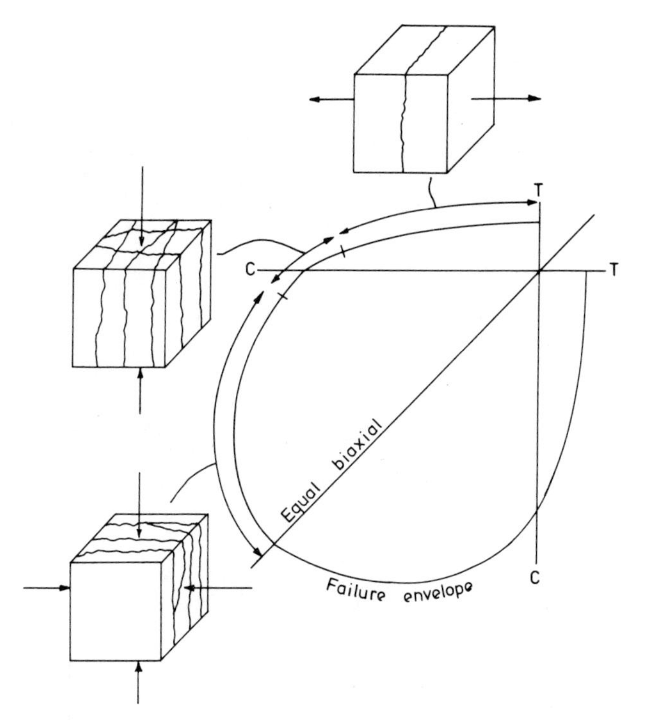

FIG. 8. Idealisation of three main modes of fracture under biaxial stress states.

principal tensile stress and that there is a special case in the vicinity of uniaxial compression where the planes of failure can be aligned in any orthogonal direction.

5.5 BEHAVIOUR OF CONCRETE UNDER TRIAXIAL STRESS

5.5.1 Introduction

Over the years, a number of investigations have been carried out to study the characteristics of concrete under multiaxial stress. These have concentrated primarily on the measurement of ultimate strength with deformational behaviour being neglected probably due to the difficulty of accurate measurement. The author has analysed the information from these tests in order to study the effect of the following main

FIG. 9. Views showing specimens of a mortar mix after failure in uniaxial and equal biaxial compression tests: (a) Cube, cylinder, prism and slab under uniaxial compression. (b) Slab under equal biaxial compression. (c) Cylinders under equal biaxial compression.

FIG. 10. Views showing specimens of a concrete mix after failure in uniaxial and equal biaxial compression tests: (a) cube, cylinder, prism and slab under uniaxial compression. (b) Slab under equal biaxial compression. (c) Cylinders under equal biaxial compression.

parameters on the characteristics of the strength envelopes:

Condition of specimen at test (i.e. saturated or dry).
Maximum size of aggregate.
Type of aggregate (i.e. lightweight or natural, etc.).
Volume of fraction of aggregate
Water/cement ratio.

Only data from tests under states of stress $\sigma_1 > \sigma_2 = \sigma_3$ with all stresses compressive, performed using a hydraulic cell arrangement have been included in order to eliminate as far as possible the effects of testing techniques.

For this exercise the data were fitted to a mathematical relationship which satisfies the following requirements which, it is felt, should be considered when deciding upon any type of data presentation technique:

(1) The information should be presented in a manner conducive to easy prediction of a property for a given mix, or type of mix, by interpolation (or extrapolation if necessary).
(2) The relationships adopted should conform to the generally accepted shape characteristics of the surface and provide a good fit to the data over the required range of values.

(3) The relationship should enable comparison to be made between groups of concretes containing results obtained by different workers.

The relationship found best in comparison with others was of normalised parabolic form:

$$\sigma_{\text{axial}}/\sigma_{\text{c}} = |A(\sigma_{\text{conf}}/\sigma_{\text{c}}) + 1|^{B}$$

where σ_{axial} = axial stress at ultimate (not corrected for changes in cross-sectional area of specimen); σ_{conf} = hydraulic confining pressure; σ_{c} = axial stress at ultimate for zero confining pressure (uniaxial strength); and A and B = constants depending on concrete mix characteristics.

For each comparison made it was attempted to keep all parameters approximately constant except the one being investigated. Typical results of this analysis, which includes information from 123 different mixes tested during the period 1928–1972, are shown in Figs. 11–13.

In drawing conclusions from this analysis it must be emphasised that data has been extracted from a large number of sources. A variety of types of loading apparatus and test procedures have been used and the mix parameters and curing procedures have, in many cases, been varied simultaneously. It has been assumed that the effects of testing are minimised by normalising the data with respect to the uniaxial compressive strength. The effects of the mix parameters, however, remain difficult to isolate.

Even accepting the above restrictions the analyses can provide an indication of likely trends, and the main conclusions obtained are as follows:

(1) Normalising the ultimate strength envelopes with respect to the uniaxial compressive strength of each mix appears to effectively eliminate the influence of water/cement ratio for practical mixes containing natural coarse aggregate.
(2) For the range of stresses applied the strength envelope for all concretes containing aggregate is *open-ended* in stress space. For cement paste and lightweight concretes, however, the envelope tends towards an intersection with the hydrostatic axis.
(3) The ultimate strength data for natural aggregate concretes can be adequately represented by a *parabolic* or *hyperbolic* function

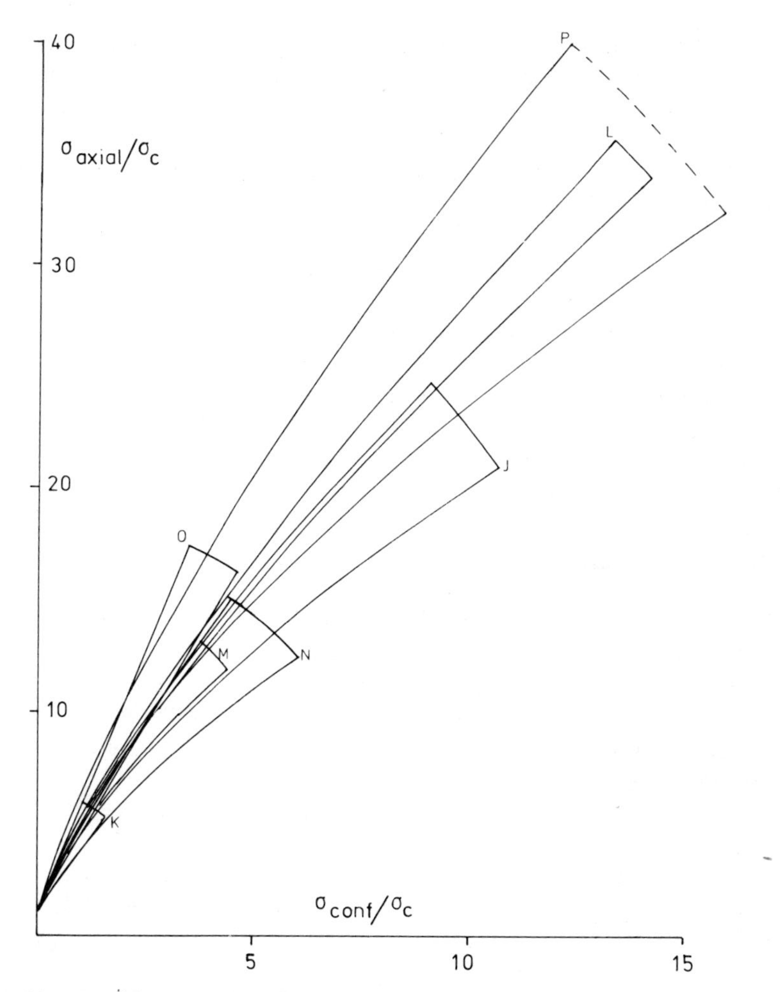

FIG. 11. Confidence regions for parabolic relationships defining normalised ultimate strength envelopes for concretes containing 20 mm natural aggregate with various ranges of W/C ratio: 0·30–0·39 (J), 0·40–0·49 (K), 0·50–0·59 (L), 0·60–0·69 (M), 0·70–0·79 (N), 0·80–0·89 (O), >0·89 (P).

capable of allowing valid interpolation to be made for stress states in the proximity of uniaxial compression, as well as for higher confining pressures.

(4) Except for cement pastes, the conditions under which the specimen is cured prior to test appear to have an insignificant

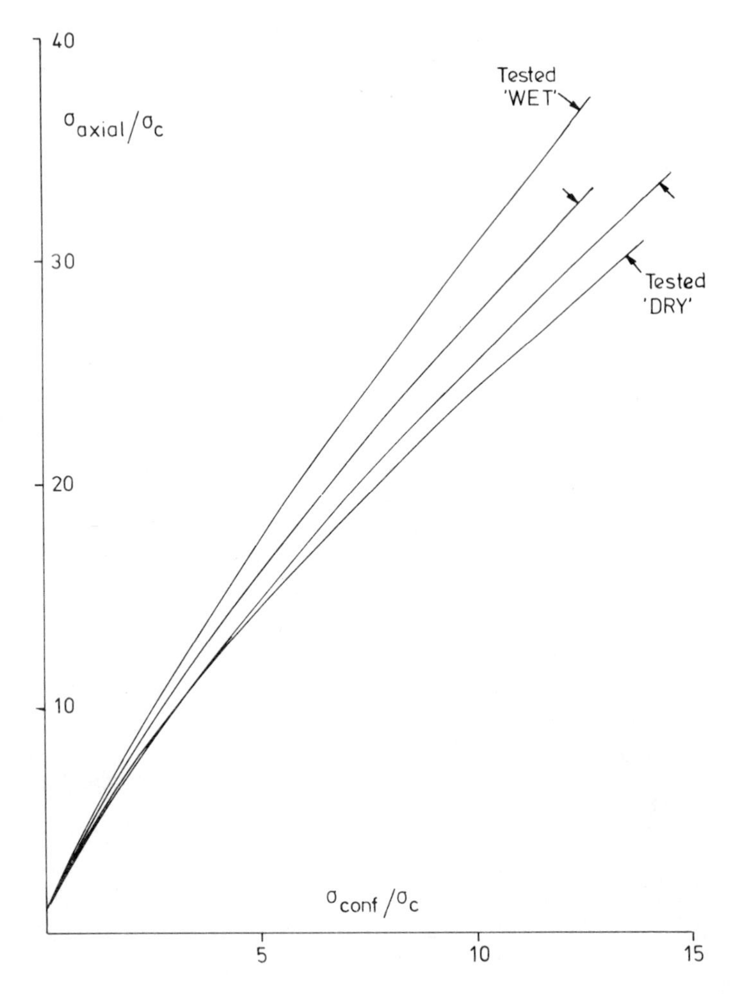

FIG. 12. Confidence regions for parabolic relationships defining normalised ultimate strength envelopes for concretes containing 20 mm natural aggregate tested in the 'wet' or 'dry' condition.

effect on the shape of the normalised ultimate strength envelope. For pastes, however, the strength envelope is significantly affected by the extent to which the paste is dried prior to test.

(5) For a maximum aggregate size above 5 mm ($\frac{3}{16}$ in), the shape of the ultimate strength envelope is largely independent of the size of aggregate.

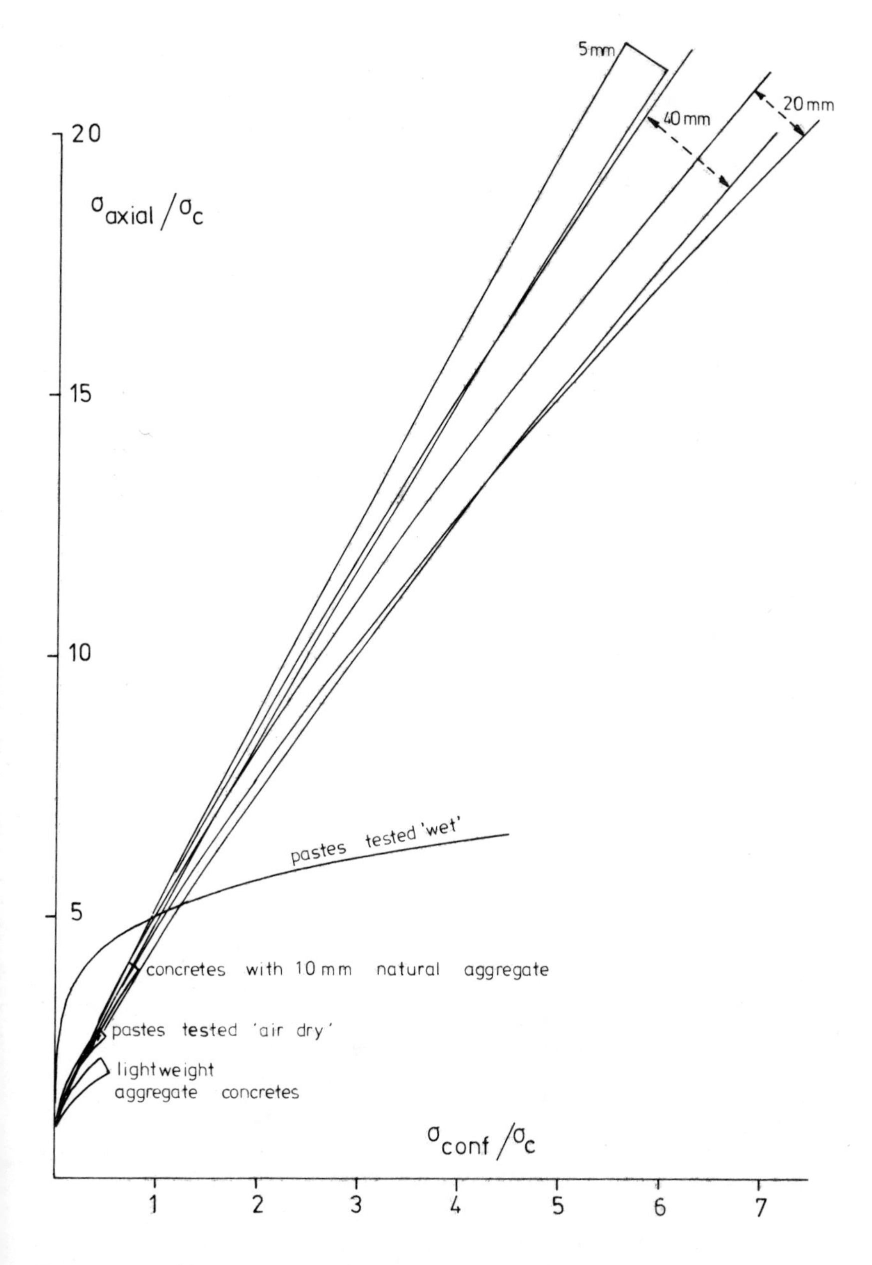

FIG. 13. Confidence regions for parabolic relationships defining normalised ultimate strength envelopes for various types of concrete.

(6) Lean concretes (i.e. those with a high volume fraction of aggregate), with correspondingly high water/cement ratios, give larger strength increases with increasing confining pressure than rich mixes with low water/cement ratios.
(7) Following from (6) and (1) it would appear that a change in the aggregate content produces a greater effect on the normalised strength envelope than a change in the water/cement ratio. Since the two variables could not be isolated in the analysis, this conclusion is tentative.

Having discussed in general the type of strength behaviour exhibited by concretes under biaxial and triaxial compressive stress states, it is appropriate now to consider deformational characteristics of concrete under these and other stress conditions with particular reference to the internal structural changes which the material undergoes. These changes will be illustrated by some data obtained to date from an ongoing comprehensive investigation to establish the criteria of concrete strength.[9,27,28,29]

5.5.2 Tests Conducted at Imperial College

The tests have been performed using a hydraulic triaxial cell capable of applying the following states of stress[30] following any axisymmetric stress path in stress space:

$\sigma_1 > \sigma_2 = \sigma_3$ with all stresses compressive (triaxial 'compression').
$\sigma_2 = \sigma_3 > \sigma_1$ with all stresses compressive (triaxial 'extension').
$\sigma_2 = \sigma_3 > \sigma_1$ with σ_2 and σ_3 compressive and σ_1 tensile (compression–compression–tension).

In the test programme much emphasis has been placed on obtaining the complete deformational behaviour of the concrete with techniques sufficiently sensitive to detect significant changes in specimen response. As a prelude to the main series of tests, idealised models of concrete comprising sand/cement mortar with single-sized roughened glass spheres as aggregate were studied in order to evaluate trends in behaviour such that the real concrete series could be more efficiently formulated. However, only the tests on real concretes will be discussed with the tests performed in the conventional manner, namely, by first applying a hydrostatic cell pressure (confining pressure) and then either increasing or decreasing the axial (deviator) stress to failure.

In order to understand more clearly the stress paths used in conventional triaxial testing Fig. 14 gives a diagrammatic

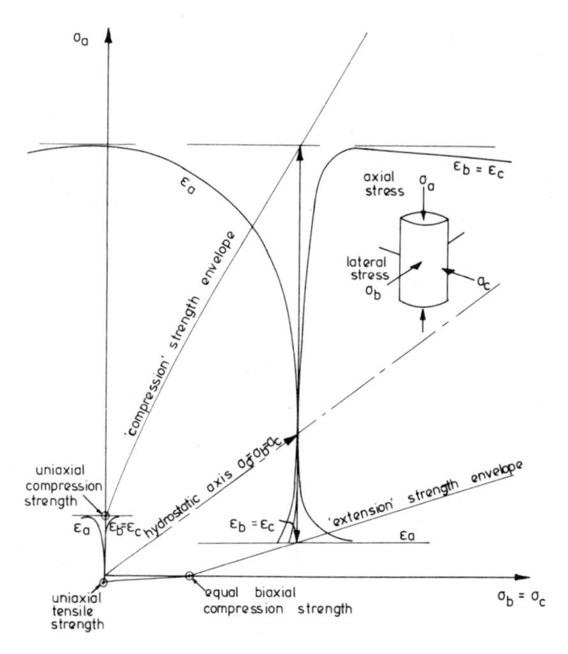

FIG. 14. Diagram showing ultimate strength envelopes on plane $\sigma_b = \sigma_c$ and typical stress–strain relationships for concrete under symmetrical triaxial stress in the directions σ_a, σ_b and σ_c respectively.

representation of the form of failure envelopes in stress space. Also, shown in the figure are typical forms of stress–strain relationship exhibited from the start of deviator loading. It is usual to describe all tests in which the axial stress σ_a is increased compressively above the confining pressure $\sigma_b = \sigma_c$ as 'compression' tests, since failure occurs in the compression mode. Tests, in which the axial stress is decreased to failure are termed 'extension' tests with failure occurring in the tensile mode even if the resultant axial stress is compressive. A special case of this latter type of test is when the specimen is suitably shaped, usually by 'waisting', such that the confining pressure can induce a tensile stress in the axial direction. This test is termed a compression–compression–tension (C/C/T) test.

5.5.2.1 *Stress–Strain Relationships*
Figures 15–21 show the variation of axial and lateral strains for various concrete mixes of which details are given in Table 2. Each relationship covers the entire loading regime, including hydrostatic as well as deviatoric, for a particular maximum level of confining pressure applied. The following main points can be noted:

(1) For a given mix and confining pressure the axial stress (σ_a)–axial strain (ε_a) relationship for 'compression' tests gradually increases in curvature with increasing stress. The corresponding axial stress (σ_a)–lateral strain ($\varepsilon_b = \varepsilon_c$) relationship, however, exhibits marked increases in curvature only as ultimate failure is approached. This phenomenon can be attributed to the process of cracking and the alighment of the cracks for this type of loading. Similar behaviour is observed under 'extension' and C/C/T stress states.

One important feature is that in all cases neither the axial stress–axial strain or axial stress–lateral strain relationship shows any linear portion or consequent 'limit of proportionality'.

(2) For a given mix the material exhibits a transition from relatively 'brittle' to more 'ductile' behaviour as the maximum confining pressure ($\sigma_a = \sigma_b$) for the test is increased. At low confining pressures, the slope of both the axial and lateral stress–strain curves reverses distinctly at ultimate. At higher confining pressures, this reversal is less obvious, particularly for the mixes containing low volume fractions of aggregate, when the relationships are aligned almost parallel to the stress axis indicating a high degree of plasticity. Such behaviour has been observed by other workers both for concrete and rocks for tests in which the variation in the overall axial strain with the nominal applied stress has been measured. It has been suggested that the phenomenon may be due to the slipping of particles of material against each other during the loading process.

Although these characteristics are produced by all stress states they are more pronounced for triaxial 'compression'.

(3) For a given confining pressure, increases in the volume fraction of aggregate cause the axial stress–axial strain relationship to become steeper, indicating increased stiffness. Hobbs[31] also has detected a similar trend. The variation of axial strain indicates that the relative increase in stiffness with volume fraction of aggregate is greater for tests under applied confining pressure than for the uniaxial compressive case. This behaviour is less obvious for 'extension' stress states.

The variation of lateral strain, however, does not show such a trend

TABLE 2

DETAILS OF MORTAR AND CONCRETE MIXES

Details of mortar and concrete mixes	*Mix reference number*					
	M1	C1	M2	C2	C3	C4
	Aggregate type					
	Irregular Thames valley river gravel				*Crushed limestone*	*Crushed granite*
Proportions by weight						
Cement (OPC)	1·00	1·00	1·00	1·00	1·00	1·00
Sand (70% coarse, 30% fine)	1·40	1·40	2·96	2·96	2·96	2·96
9·5 mm (3/8 in) aggregate	0·00	0·70	0·00	1·48	1·48	1·48
19 mm (3/4 in) aggregate	0·00	1·40	0·00	1·96	2·96	2·96
'Effective' water	0·35	0·35	0·60	0·60	0·60	0·60
Total water	0·38	0·44	0·67	0·79	0·62	0·62
Volume fractions (%)						
Fine aggregate	42·82	25·33	52·96	28·55	29·80	31·17
Coarse aggregate	0	38·14	0	42·98	44·70	43·96
Total aggregate	42·82	63·47	52·96	71·53	74·49	75·13
Properties in fresh state						
Slump (mm)	20	5	Collapse	20	Nil	Nil
Compacting factor	0·98	0·85	0·99	0·93	0·85	0·73
VB(s)	6·0	5·0	Collapse	3·0	6·0	6·0
Air content(%)	3·0	1·75	1·6	1·4	1·4	1·75
Properties in hardened state (mean of 3 specimens)						
Wet density (kg/m^3)	2260	2349	2125	2343	2468	2431
28-day cube strength (N/mm^2)	69·17	60·52	34·69	22·08	50·23	40·58
Cube strength at time of testing (N/mm^2)	91·19	73·30	41·95	23·23	63·18	53·59

Note: All tests on fresh and hardened concrete carried out in accordance with BS 1881[12]

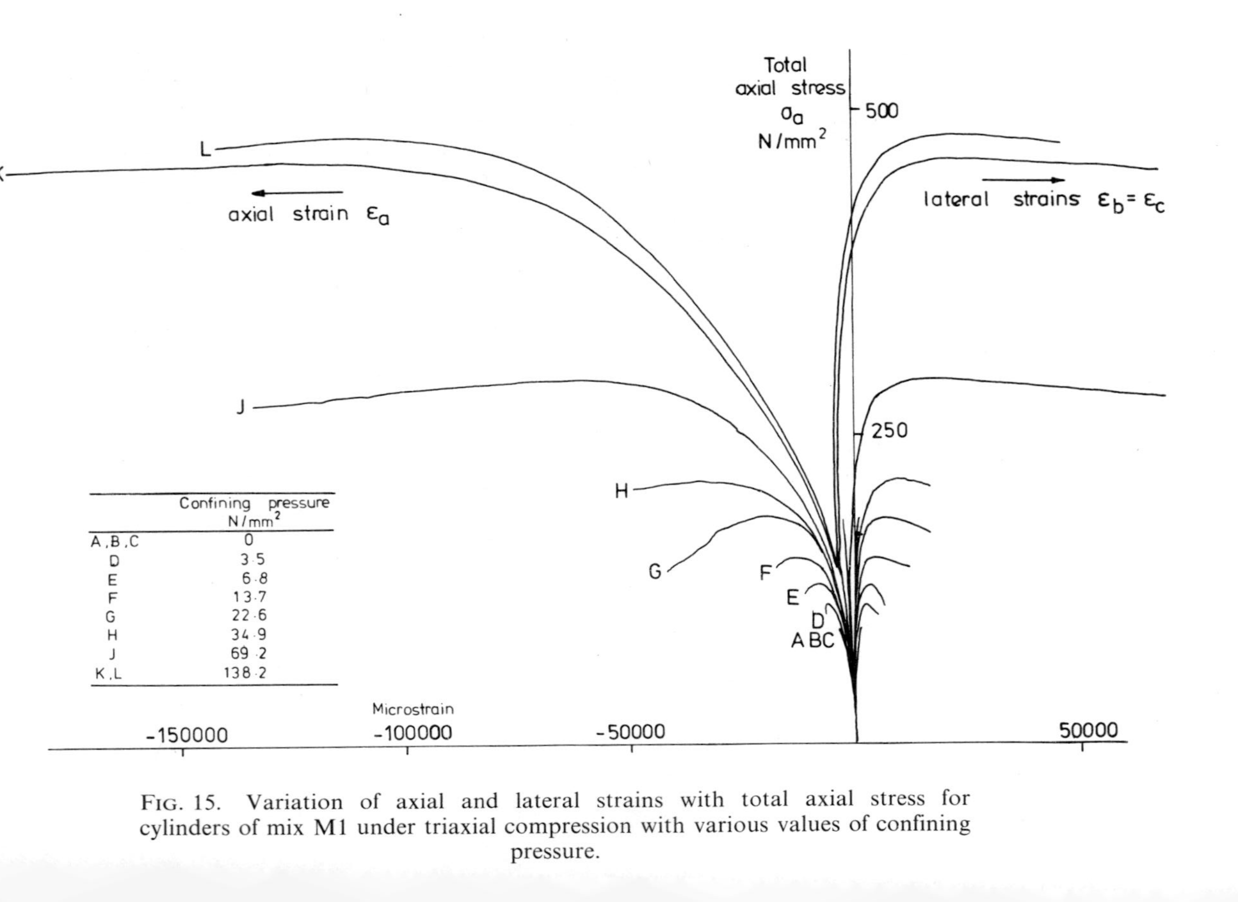

FIG. 15. Variation of axial and lateral strains with total axial stress for cylinders of mix M1 under triaxial compression with various values of confining pressure.

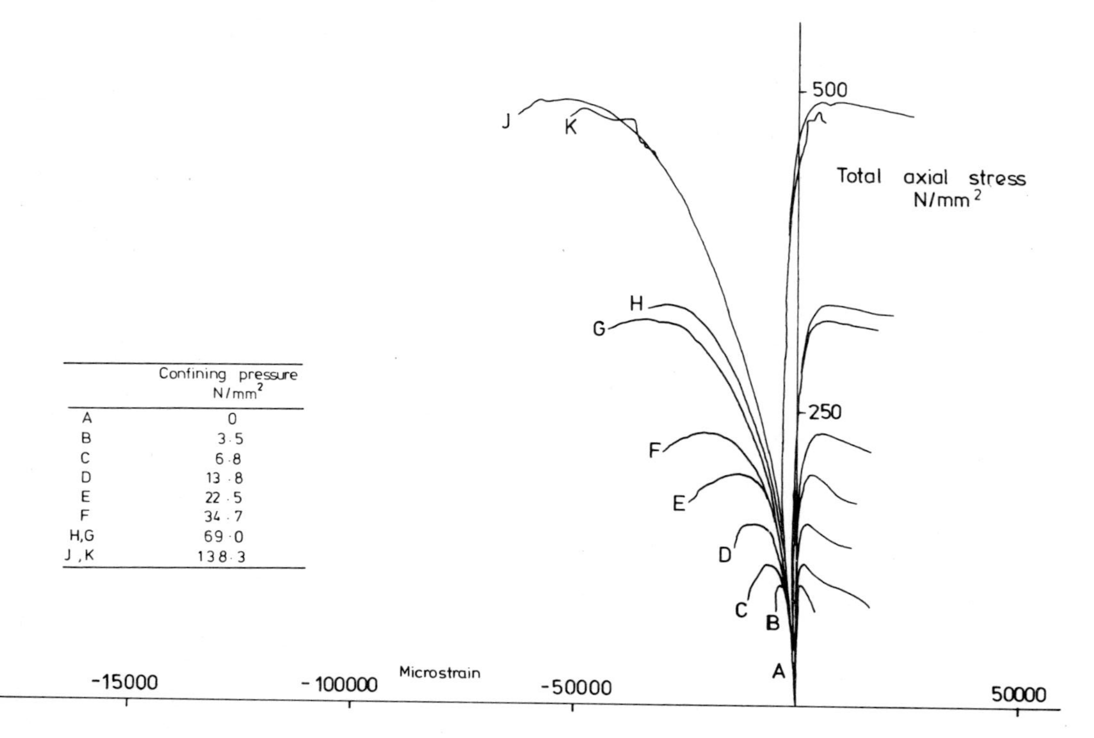

FIG. 16. Variation of axial and lateral strains with total axial stress for cylinders of mix C1 under triaxial compression with various values of confining pressure.

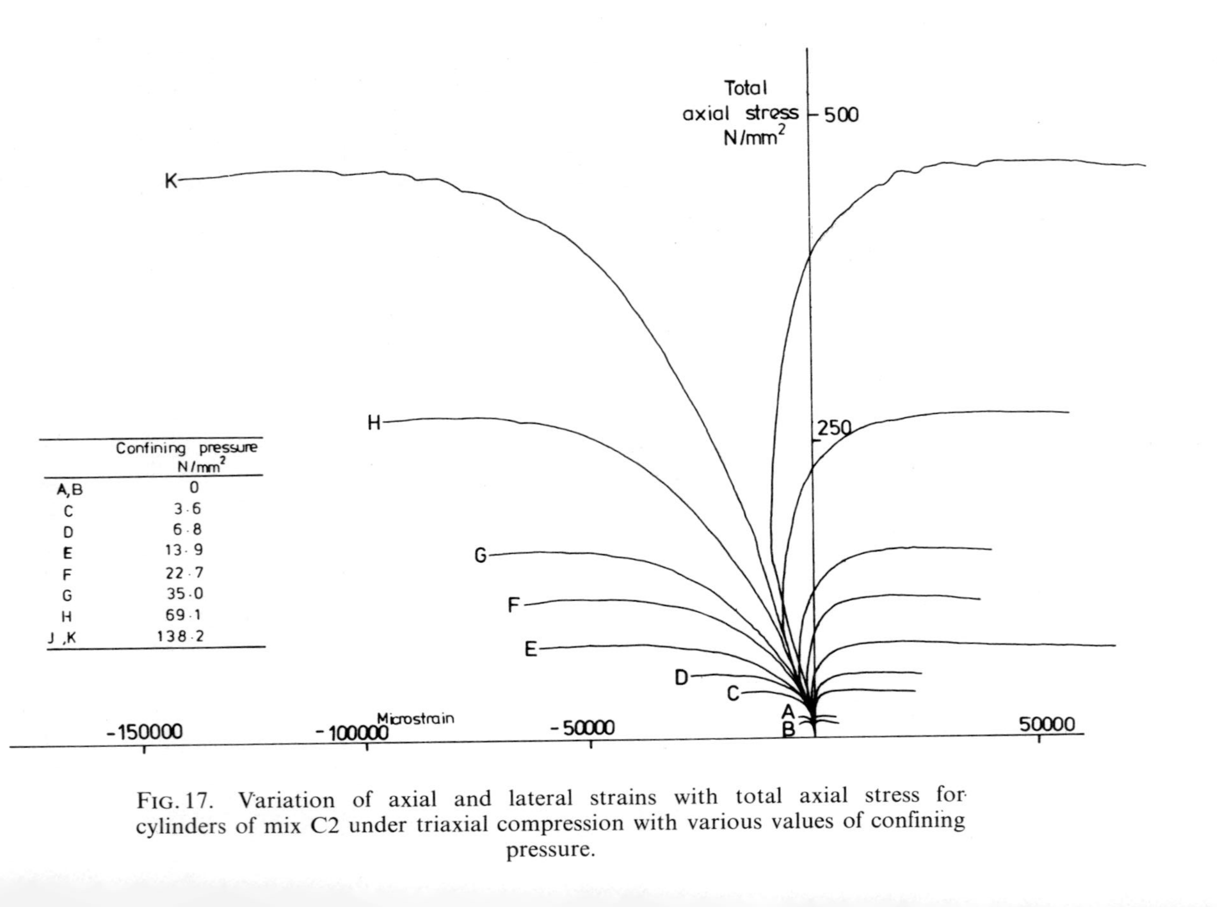

	Confining pressure N/mm²
A,B	0
C	3·6
D	6·8
E	13·9
F	22·7
G	35·0
H	69·1
J,K	138·2

FIG. 17. Variation of axial and lateral strains with total axial stress for cylinders of mix C2 under triaxial compression with various values of confining pressure.

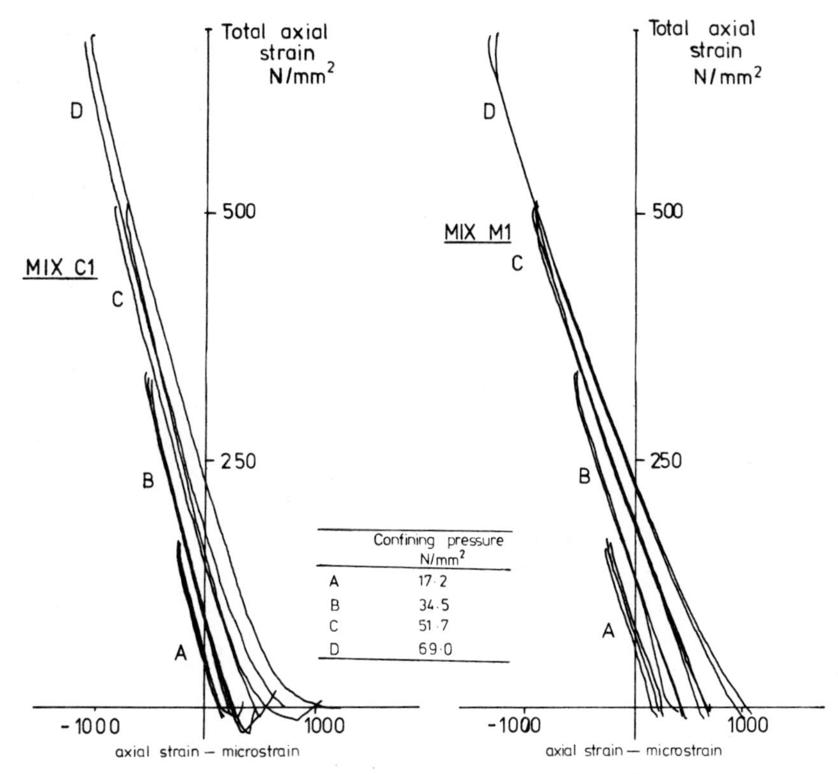

FIG. 18. Variation of axial strain with total axial stress for concrete mixes tested in triaxial 'extension' and C/C/T under various maximum confining pressures expressed in terms of 95 % confidence limits for the means.

until approaching ultimate. In fact, for a given applied confining pressure, the curves for the various mixes appear almost coincident during the initial stages of applied deviator stress.

The irregular behaviour at, or approaching, ultimate for the mixes containing higher volume fractions of aggregate subjected to relatively high confining pressures suggests disruption of individual aggregate particles. This is supported by the sounds which emanate from the specimen at the same time as the sharp changes in shape of the stress–strain curve.

(4) Comparing the relationships of mix M1 with M2 and mix C1 with C2 clearly demonstrates the considerable decrease in 'stiffness' and increase in 'ductility' with an increase of 'effective' water/cement ratio for all values of confining pressure. It should be noted that the

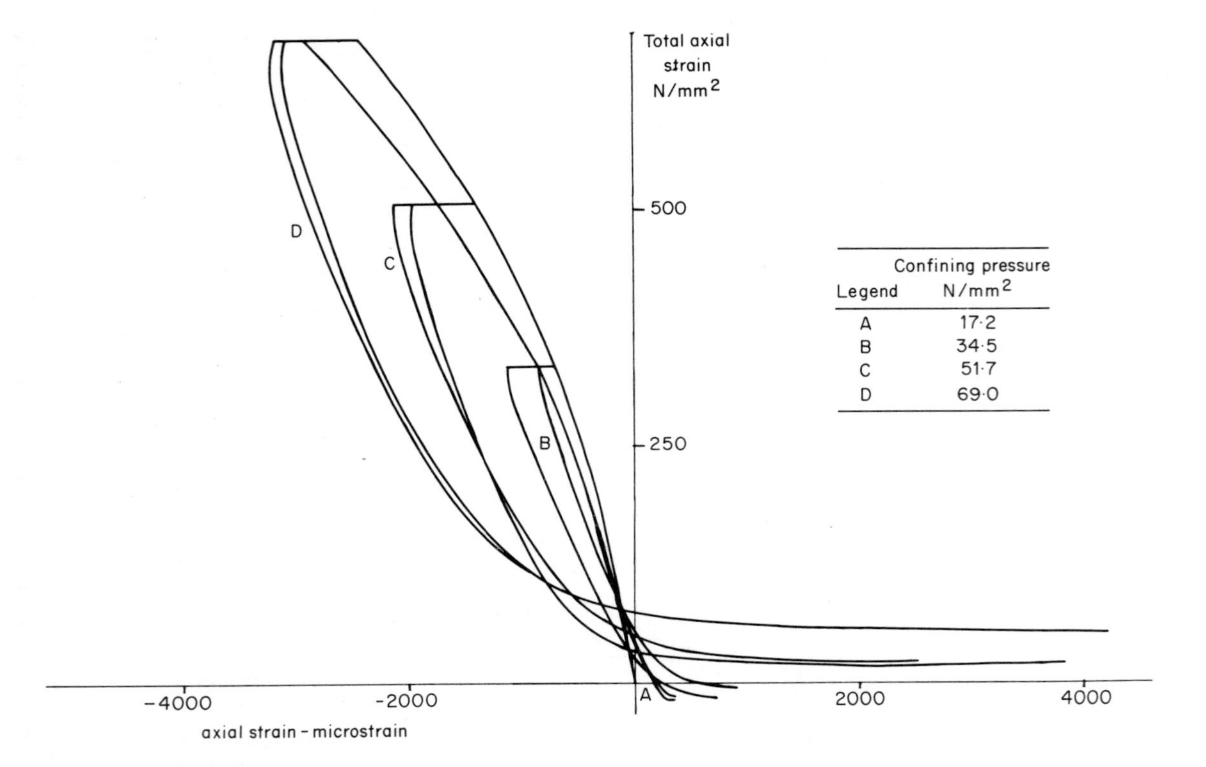

FIG. 19. Variation of axial strain with total axial stress for concrete mix C2 tested in triaxial 'extension' and C/C/T under various maximum confining pressures expressed in terms of 95 % confidence limits for the means.

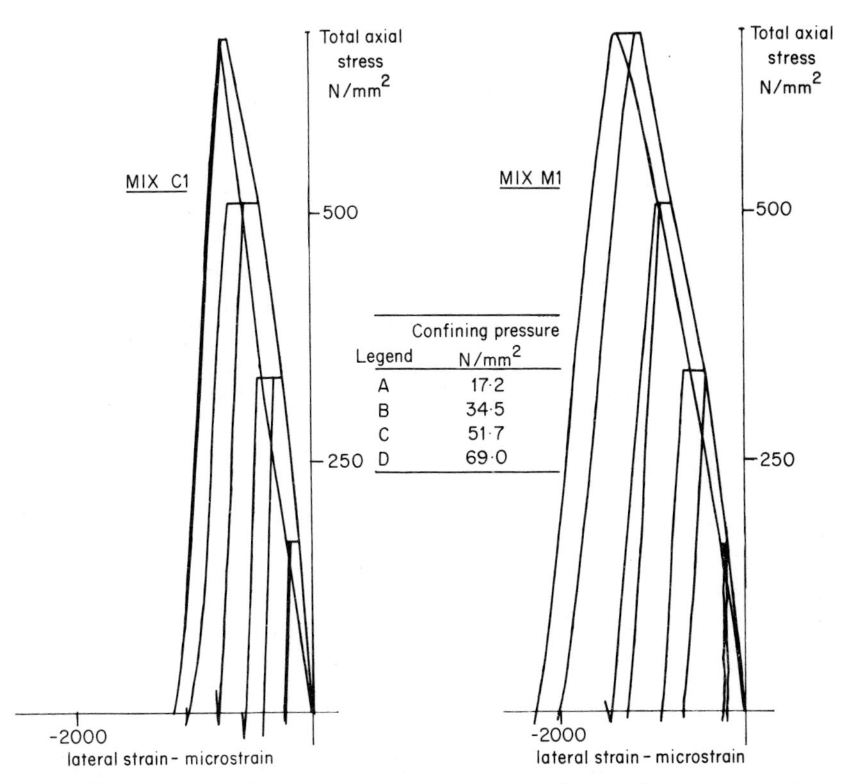

Legend	Confining pressure N/mm^2
A	17·2
B	34·5
C	51·7
D	69·0

FIG. 20. Variation of lateral strain with total axial stress for concrete mixes tested in triaxial 'extension' and C/C/T under various maximum confining pressures expressed in terms of 95% confidence limits for the means.

comparisons are not strictly valid since the volume fraction of aggregate is not constant in each case.

(5) Comparing the curves for mix C3 (limestone) and mix C4 (granite) with those for C2 (river gravel) under 'compression' stress states, shows the change in characteristics associated with varying the aggregate type, for mixes with the same 'effective' water/cement ratio containing the same volume fraction and grading of aggregate.

For given values of confining pressure the granite and limestone mixes are stiffer than those containing river gravel aggregate. This may be due to differences in aggregate stiffness and angularity which are both greater for limestone and granite than gravel, but is more likely to be due to the increased water absorption capacity of the gravel and its

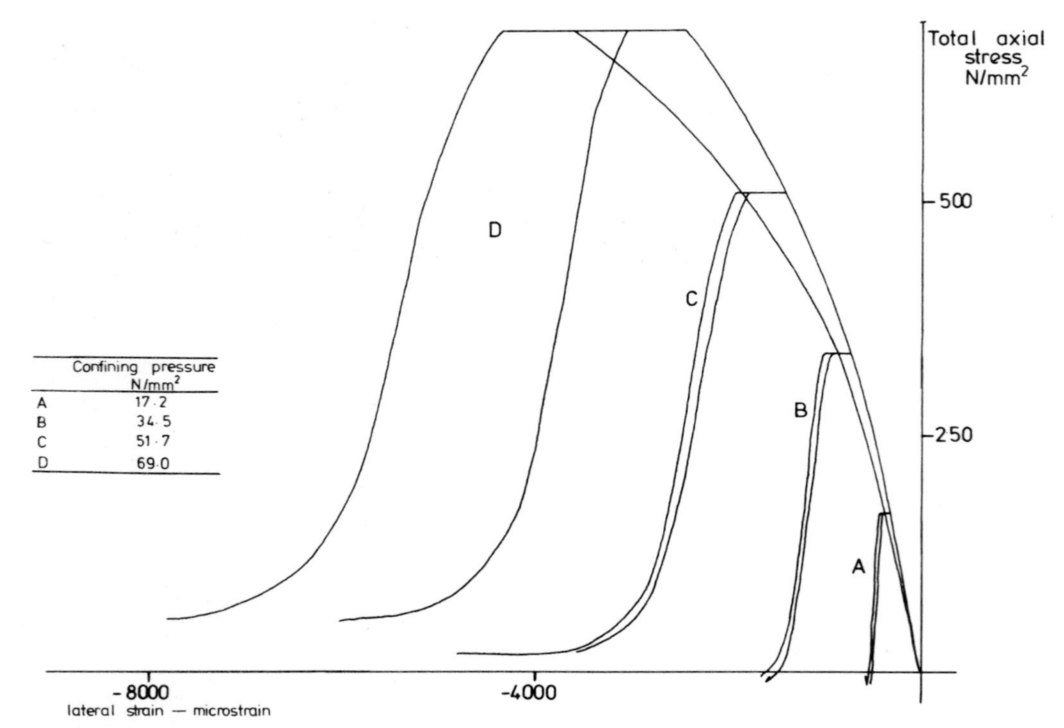

FIG. 21. Variation of lateral strain with total axial stress for concrete mix C2 tested in triaxial 'extension' and C/C/T under various levels of maximum confining pressure expressed in terms of 95 % confidence limits for the means.

effect in increasing the total water/cement ratio necessary to attain a given 'effective' water/cement ratio.

5.5.2.2 *Volume Changes*

Figures 22–25 show the typical relationships between the volume

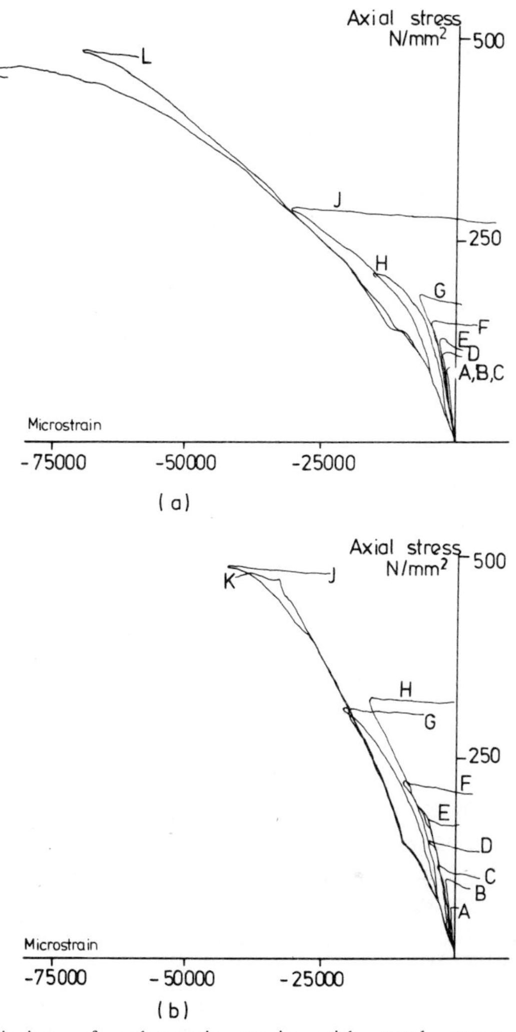

FIG. 22. Variation of volumetric strain with total stress for concretes under triaxial compression with various values of confining pressure. (a) Mix M1 (see Fig. 15). (b) Mix C1 (see Fig. 16).

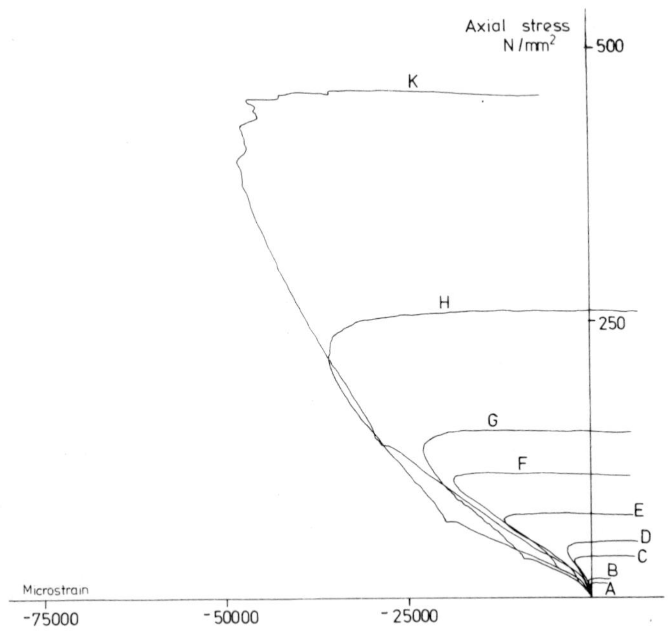

FIG. 23. Variation of volumetric strain with total axial stress for mix C2 under triaxial compression with various values of confining pressure (see Fig. 17).

changes and total axial stress at various confining pressures. Under 'compression' stress states the curves exhibit the following trends:

(1) A non-linear volume decrease for an increase in hydrostatic pressure followed by
(2) A further decrease in volume on application of deviator stress with mixes of a given 'effective' water/cement ratio containing a higher volume fraction of aggregate, giving rise to a smaller volume change at a given level of confining pressure and volume change being greater for mixes with a higher water/cement ratio.
(3) A reversal of the volume change at a stress approaching ultimate such that the volume attains a minimum value before increasing to failure.

Under 'extension' states the volume shows a small but progressive increase on application of the deviator stress and a rapid increase as the stress approaches ultimate.

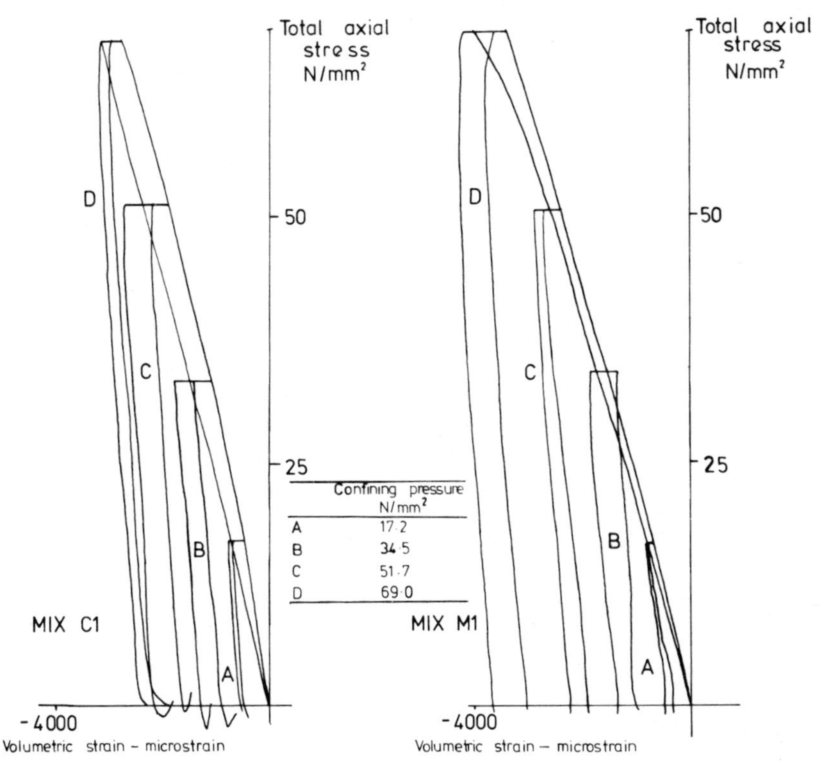

FIG. 24. Variation of volumetric strain with total axial stress for concrete mixes tested in triaxial 'extension' and C/C/T under various levels of maximum confining pressure expressed in terms of 95 % confidence limits for the means.

5.5.2.3 *Cracking*

As discussed in Section 5.4 studies of the fracture mechanisms of brittle materials (rocks and concrete) have indicated that such materials fail by crack extension in a direction parallel to the maximum principal compressive stress, or in other words, orthogonal to the maximum principal tensile stress. For uniaxial and biaxial compression and tension this behaviour can be easily identified, as it can for triaxial 'extension' tests or 'compression' tests under low confining pressures. For triaxial 'compression' at high confining pressures the mode of failure changes from 'brittle' to 'ductile' as evidenced by the stress–strain relationships (see Section 5.5.2.1) and no clear fracture mechanism is exhibited since the cracks are extremely small and localised.

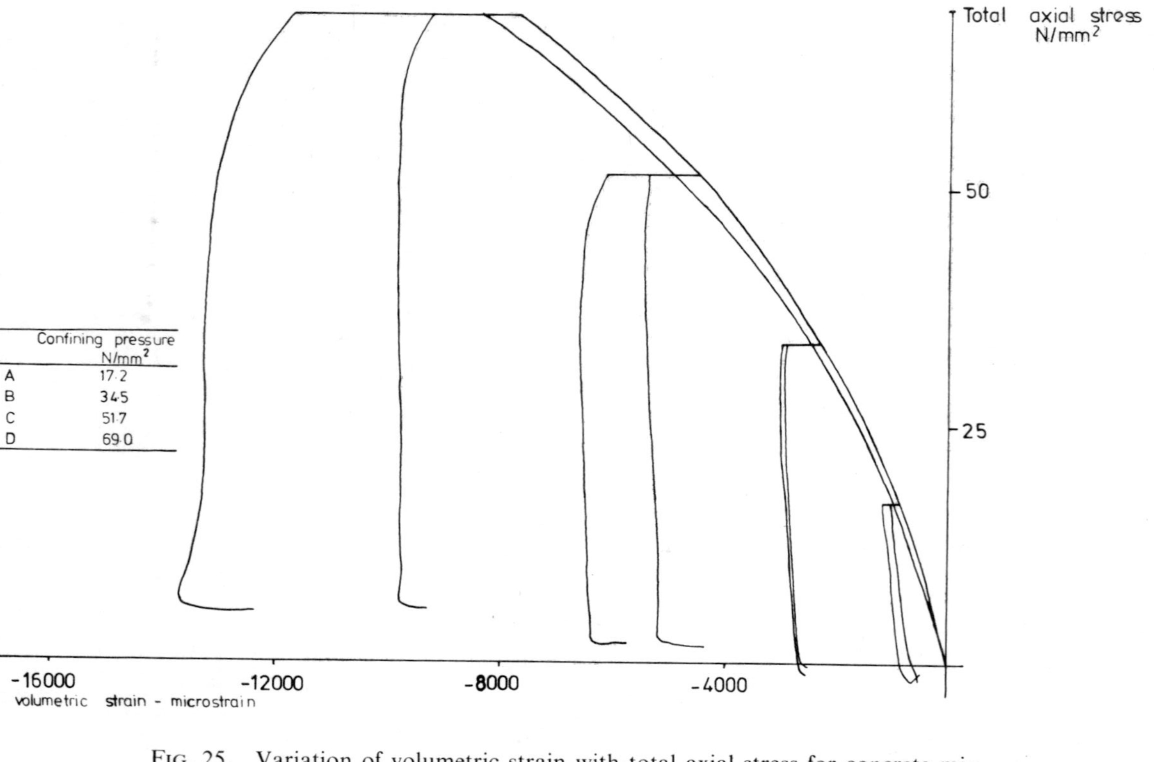

FIG. 25. Variation of volumetric strain with total axial stress for concrete mix C2 tested in triaxial 'extension' and C/C/T under various levels of maximum confining pressure expressed in terms of 95 % confidence limits for the means.

Unfortunately, the stress–strain relationships do not provide any indication of the extent of the cracking process under load since the relationships are essentially continuous and exhibit no 'yield point' prior to ultimate. However, some indication of the level at which cracking becomes significant can be provided by examining the relationships between the strains in the directions of maximum and minimum principal stresses, since such cracking will influence the strain in the direction of the minimum compressive stress more than in the direction of the maximum compressive stress. Therefore, as the cracks extend, the ratio of the strain in the direction of minimum compressive stress to the strain in the direction of maximum compressive stress should increase numerically. A typical variation of this relationship for two mixes under triaxial 'compression' is shown in Fig. 26. Only a small portion of the relationship is shown in each case up to and just beyond the departure from linearity, which is also indicated on the relationships, and this has been defined as the 'initial cracking level'[9] since it marks the start of the significant influence of cracking on the deformational behaviour of the material. For triaxial extension, typical lateral strain–axial strain relationships are shown in Figs. 27 and 28 for a mortar and concrete respectively. The initial cracking level is not normally assessed from overall strain–strain plots but rather from the changes in the strain ratio for small increments of stress, where the linear portion of each relationship is clearly seen (Fig. 29). Another clearly defined and more advanced level of the cracking process under triaxial 'compression' stress states is that defined by the stage at which the volume change relationships attain a minimum value before dilating to failure. For maximum 'extension', where no minimum volume is indicated, the point at which the volume increases rapidly is used. This level has been defined as 'final breakdown[9] and, together with 'initial cracking' can be adopted as design criteria[9] since the stages denoted by these and the ultimate level correspond to stable, quasi-stable and unstable behaviour under fluctuating loads and to definite behavioural patterns under sustained load.

5.5.2.4 *Stresses at Initial Cracking, Final Breakdown and Ultimate*
Figures 30–32 show envelopes denoting the variation of axial stress with confining pressure for various concrete mixes. For initial cracking the curves all show the same trend namely, that, as the confining pressure of the test is increased for a given mix the envelope tends

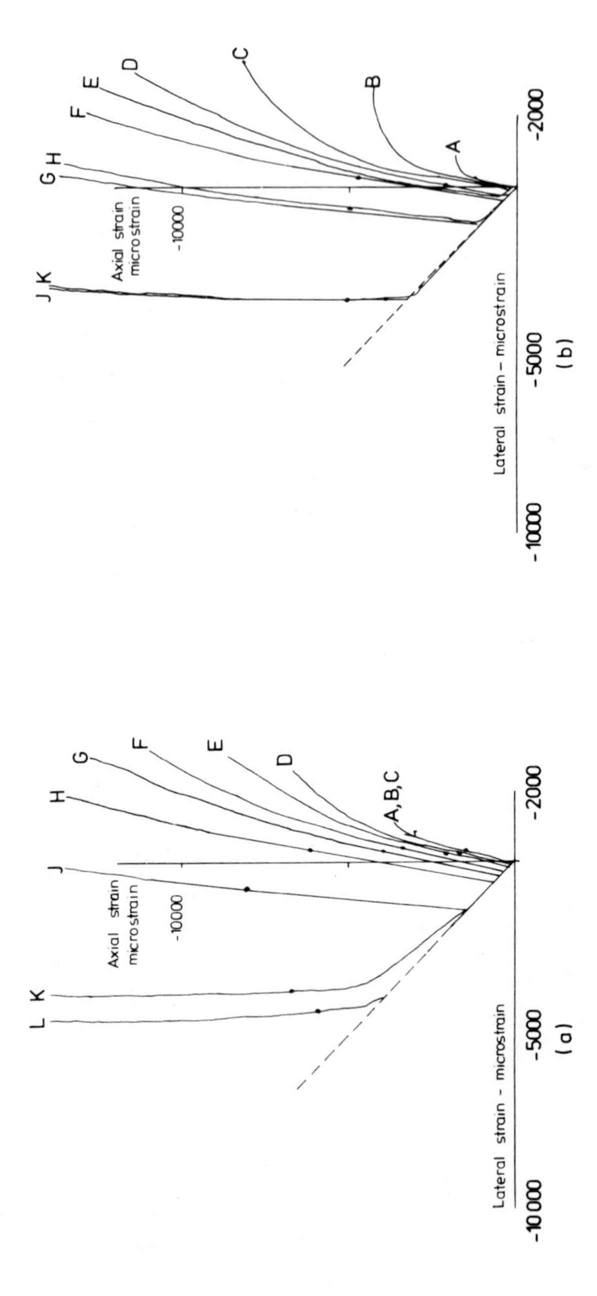

FIG. 26. Relationships between axial and lateral strains for real concretes under triaxial 'compression' with various values of confining pressure (see Figs. 15 and 16). ●, initial cracking level. (a) Mix M1. (b) Mix C1.

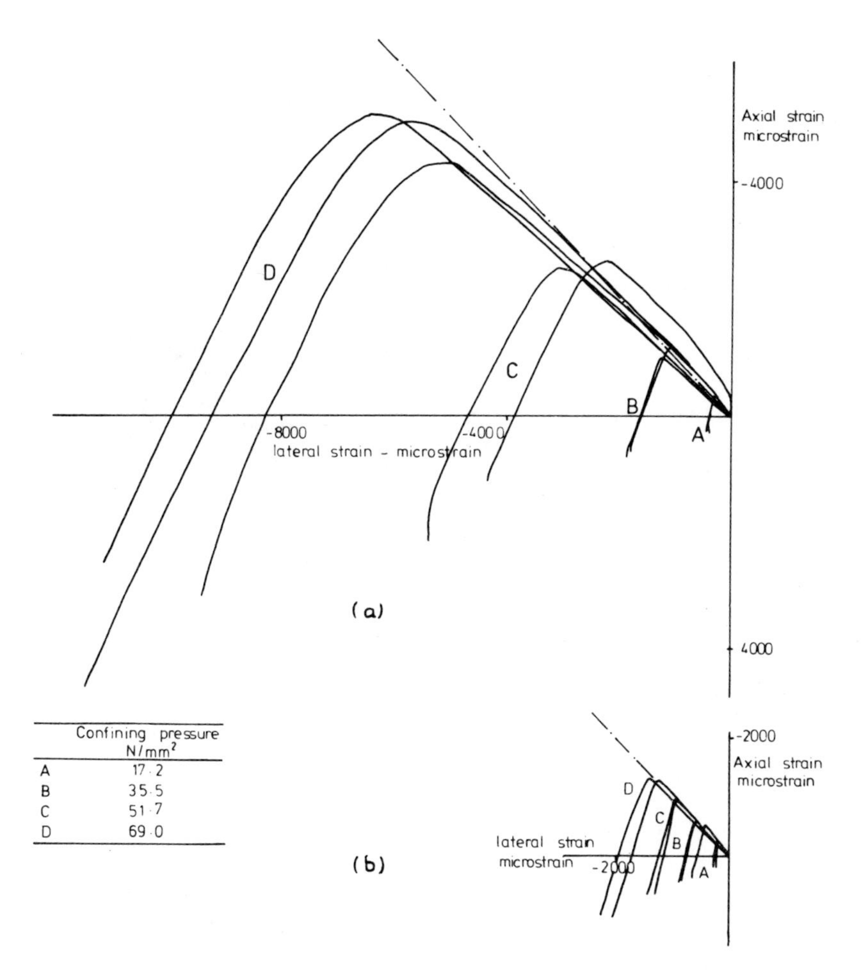

	Confining pressure N/mm^2
A	17·2
B	35·5
C	51·7
D	69·0

FIG. 27. Variation of lateral strain with axial strain for concrete mixes tested in triaxial 'extension' and C/C/T under various maximum confining pressures. (a) Mix M2. (b) Mix M1.

towards intersection with the hydrostatic axis such that a closed portion of stress-space is defined.

The author has postulated that such behaviour may be due to both local stress or strain concentrations in the heterogeneous material under hydrostatic compression and collapse of the cement paste phase. This is supported by evidence of failure mechanisms found for concretes

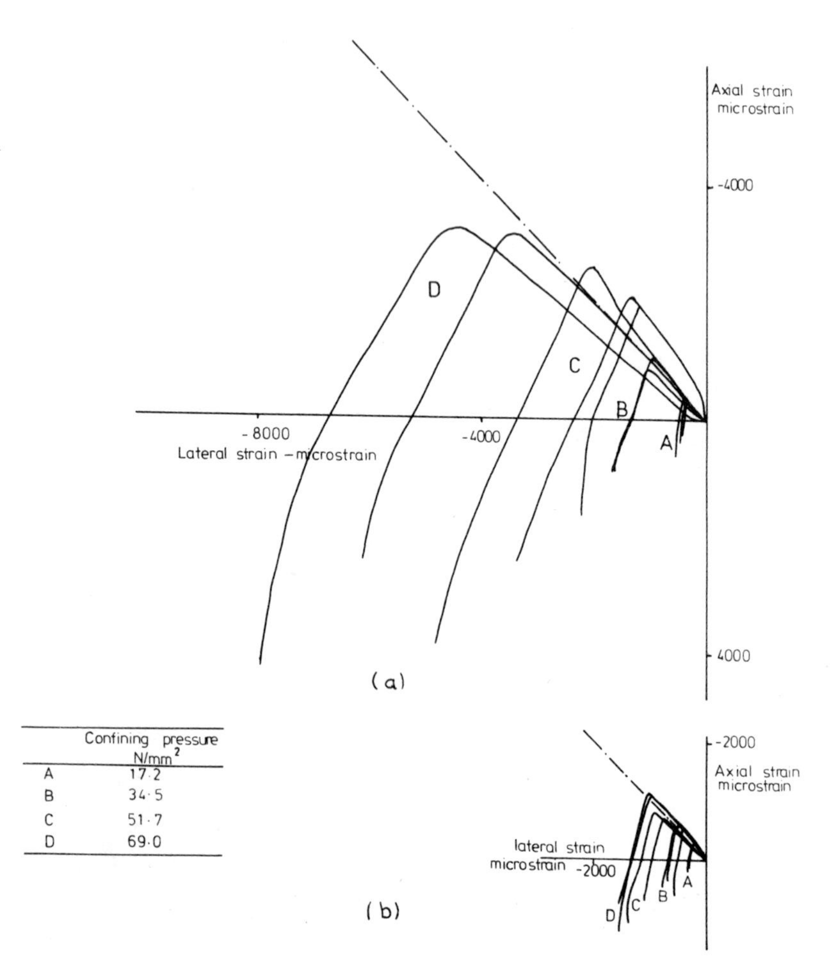

	Confining pressure N/mm^2
A	17·2
B	34·5
C	51·7
D	69·0

FIG. 28. Variation of lateral strain with axial strain for concrete mixes tested in triaxial 'extension' and C/C/T under various maximum confining pressures. (a) Mix C2. (b) Mix C1.

under hydrostatic pressure and failure envelopes for pastes,[32] lightweight concretes[33,34] and other porous materials.[35]

This behaviour is to be contrasted with the ever-increasing stress at final breakdown and ultimate with increasing compressive confining pressure which gives rise to open-ended envelopes, at least within that portion of stress space covered by the tests.

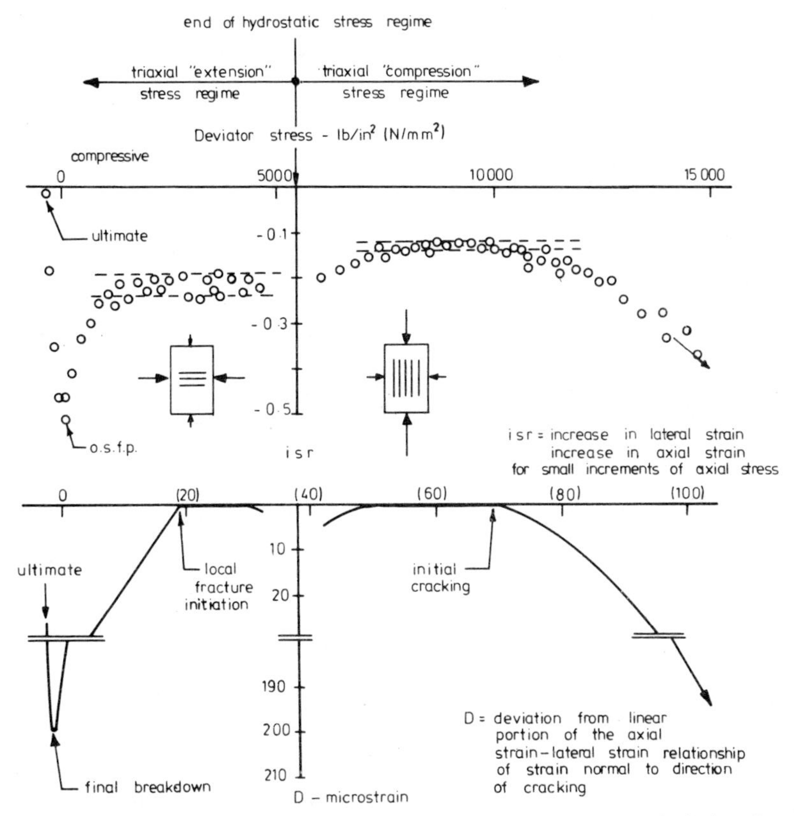

FIG. 29. Typical variation of incremental strain ratio (isr) and deviation from linearity (D) for a gravel concrete subjected to triaxial 'compression' and 'extension' stress states.

5.5.2.5 *Failure Modes*

The cracking of concrete under multiaxial stress discussed in Section 5.5.2.3 results in the complete disruption of specimens at, or beyond, the ultimate stress level in a manner which is dependent on the type of stress state imposed. For the cases of uniaxial and biaxial stress states the failure modes are covered in Section 5.4.4. For triaxial 'compression', where each principal stress is maintained compressive or increased in compression, under low values of volumetric stress, failure can be observed to occur along planes parallel to the maximum compressive stress in a similar manner to uniaxial compression.

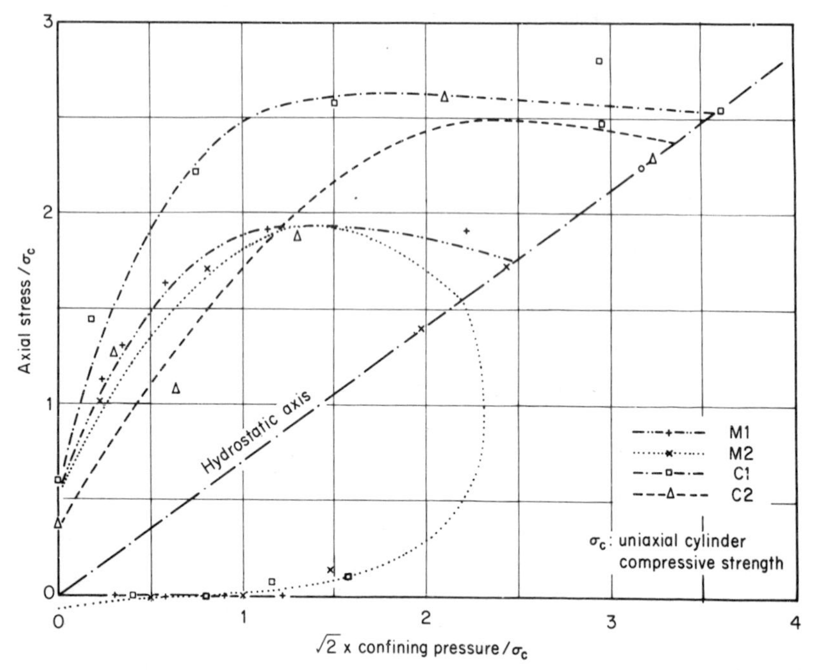

FIG. 30. Variation of total axial stress at initial cracking level with confining pressure for gravel concretes, and mortars under triaxial 'compression' and 'extension' stress states. σ_c—uniaxial cylinder compressive strength.

However, for larger values of volumetric stress the failure planes become less distinct as the behaviour becomes more ductile. If a conventional hydraulic coaxial cell is used to apply loads then this behaviour gives rise to the 'bulged' profiles of specimens under high confining pressures allied to the considerable shortening in the axial direction. For the 'extension' or compression–compression–tension types of test failure occurs abruptly on planes orthogonal to the minimum compressive stress.

Examples of modes of failure under triaxial states of stress applied in a triaxial cell are shown in Figs. 33 and 34.

5.6 REPRESENTATION OF MULTIAXIAL TEST DATA

5.6.1 Critical Stress Envelopes

In preceding sections strength data has been shown by plotting derived results in two dimensions using natural scales. Such a method of

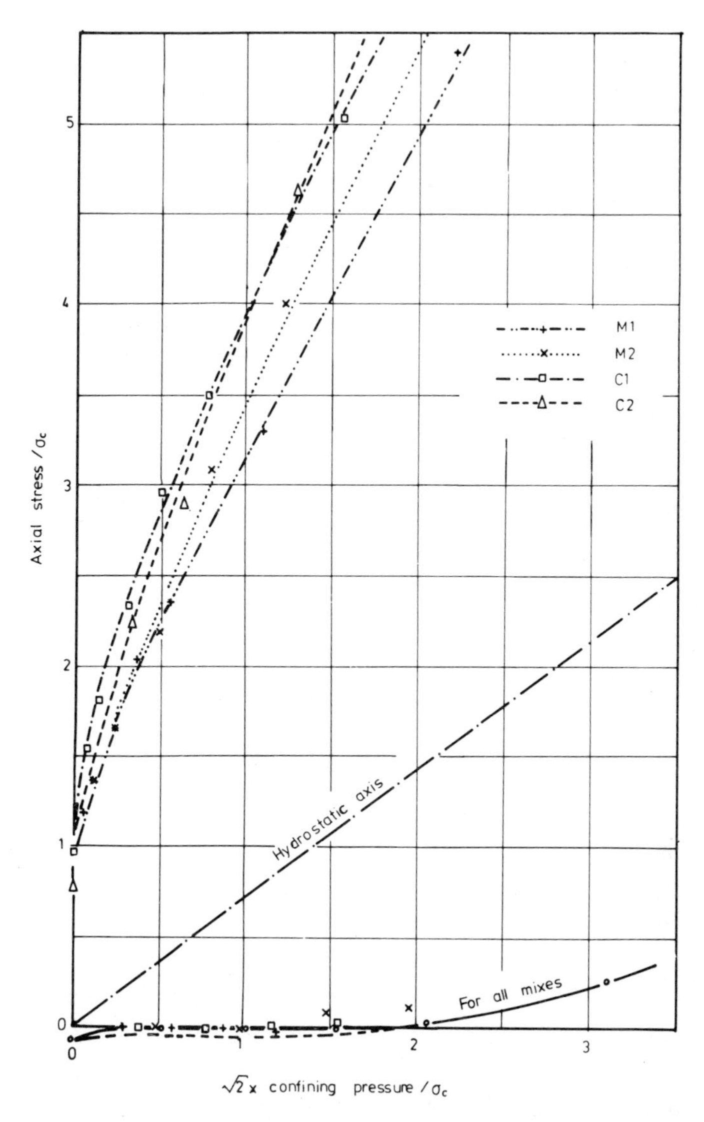

FIG. 31. Variation of total axial stress at final breakdown level with confining pressure for gravel concretes and mortars under triaxial 'compression' and 'extension' stress states. σ_c—uniaxial cylinder compressive strength.

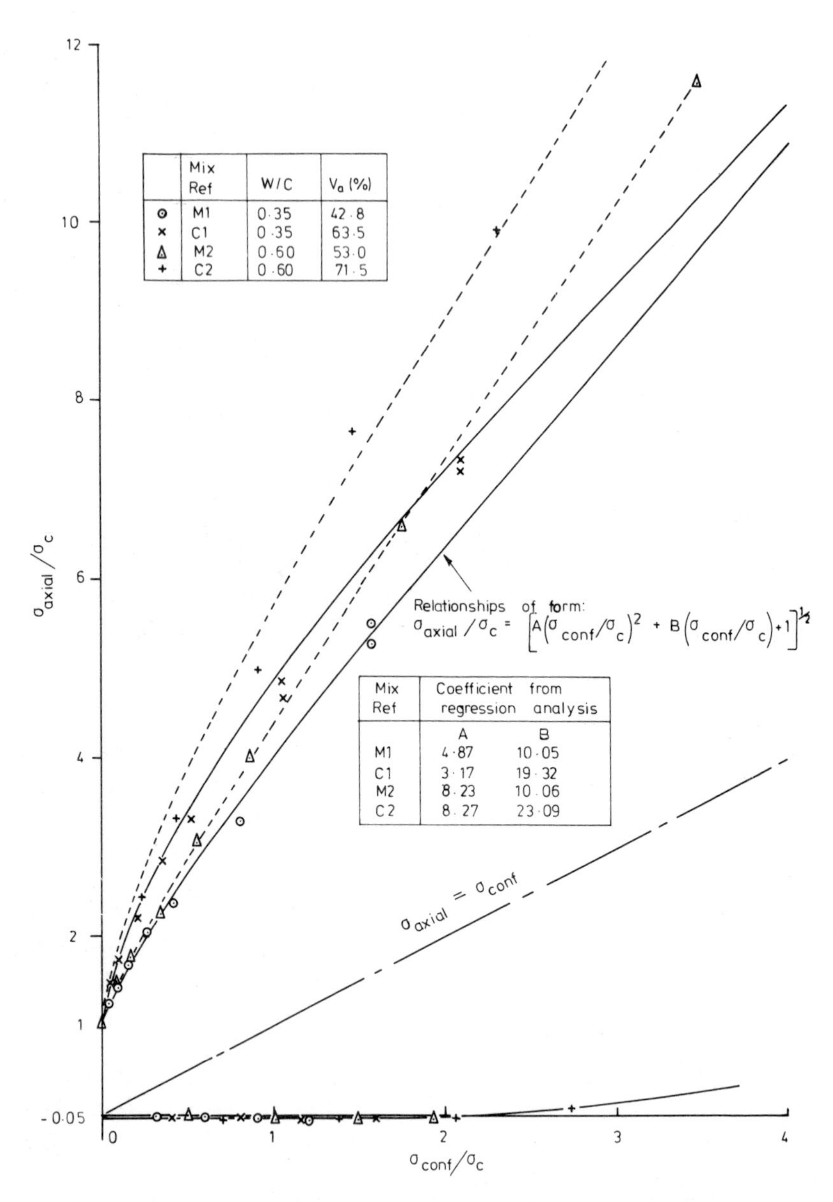

FIG. 32. Relationships between normalised axial stress and confining pressure at ultimate under triaxial 'compression' and 'extension' for gravel mortars and concretes.

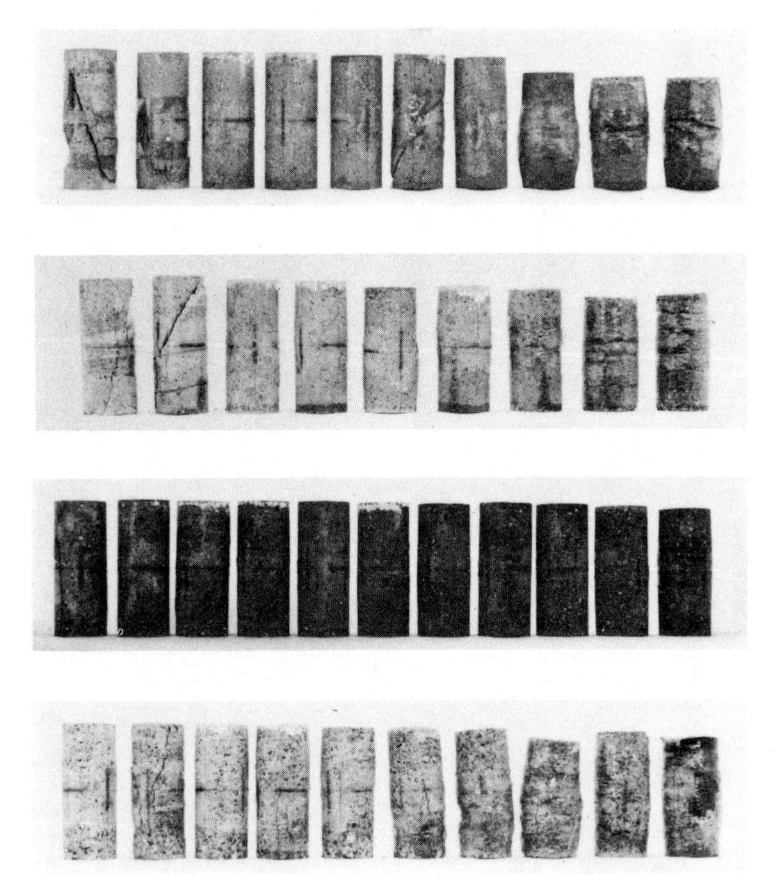

FIG. 33. Specimens of mortars and concretes after failure under triaxial 'compression': (a) M1. (b) BC1. (c) M2. (d) C2. With confining pressure increasing from left to right.

representation allows immediate visual comparisons to be made and is ideal for plane stress tests or tests in which the stresses are applied axisymmetrically. However, it suffers from the disadvantage that the criticality of more generalised stress states, i.e. where $\sigma_1 \neq \sigma_2 \neq \sigma_3$, cannot be assessed easily, and if the data are to be used as criteria for design we need to describe three-dimensional envelopes in stress space either graphically or mathematically.

Examples of graphical representation of strength envelopes have been

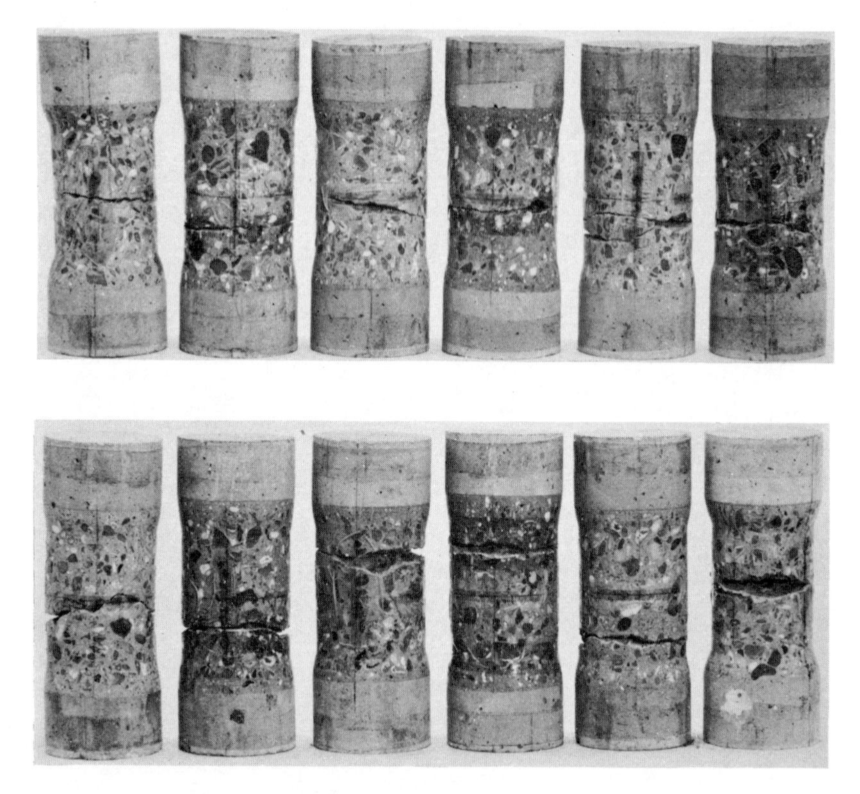

FIG. 34. Specimens of a concrete mix after failure under triaxial 'extension' and C/C/T.

given by Newman and Newman[36] and it has been pointed out that due to the inherent variability of results from concretes and other brittle materials any theory of failure deduced from a given plot can be dependent on the methods of plotting. Thus it must be ensured that the choice of a method for describing data is given careful consideration to ensure that incorrect conclusions are not drawn with regard to material behaviour.

Failure envelopes are usually expressed directly in terms of principal stresses and, owing to the difficulty of mathematically describing their shapes for generalised stress states, resort is usually made to graphical methods by decomposing stress states into volumetric and deviatoric components[36,37] where

volumetric (hydrostatic) component of stress $\sigma_v = (\sigma_1 + \sigma_2 + \sigma_3)/3$

deviatoric components $\sigma_{1D} = \sigma_1 - \sigma_v$

$$\sigma_{2D} = \sigma_2 - \sigma_v$$

$$\sigma_{3D} = \sigma_3 - \sigma_v$$

By this technique the criticality of a particular stress state $\sigma_1\sigma_2\sigma_3$ is examined by plotting the deviatoric components of that stress state on the envelope for the appropriate volumetric stress plane and checking whether the point lies within or outside the envelope.

This technique has been employed by the author in deriving design criteria for the cracking and ultimate strength of various concretes under generalised stress states[38] and envelopes for a typical concrete are given in Figs. 35 and 36.

Another approach to the formulation of design criteria is to collect together all relevant information on the strength of concrete under complex states of stress and derive safe lower bound envelopes to the data which can then be described mathematically. This technique has been used by Hannant[37] to produce ultimate failure envelopes and by Hobbs *et al.*[39] to produce both ultimate and serviceability envelopes suitable for use with limit state design methods.

5.6.2 Stress–Strain Relationships

Most investigations into the behaviour of concrete under multiaxial states of stress have been primarily concerned with the determination of the strength characteristics of concrete and many of the results obtained have been combined in nomograms which may be used for design purposes (see Section 5.6.1).

However, although some data for the deformational properties of concrete under various loading regimes are available, it is difficult to present this in a suitable form for use with computer-based techniques. Indeed, the accuracy of some data which are available is often in question due to uncertainties concerning the efficiency of the various testing systems used to achieve given states of stress within specimens.

Nevertheless, attempts have been made to produce a mathematical description of the deformational behaviour of concrete under various

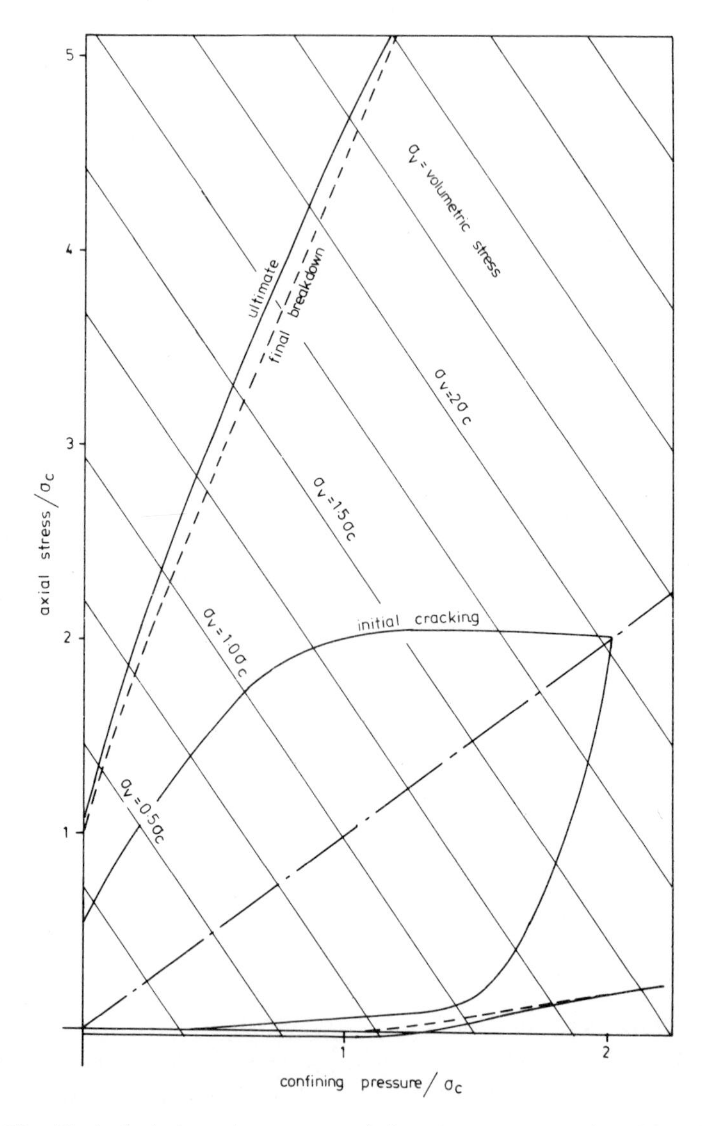

Fig. 35. Typical design chart on rendulic plane ($\sigma_2 = \sigma_3$) for determining combination of normalised stress at initial cracking, final breakdown and ultimate for a gravel concrete.

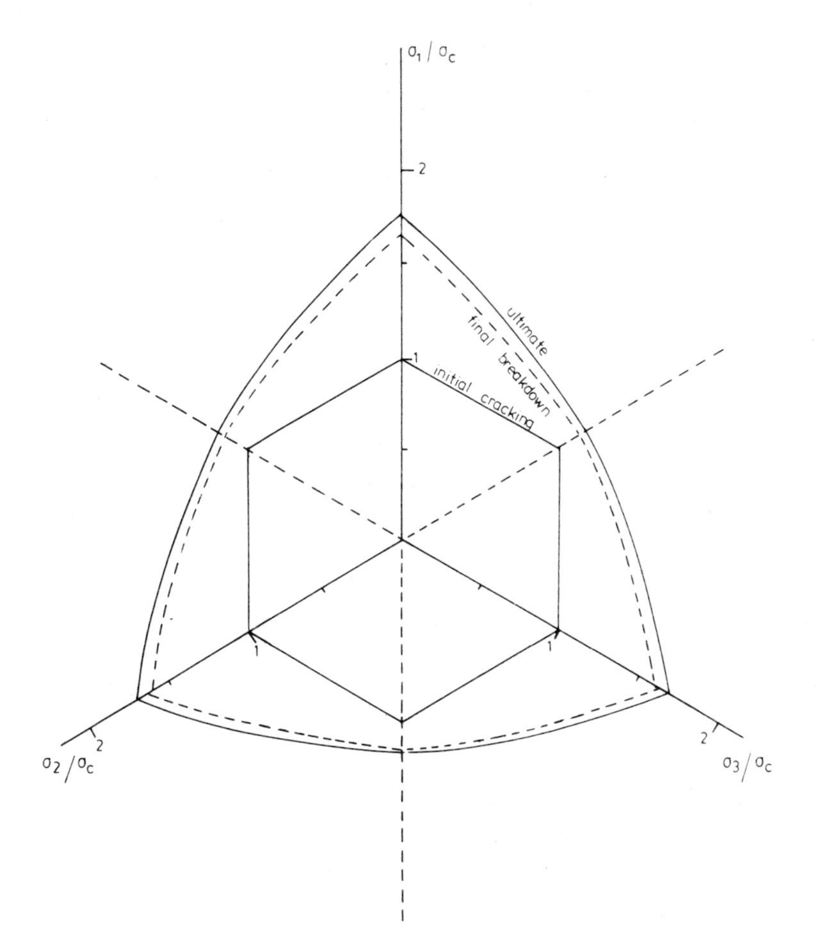

FIG. 36. Envelopes on equivolumetric stress plane $\sigma_v = 0{\cdot}75\,\sigma_c$ defining initial cracking, final breakdown and ultimate for a gravel concrete (based on Fig. 35).

types of loading conditions. In these attempts the following methods have been employed for the derivation of mathematical expressions:

(1) *Curve fitting.* This method has been used predominantly to describe uniaxial stress–strain data.[40,41] Its basic disadvantage is the lack of an underlying theory which would allow extrapolation procedures.

(2) *Physical models of the microstructure.* Examples of such models are (a) multiphase models,[42,43] (b) truss analogies,[44,45] and (c) models representing the fracture behaviour.[42,43] One serious drawback to the use of current models is the expense of incorporating them in computer programs and another is the fact that they are mainly qualitative and require the addition of empirical factors.
(3) *Characterisations based on measures of 'damage'.*[46] In such methods the material properties are expressed as a function of 'damage' or change of state experienced by the material. The main problem of this approach is the determination of a suitable measure of 'damage'.
(4) *Nonlinear viscoelasticity.* Examples of this approach are constitutive equations derived from theoretical considerations based on nonlinear elasticity,[47–51] hypoelasticity[52] and 'equivalent time' characterisations.[53,54] In this latter technique the nonlinearities are accounted for by making the 'equivalent time'[47] a function of the strain history; to date this approach, however, ignores 'damage'-induced anisotropy.
(5) *Modified plasticity theory.*[55] This concept has been extended from metals to concrete and it is based on principles of the classical theory of plasticity such as the normality principle, similarity of shapes of yield surfaces in stress space, etc., which are not valid for concrete.

In all the above methods the deformational behaviour of concrete has been expressed in a mathematical form including parameters which can only be evaluated by analysing experimental values. The goodness of fit of the derived expressions to the true stress–strain relationship of concrete, therefore, depends not only upon the underlying theory regarding the concrete behaviour but also the validity of the experimental values used.

An attempt to characterise the stress–strain relationship of concrete more effectively has been made by developing a model in conjunction with approaches of type (3), (4) or (5) above. In this way it is hoped to predict the complete deformational behaviour of the materials and to conclude the effects of 'damage'-induced anisotropy.

This approach has led to a mathematical description of the deformational properties up to ultimate strength of concretes with uniaxial cylinder strengths of from 20–70 N/mm^2 under any type of short-term static loading condition, in a form suitable for use with computer-based methods of structural analysis.[56] This description is currently being refined to predict the post ultimate strength behaviour of the same range of concretes.

This derivation of a mathematical expression is based on an analysis of experimental data obtained at Imperial College, London in previous investigations and makes use of the assumption that any nonlinear material behaviour is due to the fracture processes which occur under increasing strain. Figures 37 and 38 show typical comparisons between the predicted and experimentally determined stress–strain relationships for a concrete under triaxial 'compression' and 'extension' stress states. The experimental data used in this investigation have been validated by comparing them with those obtained in a recently completed international cooperative programme of research which has been concerned with the effect of testing techniques on the behaviour of concrete under biaxial and triaxial stress states.[23]

5.7 CONCLUDING REMARKS

It has been shown in this chapter that the behaviour of an element of concrete under load depends on the stress or strain state applied to the boundary of that element. It is possible to generate an infinite variety of boundary conditions using different loading techniques which can vary between the following stress conditions:

(1) 'Compression' stress states

where $\sigma_1 > \sigma_2 = \sigma_3 > 0$

A special case is

uniaxial compression $\sigma_1 > 0$, $\sigma_2 = \sigma_3 = 0$

(2) 'Extension' stress states $\sigma_1 = \sigma_2 > \sigma_3$

where σ_3 can be compressive or tensile.

A special case is

biaxial compression $\sigma_1 = \sigma_2 > 0$, $\sigma_3 = 0$

Between the above stress states the intermediate stress σ_2 can vary between σ_3 and σ_1.

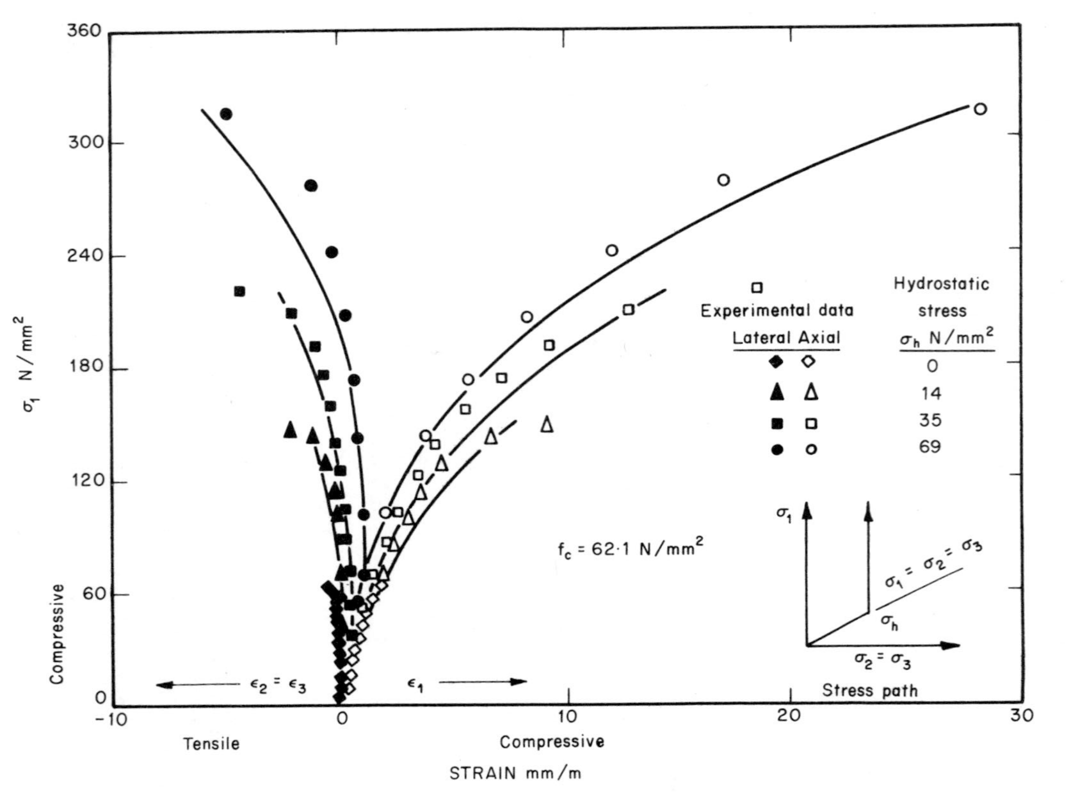

FIG. 37. Comparison between predicted and measured stress–strain relationships for a concrete under triaxial 'compression' stress states. —, predicted relationships. $f_c = 62{\cdot}1$ N/mm².

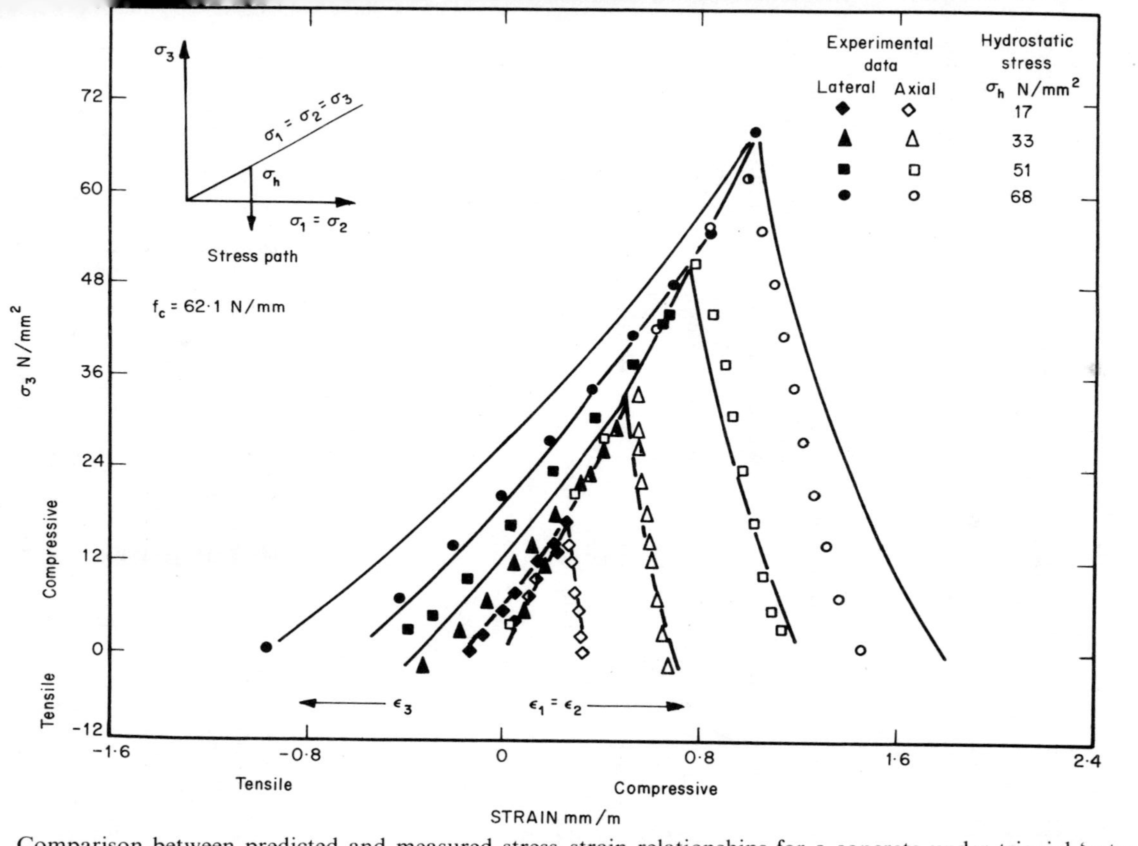

FIG. 38. Comparison between predicted and measured stress–strain relationships for a concrete under triaxial 'extension' stress states, —, predicted relationships.

For triaxial 'compression' stress states where $\sigma_2 = \sigma_3 > 0$, concrete can resist stresses in excess of those capable of being carried in simple (uniaxial) or biaxial compression, and this enhancement can be exploited in structural design by introducing secondary active or passive restraints. One example of an active restraint is the wire-wound or prestressed concrete pressure vessel[57] while passive restraint can be provided by reinforcement or surrounding materials. The failure mechanism under such stress states is by crack extension in the direction of maximum principal compressive stress with distinct stages of fracture initiation, propagation and ultimate breakdown being identifiable as the stress is increased. The strains at ultimate increase dramatically as σ_2 and σ_3 increase to give characteristics of ductility.

However, for the 'extension' type of stress state complete disruption can occur at much lower stresses particularly if one or more stresses are tensile. As for 'compression' the failure mechanism is one of crack propagation in the direction of the most compressive principal stress, but in this case the initiation and propagation processes occur almost simultaneously and lead to an abrupt loss of load-carrying capacity at ultimate.

In view of the above differences in behaviour it is necessary to describe the deformational and strength characteristics of concrete under generalised stress conditions so that realistic analyses of complex structures can be performed and the properties of the material exploited to the full.

REFERENCES

1. MARIN, J. (1935). Failure theories of materials subjected to combined stresses. *Trans. Am. Soc. Civ. Eng.*, **101,** 1162–78.
2. NADAI, A. (1950). *Theory of Flow and Fracture of Solids, Vol. I.* McGraw-Hill, New York. 572 pp.
3. TIMOSHENKO, S. P. (1953). *History of Strength of Materials.* McGraw-Hill, New York. 445 pp.
4. RICHART, F. E., BRANDTZAEG, A. and BROWN, R. L. (1928). A study of the failure of concrete under combined compressive stresses. *Eng. Expt. Stat. Bulletin No. 185*, University of Illinois. pp. 1–102.
5. RICHART, F. E., BRANDTZAEG, A. and BROWN, R. L. (1929). The failure of plain and spirally reinforced concrete in compression, *Eng. Expt. Stat. Bulletin No. 190*, University of Illinois.
6. KUPFER, H., HILSDORF, H. K. and RUSCH, H. (1969). Behaviour of concrete under biaxial stresses. *J. Am. Concrete Inst., Proc.*, **66** (8), 656–66.

7. ANDENAES, E., GERSTLE, K. and KO, H.-Y. (1977). Response of mortar and concrete to biaxial compression, *J. Eng. Mech. Div. (ASCE)* **103** (EM4), 515–25
8. SCHICKERT, G. (1973). On the influence of different load application techniques on the lateral strain and fracture of concrete specimens. *Cement & Concrete Research*, **3,** 487–94.
9. NEWMAN, J. B. (1973). Deformational behaviour, failure mechanisms and design criteria for concretes under combinations of stress. *CIRIA Report*, December. 583 pp.
10. PANDIT, G. S., ZIMMERMAN, R. M., KUPFER, H., HILSDORF, H. K. and RUSCH, H. (1970). Discussion of Reference 6, *J. Am. Concrete Inst., Proc.*, **67** (2), 194–7.
11. BELLAMY, C. J. (1961). Strength of concrete under combined stress. *J. Am. Concrete Inst., Proc.*, **58** (4), 367–82.
12. BRITISH STANDARDS INSTITUTION (1970). BS 1881, Part 4: Methods of testing concrete for strength. BSI, London.
13. CHINN, J. and ZIMMERMAN, R. M. (1965). Behaviour of plain concrete under various high triaxial compression loading conditions. *Tech. Rep. No. WLTR 64–163.* Res. and Tech. Div. Air Force Weapons Lab. Air Force Systems Command, Kirtland A.F.B., New Mexico, USA. 138 pp.
14. VILE, G. W. D. (1965). *Behaviour of concrete under simple and combined stresses*, Ph.D. Thesis, University of London. 559 pp.
15. WASTLUND, G. (1937). Nya ron Angaende Betongens Grundlaggande Hall fast Hetsegenskaper (New Evidence regarding the Basic Strength Properties of Concrete). *Betong*, Heft 3, 189–205.
16. WEIGLER, H. and BECKER, B. (1961). Uber das Bruck-und Verformungsverhalten von Beton bei Nehrachsiger Beanspruchung. *Bauingenieur*, **36,** Heft 10, pp. 390–6.
17. WEIGLER, H. and BECKER, B. (1963). Untersuchungen uber das Bruch-und Verformungsoerhalten von beton bei Zweichsiger Beanspruchung. (Investigation into Strength and Deformation Properties of Concrete subjected to Biaxial Stresses). *Deutscher Ausschus fur Stahlbeton*, Heft 157, Berlin.
18. OPITZ, H. (1968). Festigkeit und Verformungseigenschaftendes Betons bei Zweiachsiger Druckbeanspruchung (Strength and Deformation Properties of Concrete under Biaxial Compressive Loading). *Proc. of Inter. Colloquium Festigkeitsprobleme des Betons, Dresden, June 1968*. pp. 1520–1523.
19. KUPFER, H., HILSDORF, H. K. and RUSCH, H. (1969). Behaviour of concrete under biaxial stresses. *J. Am. Concrete Inst., Proc.*, **66** (8), 656–66.
20. MILLS, L. L. and ZIMMERMAN, R. M. (1970). Compressive strength of plain concrete under multiaxial loading conditions. *J. Am. Concrete Inst., Proc.*, **67** (10), 802–7.
21. KOBAYASHI, S. and KOYANAGI, W. (1972). Fracture criteria of cement paste, mortar and concrete subjected to multiaxial compressive stresses. *Proc. RILEM Int. Symp., The Deformation and the Rupture of Solids Subjected to Multiaxial Stresses, Cannes, Oct. 1972, Vol. I.* pp. 131–48.

22. STEGBAUER A. and LINSE, D. (1972). Comparison of stress–strain behaviour of concrete and other materials under biaxial loading. *Proc. RILEM Int. Symp., The Deformation and the Rupture of Solids Subjected to Multiaxial Stresses, Cannes, Oct. 1972, Vol. I.* pp. 229–43.
23. GERSTLE, K. *et al.* (1978). *Strength of concrete under multiaxial stress states.* Paper presented at McHenry Symposium, Mexico City, October 1976. ACI Special Publication SP-55, pp. 103–31.
24. MCHENRY, D. and KARNI, J. (1958). Strength of concrete under combined tensile and compressive stress. *J. Am. Concrete Inst., Proc.*, **54** (10), 829–40.
25. BRESLER, B. and PISTER, K. S. (1957). Failure of plain concrete under combined stresses. *Trans. ASCE*, **122,** *Paper No.* 2897, 1049–59.
26. LINSE, D. and ASCHL, H. (1976). *Investigations on the behaviour of concrete under multiaxial loading.* Report of part of joint research programme, Tech. Univ. Munich. Personal communication.
27. NEWMAN, J. B. (1973). *Criteria for concrete strength*, Ph.D. Thesis, University of London. 583 pp.
28. KOTSOVOS, M. D. (1974). *Failure criteria for concrete under generalised stress states*, Ph.D. Thesis, University of London. 284 pp.
29. KOTSOVOS, M. D. and NEWMAN, J. B. (1977). Behaviour of concrete under multiaxial stress. *J. Am. Concrete Inst., Proc.*, **74** (9), 443–6.
30. NEWMAN, J. B. (1974). Apparatus for testing concrete under multiaxial states of stress. *Mag. of Con. Res.*, **26** (89), 229–38.
31. HOBBS, D. W. (1972). The strength and deformation properties of plain concrete under combined stress. *Proc. RILEM Int. Symp., The Deformation and the Rupture of Solids Subjected to Multiaxial Stresses, Cannes, Oct. 1972, Vol. I.* pp. 97–111.
32. GVOZDEV, A. A. (1959). Some mechanical properties of concrete particularly important for the structural mechanics of reinforced concrete structures. *Trans. of Scientific Res. Inst. for Reinf. Conc.* (*N.I.I. Zh.B.*), No. 5, Gosstroiizdat, Moscow (in Russian).
33. SHIDELER, J. J. (1957). Lightweight aggregate concrete for structural use. *J. Am. Concrete Inst.*, **29** (4), 299–329.
34. HANSON, J. A. (1963). Strength of structural lightweight concrete under combined stress. *Journ. PCA*, **5** (1), 39–46.
35. DERUNTZ, J. A. and HOFFMAN, O. (1969). The static strength of syntactic foam. *J. Appl. Mech.*, **36E** (3), 551–7.
36. NEWMAN, K. and NEWMAN, J. B. (1971). Failure theories and design criteria for plain concrete. In, *Structure, Solid Mechanics and Engineering Design.* Ed. M. Te'Eni, Wiley-Interscience, London. 1484 pp.
37. HANNANT, D. J. (1974). Nomograms for the failure of plain concrete subjected to multiaxial stresses. *The Structural Engineer*, **52** (5), 151–65.
38. NEWMAN, J. B. and NEWMAN, K. (1978). Development of design criteria for concrete under combined states of stress. *CIRIA Technical Note No. 93.* 52 pp.
39. HOBBS, D. W., POMEROY, C. D. and NEWMAN, J. B. (1977). Design stresses for concrete structures subject to multi-axial stresses. *The Structural Engineer*, **55,** 151–64.
40. POPOVICS, S. (1970). A review of stress-strain relationships for concrete. *J. Am. Concrete Inst., Proc.*, **67** (3), 243–8.

41. POPOVICS, S. (1973). A numerical approach to the complete stress–strain curve of concrete. *Cement & Concrete Research*, **3** (5), 583–99.
42. BUYOZTURK, O., NILSON, A. H. and SLATE, F. O. (1971). Stress–strain response and fracture of a concrete model in biaxial loading. *J. Am. Concrete Inst., Proc.*, **68** (8), 590–9.
43. BUYOZTURK, O., NILSON, A. H. and SLATE, F. O. (1972). Deformation and fracture of particulate composite. *J. Eng. Mech. Div.* (ASCE), **98** (EM.3), Proc. Paper 8970, 581–93.
44. ANSON, M. (1964). An investigation into a hypothetical deformation and failure mechanism for concrete. *Mag. Conc. Res.*, **16** (47), 73–82.
45. ENDEBROOK, E. G. and TRAINA, L. A. (1972). Static concrete constitutive relations based on cubical specimens. *Tech. Rep. No. AFWL-TR-72-59*, **1,** Air Force Weapons Lab. Air Force Systems Command. Kirtland A.F.B., New Mexico, USA. 204 pp.
46. ROMSTAD, K. M., TAYLOR, M. A. and HERRMANN, L. R. (1974). Numerical biaxial characterisation for concrete. *J. Eng. Mech. Div.* (ASCE), **100** (EM 3), Proc. Paper 10879, 935–48.
47. KUPFER, H. B. and GERSTLE, K. H. (1973). Behaviour of concrete under biaxial stresses. *J. Eng. Mech. Div.* (ASCE), **99** (EM 4), Proc. Paper 9917, 852–66.
48. LIU, T. C. Y., NILSON, A. H. and SLATE, F. O. (1972). Biaxial stress–strain relations for concrete. *J. Struct. Div.* (*ASCE*), **98** (ST 5), Proc. Paper 8095, 1025–34.
49. LIU, T. C. Y., NILSON, A. H. and SLATE, F. O. (1972). Stress–strain response and fracture of concrete in uniaxial and biaxial compression. *J. Am. Concrete Inst., Proc.*, **A69** (5), 291–5.
50. DALANISWAMY, R. and SHAH, S. P., (1974). Fracture and stress–strain relationship of concrete under triaxial compression. *J. Struct. Div.* (*ASCE*), **100** (ST 5), Proc. Paper 10547, 901–16.
51. CEDOLIN, L., CRUTZEN, Y. R. J. and POLI, S. D. (1977). Triaxial stress–strain relationship for concrete. *J. Eng. Mech. Div.* (*ASCE*), **103** (EM 3), Proc. Paper 12969, 432–9.
52. COON, M. D. and EVANS, R. J. (1972). Incremental constitutive laws and their associated failure criteria with application to plain concrete. *Int. Journ. of Solids and Structures*, **8,** 1169–83.
53. BAZANT, Z. P. and BHAT, P. D. (1976). Endochronic theory of inelasticity and failure of concrete. *J. Eng. Mech. Div.* (ASCE), **102** (EM 4), Proc. Paper 12360, 331–44.
54. BAZANT, Z. P. and SHIEH, C. L. (1978). Endochronic model for nonlinear triaxial behaviour of concrete. *Nuc. Eng. and Des.* **47,** 305–15.
55. CHEN, A. C. T. and CHEN, W. F. (1975). Constitutive relations for concrete. *J. Eng. Mech. Div.* (ASCE), **101** (EM 4), Proc. Paper 11529, 465–81.
56. KOTSOVOS, M. D. and NEWMAN, J. B. (1972). Generalised stress–strain relation for concrete. *J. Eng. Mech. Div.* (ASCE).
57. *Proceedings of Conference on Prestressed Concrete Pressure Vessels Institution of Civil Engineers*, London, 1968.

Chapter 6

FRACTURE MECHANICS APPLIED TO CONCRETE

R. NARAYAN SWAMY

Department of Civil and Structural Engineering, University of Sheffield, UK

SUMMARY

Most structural materials and fabricated structures contain flaws or cracks that are introduced during fabrication or whilst in service. Fracture mechanics offers a practical and exciting development in modern engineering design which relates the fracture strength of a material or structure under a given applied stress and the critical flaw sizes in the material. Although fracture mechanics concepts have been extensively applied to metals, ceramics and polymers, their use in concrete materials and concrete structures is relatively new. In this chapter, the basic fracture mechanics concepts are reviewed and the published fracture toughness parameters relating to concrete materials critically assessed. Current developments taking place in the field of fracture mechanics are described and possible future lines of application indicated.

6.1 INTRODUCTION

It is well known that fabricated structures and most structural materials contain small crack-like flaws which are either inherent in the basic materials or introduced during fabrication or service. The service performance, operational life and the fracture behaviour of the structure and material will then depend on the applied stress level, the initial size of the flaw, material properties and the mechanisms by which the cracks propagate and proceed to fracture. One of the most exciting developments in modern engineering design is the use of methods of fracture mechanics which provide a quantitative relationship between the size of defect and the applied working stress.

Fracture mechanics is concerned with the problem and conditions of cracks propagating through a material and the mechanism of fracture in the presence of these cracks and crack-like defects. Associated with the crack propagation problem is the problem of determining the stress state at a crack tip. The property which gives structural materials adequate strength in the presence of cracks and crack-like defects is termed 'fracture toughness'. Toughness is a measure of the energy or work required to produce fracture and hence increases with the ductility of a material, which is a measure of the strain required to produce fracture.

From an engineering point of view, the main objective in developing fracture mechanics concepts is to relate the development of fracture, its orientation and ultimate failure to the applied state of stress. Exact relations between flaw size and fracture toughness have been developed for many types of stress systems. These relations make it possible to calculate, with a reasonable degree of accuracy, the fracture strength of a body and to predict the life of a structure when the applied stresses, fracture toughness and effective or critical flaw sizes are known. Such data on flaw sizes also enable a rational approach to the use of non-destructive testing techniques.

In complex structural systems, and in structures subjected to complex load–time–temperature histories, the problem of fracture strength prediction and life estimation is considerably more difficult. Even then, the fracture mechanics concepts hold great promise. There is considerable evidence to show that prediction of service performance and service life that assumes no initial crack in the structure, cannot only be unrealistic but also dangerous.

Although fracture mechanics concepts have proved to be invaluable tools in characterising the failure processes in metals, polymers and ceramics, there is virtually little or no application of these ideas to concrete and concrete structures. The reasons are not difficult to understand. Both the heterogeneity of concrete systems and their fracture processes are extremely complex, and in any case, the relationships between fracture theory and engineering design are not only not clearly defined, but are also not fully appreciated by engineers and researchers.

The main purpose of this chapter is to describe the basic concepts of fracture mechanics and to show how these concepts have been applied to concrete materials. Emphasis is laid on current new developments in fracture mechanics ideas and their possible future applications.

6.2 THEORETICAL STRENGTH OF MATERIALS

The theory of completely brittle cracks has been well understood for a long time. The theoretical strength of perfectly homogeneous, flawless, elastic materials is determined by the molecular cohesive forces within the material which, in turn, depend on the interatomic spacing, a (Fig. 1(a)). A rough estimate of the maximum tensile stress required to

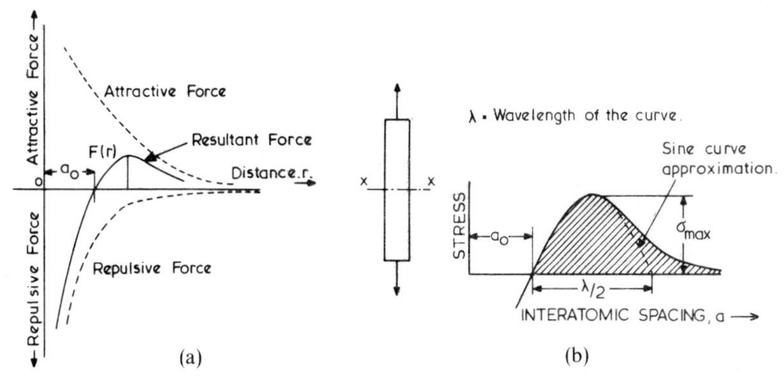

FIG. 1. (a) Variation of interatomic forces between atoms with the separation distance. (b) Sine wave approximation for atomic stress–strain curve.

overcome the molecular cohesion can be obtained by considering the work done by the interatomic forces acting across the plane of fracture and the energy of the two new fracture surfaces produced at rupture. Considering the resultant curve to be a sine curve (Fig. 1(b)), the 'molecular strength' or 'theoretical strength' σ_m† is given by

$$\sigma_m = (E\gamma/a_0)^{1/2} \tag{1}$$

†Orowan[3] has also derived this equation by considering the density of elastic energy between two atomic planes and shown that $\sigma_m = (2E\gamma/a_0)^{1/2}$. This will modify eqn. (5).

where E is the Young's modulus of the material, γ,† the surface energy and a_0 is the equilibrium atomic separation.

Numerous theoretical calculations have indicated that the strength of the atomic forces holding the material together should be about one-tenth of the material's elastic modulus. However, because of the presence of flaws and cracks, measured strengths vary between $E/10^3$ and $E/10^2$. A good measure of the surface energy for most inorganic materials is $Ea_0/20$.

Size effects, as evidenced by the tests of Leonardo da Vinci carried out almost five centuries ago, and the presence of very small microcracks or flaws, which cause stress concentration around them, have long been recognised to account for the discrepancy between the theoretically estimated and the observed values of the breaking strength of materials. Pre-existing cracks are thus the precursors to fracture; by eliminating such cracks, the tensile strengths can be made to approach the theoretical estimates as shown by Ernsberger for glass.

6.3 GRIFFITH'S THEORY

Inglis[1] was the first to confirm this hypothesis by calculating the stress distribution around a flat elliptical hole of major axis $2c$, which could be degenerated into a crack, in a homogeneous, elastic plate (Fig. 2(a)). If the plate is subjected to an average tensile stress, σ, perpendicular to the major axis, the stress concentration which occurs at the crack tip is given by

$$\sigma_c = \sigma\, 2(c/\rho)^{1/2} \tag{2}$$

where ρ is the radius of curvature at the ends of the major axis. As the crack tip becomes infinitely sharp, i.e. as $\rho \to 0$, $\sigma_c \to \infty$ and the material is expected to have negligible strength. The term $2(c/\rho)^{1/2}$ is defined as the stress concentration factor and describes the effect of crack geometry on the local crack tip stress level.

Griffith was the first to compute the relation between size of crack and the critical normal stress which will cause the crack to propagate to failure. Griffith's theory[2] for brittle fracture recognises that

†In an ideal brittle material γ will be replaced by the surface tension T of the material.

materials contain defects such as impurities, microcracks or other flaws either on the surface or within the material and that these minute flaws can propagate catastrophically once a critical stress is reached (Fig. 2(b)). Such flaws act as stress raisers so that when the applied stress is increased, the stress at the flaw reaches approximately the theoretical strength of the material. Griffith's theory, therefore, assumes that the theoretical strength is the true microscopic fracture stress which is actually reached only within a very small volume of the material, while the mean stress may remain very low. According to Griffith, the crack will become unstable and lead to fracture when the decrease in elastic energy of the stress field of the crack (i.e. the strain energy release rate) equals or exceeds the surface free energy of the new crack surfaces (Fig. 3).

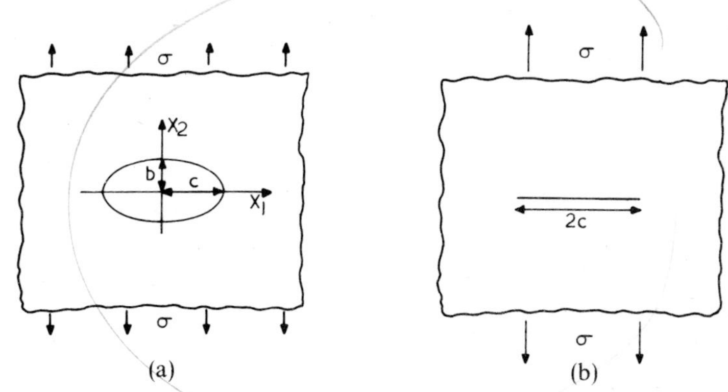

FIG. 2. (a) The Inglis Model: elliptical hole in a uniformly stressed infinite plate. (b) Griffith's model: geometrical configuration.

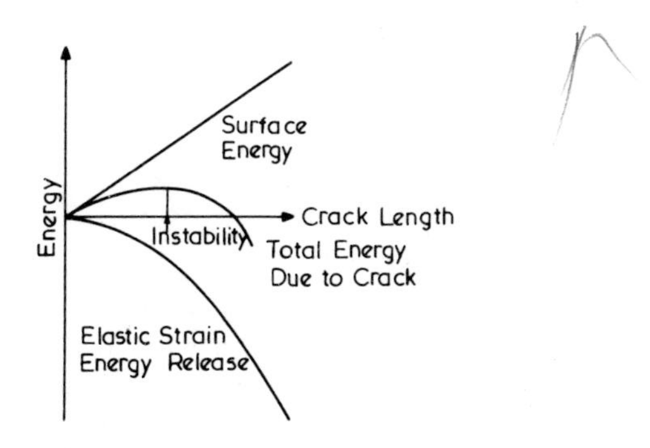

FIG. 3. Schematic representation of Griffith's energy balance in relation to crack length.

Surface cracks of depth c produce approximately the same stress concentration as internal cracks of length $2c$. Considering a two-dimensional continuum model, loaded uniformly at the infinite boundaries and containing a through crack of length $2c$ and using Inglis's solution, Griffith showed that, in the case of plane state of stress (i.e. for a thin plate where the thickness is small compared with the length $2c$ of the crack) the crack will spread and cause fracture when the stress σ applied normally to the direction of the crack exceeds the critical value (Fig. 2(b))

$$\sigma = \left(\frac{2E\gamma}{\pi c}\right)^{1/2}. \tag{3}$$

The measured strength of a body is thus related to its material properties and the length of the pre-existing crack. Many results on brittle-like materials have shown that Griffith's theory predicts the correct functional relationship between stress at fracture and flaw size. The theory also has the advantage that it predicts a theoretical cohesive strength of the defect-free material of the right order of magnitude, of about $0{\cdot}1E$, which has also been verified on single-crystal whiskers.

Since the greatest sharpness that a crack could have is the smallest distance between atoms, $\rho = a$ and eqn. (2) becomes (3),

$$\sigma_c = 2\sigma(c/a)^{1/2} \tag{4}$$

and if it is assumed that fracture occurs when the stress at the stress concentration reaches the value σ_m of the theoretical strength given by eqn. (1),

$$\sigma_c = \sigma_m = 2\sigma(c/a)^{1/2}$$
$$= (E\gamma/a)^{1/2}$$

or,

$$\sigma = (E\gamma/4c)^{1/2} \tag{5}$$

which differs from Griffith's formula (3) only by a factor of $(8/\pi)^{1/2} = 1{\cdot}6$ approximately. It will also be seen that Griffith's equation gives a value just 20 % lower than eqn. (1) if we put $c = a_0$.

For a thick plate of material having Poisson's ratio ν or for plane strain conditions, eqn. (3) becomes

$$\sigma = \left\{ \frac{2E\gamma}{\pi c(1-\nu^2)} \right\}^{1/2} \tag{6}$$

Griffith's analysis is valid only to ideal brittle elastic (i.e. non-ductile) materials containing a very sharp crack and which obey Hooke's law, in which strain energy is completely and reversibly stored up to the onset of fracture. The only work required for such a fracture is that needed to overcome the cohesive forces between the atoms on either side of the cleavage crack and no other energy dissipating mechanism is present. This work is so small that it can be supplied from the elastic energy stored in the specimen and testing machine at the moment of fracture. Griffith's theory also assumes that no crack-stopping process is present and that the critical stress is not reduced by local inelastic deformation at the crack tip. The high stress concentration at the edges of the crack will induce a certain amount of plastic flow; however, this is confined to a small region and the stress distribution remote from the crack will not be affected.

The Griffith equation represents not only a necessary, but also a sufficient condition of fracture in a fully brittle specimen involving very low energy absorption. The theory can only, however, be applied if plastic deformation is absent or confined to the immediate vicinity of the crack walls so that the bulk of the body is still elastic.

6.4 EXTENSION OF GRIFFITH'S THEORY

Griffith's theory of rupture has also been extended to three dimensions.[4,5] For a three-dimensional body (infinite in size with respect to crack size) of elastic modulus E and Poisson's ratio ν and subjected to tensile stresses, the fracture equation defining the critical applied stress σ (normal to the crack) required to cause an internal penny-shaped crack— a flat oblate ellipsoid or a plane crack bounded by a circle— to become unstable is given by

$$\sigma \geqq \left\{ \frac{\pi E\gamma}{2c(1-\nu^2)} \right\}^{1/2} \tag{7}$$

The three-dimensional model thus gives a critical stress which differs from the two-dimensional value (eqn. (3)) by a factor $\pi/\{2(1-\nu^2)^{1/2}\}$ and from eqn. (6) by a factor $\pi/2$. There is thus hardly any difference between the critical stress for a central crack of length $2c$ in a sheet and a penny-shaped crack of radius c. In fact, for a wide variety of discontinuity problems, the application of an energy balance criterion leads to quantitatively similar Griffith-type results. (Table 1.)

TABLE 1

CRITICAL GRIFFITH STRESSES

Geometry	*Two-dimensional crack*	*Cylindrical cavity*[a] (a/b→0)	*Three-dimensional crack*	*Spherical cavity*[a] (a/b→0)
Critical stress	$\left(\frac{2}{\pi}\right)^{1/2}\left(\frac{E\gamma}{c}\right)^{1/2}$	$\left(\frac{\sqrt{2}}{2}\right)\left(\frac{E\gamma}{c}\right)^{1/2}$	$\left(\frac{2\pi}{3}\right)^{1/2}\left(\frac{E\gamma}{c}\right)^{1/2}$	$\left(\frac{4}{3}\right)\left(\frac{E\gamma}{c}\right)^{1/}$
Loading	Uniaxial	Biaxial	Uniaxial	Triaxial

[a] a = internal radius, b = external radius.

The energy balance concept is the essential cornerstone of fracture mechanics. The Griffith's concept indicates that fracture will occur when the change in stored energy equals or exceeds the energy required to form the new surfaces. Once this condition is satisfied, fracture ensues and the excess strain energy within the body relative to the energy required to form the new surfaces will dissipate itself as mechanical losses in the fracture process such as heat and kinetic energy.

The concept itself can then be considered to be independent of the material to which it is applied, provided the mechanism of energy dissipation during the fracture process appropriate to the material is considered. Griffith's theory can therefore be generalised in the form

$$\sigma = k\left(\frac{E\gamma_{\rm i}}{c}\right)^{1/2} \tag{8}$$

where k is a constant for the geometry under consideration and γ_i reflects the particular dissipation processes for the material concerned. For very brittle materials, the material's surface energy, γ, is the only mode of energy dissipation. For ductile and viscoelastic fracture processes, γ_i should include all mechanisms concurrent with crack extension that dissipate energy. At high velocities of crack propagation, a kinetic energy term must also be included, and the limiting crack velocity is largely determined by the kinetic energy term.

6.5 EXTENSION TO NON-BRITTLE MATERIALS

Since its original application to brittle materials, the energy balance concept has been extended to ductile materials,[6–9] viscoelastic fracture,[10] and rocks.[11]

For semi-brittle materials, such as metallic materials and polymers, the original Griffith equation is not valid without modification because plastic deformation around the crack accompanies crack propagation, and the fracture energy is found to be several orders of magnitude (sometimes by as much as 10^3) greater than the surface energy of the material. The extremely high elastic stress concentrations assumed in the Griffith theory cannot, therefore, occur in a ductile material with plastic flow; even after the formation of the initial crack, failure cannot occur without further plastic deformation, and the crack opens up considerably before failure occurs.

Both Orowan[3,6,7] and Irwin[8,9] have shown that Griffith's theory could be applied to ductile materials provided that a plastic term is included in the energy term. Accordingly

$$\sigma = \left\{ \frac{2E(\gamma + \gamma_p)}{\pi c} \right\}^{1/2} \tag{9}$$

where γ_p represents the energy expended in the plastic work necessary to produce unstable crack propagation.

Because $\gamma_p \gg \gamma$, eqn. (9) leads to

$$\sigma \simeq \left(\frac{2E\gamma_p}{\pi c} \right)^{1/2} \tag{10}$$

From eqns. (1) and (2),

$$\sigma = \left(\frac{2E\gamma}{\pi c} \right)^{1/2} \left(\frac{\pi \rho}{8a_0} \right)^{1/2} \tag{11}$$

Comparing eqns. (10) and (11) suggests that plastic deformation can be related to a blunting process at the crack tip, i.e. ρ will increase with γ_p. Also, comparison of eqns. (3) and (11) indicates that $\rho \simeq 3a_0$ is a lower limit of the 'effective' radius of an elastic crack. Thus, when $\rho < (8/\pi)a_0$, the stress for unstable crack propagation is given by eqn. (3), and when $\rho > (8/\pi)a_0$, eqn. (11) becomes valid.

6.6 DEVELOPMENT OF THE FRACTURE TOUGHNESS CONCEPT

The resistance of a material to the propagation of an existing crack is known as its fracture toughness. This determines how large a crack the material can tolerate when loaded to any given stress level. The fracture toughness approach is thus essentially concerned with the role of crack propagation in brittle fracture.

6.7 ASSESSMENT OF FRACTURE TOUGHNESS

There are several methods of assessing the fracture toughness of a material. All these methods have primarily been developed for metallic materials, and only those relevant to concrete materials are discussed here in any depth.

All the various approaches of assessing fracture toughness can be broadly classified into two.

(1) Methods based on linear elastic fracture mechanics. In these methods, some property of the stress–strain distribution in the vicinity of the crack or notch is used to predict the criterion of low stress fracture. A number of different approaches ranging from critical stress to critical strain criteria have been suggested.

(2) Methods applied to cracked bodies which cannot be regarded as linearly elastic and for conditions involving both elastic and plastic deformations.

6.8. LINEAR ELASTIC FRACTURE MECHANICS

Linear elastic fracture mechanics is the study of stress and displacement fields near a crack tip in an isotropic, homogeneous, elastic material at the onset of rapid, unstable crack propagation which leads to fracture. The theory essentially provides a means of predicting the fracture stress of structures or their components containing sharp flaws or cracks of known size and location in terms of a single parameter. The concepts of the theory are most applicable to brittle materials in which the inelastic region near the crack tip is small compared to the flaw and specimen dimensions. The basic theory can be developed in terms of either an energy approach or a stress-intensity approach. Both approaches are closely related and yield identical results. Apart from metallic materials, linear elastic fracture mechanics has also been applied to anisotropic materials such as reinforced plastics and wood.

6.8.1 Energy Considerations

The energy approach to fracture instability is, of course, due to Griffith and has already been discussed. In the energy approach, the criterion to crack propagation of a crack in a body is expressed in terms of the rate of change, with respect to crack extension, of the various energy terms involved in the process.

In the Griffith brittle fracture theory, the surface energy, γ, absorbed during crack propagation is a clearly defined material property, whereas the work of plastic deformation, γ_p, at the crack tip involves mechanisms which, still, are not well understood. Although the energy absorbed in crack growth, dQ, is influenced by many unknown parameters, the available energy for crack extension, dU, depends only on the elastic properties of the specimen and the applied load. If the development of kinetic energy is negligible, then $dU = dQ$, and the rate of energy absorption at any stage in the crack growth can be determined by the instantaneous value of the rate of supply of available energy. The latter, dU/dA, where dA is the increase in crack area, is denoted by the symbol G and referred to as the *strain energy release rate*[12] measured in units of kN/m (or ergs/cm^2). G is also referred to as the *crack driving force* or *crack extension force.*

During stable crack propagation, G is entirely absorbed by the work involved in plastic flow and other energy dissipating mechanisms as well as the relatively smaller surface energy increase. Crack extension or crack instability, or the onset of fast fracture is associated with a critical value of G such that an incremental increase in G will be greater than the work dissipated and unstable crack growth will occur. The critical value of G at instability of the crack is denoted by G_c, the unstable condition for plane stress conditions.

Expressions for G have been obtained for a number of geometries and loading conditions. For the case of a crack length $2c$ in an infinite plate, the relationship is

$$G = \prod c\sigma^2/E \tag{12}$$

the strain energy release rate being proportional to the product of the square of the stress and the crack length.

The value G_c provides a convenient parameter to include all supplementary energy-dissipating terms, such as plastic flow, which could in turn produce heat or sound, in addition to the work required to fracture the atomic bonds. G_c is thus a measure of a material's resistance to fracture and became known as the material's toughness or the crack-resistance force of the material.

6.8.2 Stress Intensity Factors

When a cracked body is stressed, the crack surfaces move relative to each other. Kinematically, three modes of crack surface displacement can be identified (Fig. 4(a)); associated with each mode is a particular type of elastic crack tip stress field.[12,13] On a macroscopic scale, the opening mode is most commonly experienced in practice and the discussion here is related to the opening mode stress intensity factor.

Since the rate of supply of available energy, G, for crack propagation depends only on the elastic properties of the specimen and the applied load, it should be possible to relate G to the stress field within the specimen. Assuming the crack to be contained in a two-dimensional plane sheet of isotropic, elastic material, the stresses in the vicinity of a crack tip can be analysed. For plane stress conditions, the stress normal to the crack at a distance r from the crack tip and making an angle θ with the crack plane is given by (Fig. 4(b)).

$$\sigma_y = \frac{K}{(2\prod r)^{\frac{1}{2}}} \cos\frac{\theta}{2}\left(1 + \sin\frac{\theta}{2}\sin\frac{3\theta}{2}\right) \tag{13}$$

where K is a parameter which depends on the applied stress σ acting in the y direction, the crack length $2c$, and the dimensions of the sheet. By this method by specifying the value of K, the stress can be determined at any point (r, θ).

K is called the *stress intensity factor*. It is a convenient measure that characterises the *intensity* of the elastic stress field at all points in the vicinity of the crack tip. For any length of crack or combination of forces applied to the body, the local stress will decrease inversely with the square root of the distance, and K gives the precise magnitude of the stress by embracing in one convenient form both the crack geometry and the net contribution of the applied forces.

The stress intensity factor is thus a measure of the mechanical properties in the presence of a crack, in the same way that the term, 'stress', characterises the mechanical properties in an uncracked specimen. K has dimensions of stress $\times$ (length)$^{1/2}$ and the usual units are $MNm^{-3/2}$.

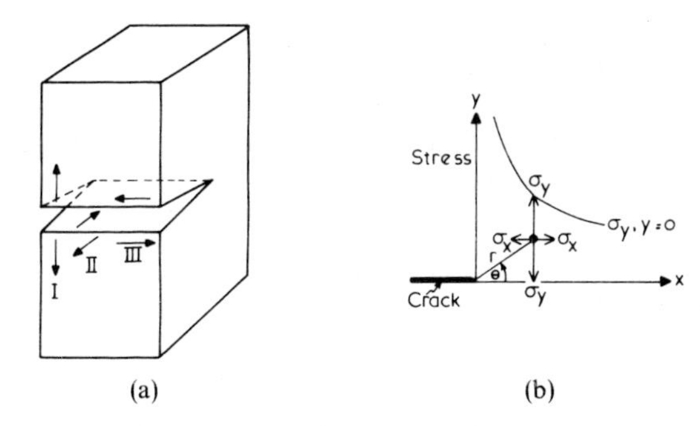

FIG. 4. (a) Basic modes of crack surface displacement. I—opening mode, II—edge sliding mode and III—shear mode. (b) Schematic representation of stress distribution at the crack tip stress field.

Small-scale non-linear effects, such as those due to yielding, microstructural irregularities, internal stresses and local irregularities in the crack surface do not affect the general character of the stress field and can therefore be neglected in a reasonable approximation.[12] The stress intensity factor therefore provides a convenient mathematical framework for the study of fracture processes.

The notion of K implies that the higher the value of K, the more severe a crack. When K reaches a critical value K_c, sufficient energy is being supplied to the crack tip, and crack propagation will occur, usually catastrophically. K_c is a useful measure of *fracture toughness* of a material or its resistance to brittle fracture. The great advantage of using K is that it provides a single parameter characterisation and its evaluation is a normal stress analysis problem involving the applied stress, crack length and specimen configuration. Stress intensity factors can be computed for various crack configurations;[13] an external flaw is generally considerably more severe than an internal flaw of the same overall size, and the depth rather than the visible length of a surface flaw controls its severity.

Although G_c became initially known as a material's fracture toughness, in recent years less emphasis has been given to the energy approach to fracture and it is now more usual to express fracture toughness in terms of K rather than G values. There are several reasons for this: First, K varies linearly with the applied load or nominal stress for a given geometry. Secondly, it is more logical and rational to consider stress–strain distributions at crack tips to determine conditions of fracture rather than energy balances. Thirdly, if failure occurs under a critical stress, then a Griffith type relationship results without considerations of any energy dissipation processes involved. Further, the K values are algebraically additive so that, for a given mode of crack surface displacement, the K value resulting from several stresses may be simply determined as the sum of the individual contributions.

From dimensional analysis of eqn. (13), for a crack length of $2c$ in an infinitely wide plate

$$K = \sigma(\prod c)^{1/2} \tag{14}$$

and in general terms,

$$K = \alpha . \sigma(\prod c)^{1/2} \tag{15}$$

where α is a constant, of the order of one, depending on geometry and loading conditions. The theory thus predicts that strength is inversely proportional to $\sqrt{c}$.

6.8.3 Relation Between G and K

For elastic loading conditions, the strain energy is uniquely defined by the stress distribution, and hence a simple relationship exists between the stress intensity factor K and the strain energy release rate, G. For plane stress,

$$K^2 = EG \text{ and } K_c^2 = EG_c \tag{16}$$

and for plane strain,

$$K_I^2 = \frac{EG}{1-\nu^2} \text{ and } K_{Ic}^2 = \frac{EG_{Ic}}{1-\nu^2} \tag{17}$$

where the Roman numerical subscript I designates the first or opening mode of crack extension. From eqns. (15) and (17).

$$\sigma_c = \left[\frac{EG_c}{\prod c(1-\nu^2)}\right]^{1/2} \tag{18}$$

which is the Griffith criterion of fracture with G_c replacing the surface energy term, and c is now the size of the critical flaw which will permit continuous crack growth. For metals, G_c is several orders of magnitude larger than the thermodynamic surface energy due to the large plastic deformation that is expended prior to the onset of unstable crack propagation. In ceramics G_c is found to vary between 2 and 100 times the surface energy. In glass, G_c is about 10 times its solid state surface energy; in concrete, G_c is about 12 times the surface energy.

Comparing eqns. (14) and (16) with eqn. (9),

$$G = 2(\gamma + \gamma_p) \tag{19}$$

Measured values of K_c or G_c are found to vary with the cross-section of the specimen used, and this variation is related to fracture surfaces. The range of values depends on the stress condition at the tip of the crack and decreases with increase of sheet thickness. As the sheet thickness increases, the state of stress in the vicinity of the crack changes from plane stress to plane strain conditions, and K_c and G_c approach minimum limiting values (Fig. 5). These minimum values are denoted by K_{Ic} and G_{ic} and are considered to be material constants.

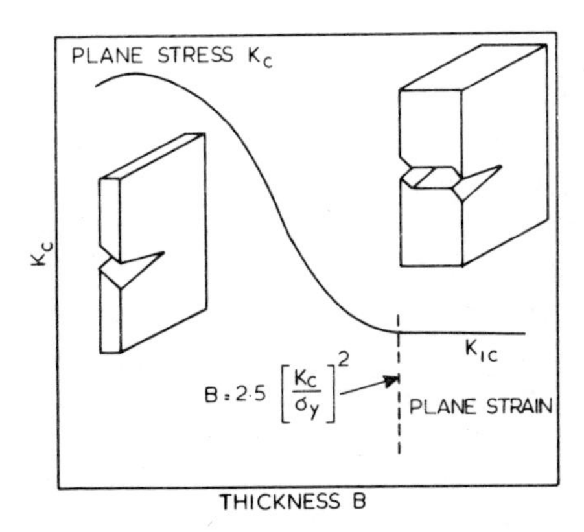

FIG. 5. Schematic representation of variation of K_c with specimen thickness.

The distinction between K_{Ic} and K_I is important and can be compared to the distinction between strength and stress. Thus, as plate thickness increases, plane strain fracture toughness is being measured in a plane stress situation.

Because of the change from plane stress to plane strain conditions (Fig. 6), K_{Ic} is often referred to as the plane strain fracture toughness, and represents a basic material property in the same sense as the proof stress measured in a tensile test on a crack-free specimen. K_{Ic} and G_{Ic}

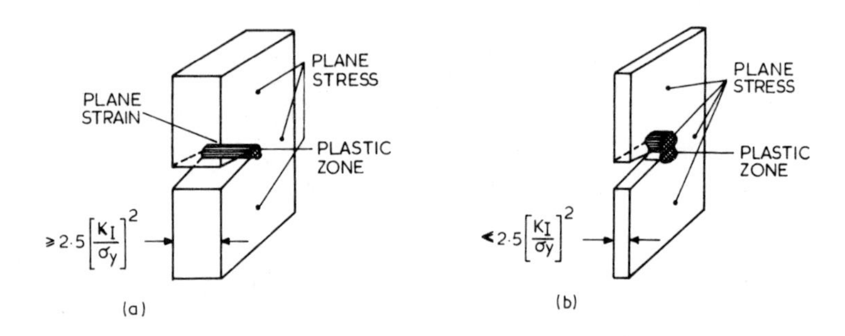

FIG. 6. State of stress in a cracked plate (a) Plane strain. (b) Plane stress.

are thus independent of specimen dimensions in contrast to K_c and G_c which depend to some extent on the initial crack length as well. K_{Ic} and G_{Ic} thus provide an invariant fracture characteristic for many materials and are therefore of more interest for general evaluation of material property than K_c and G_c.

6.8.4 Effect of Plastic Zone

The elastic stress analysis assumes perfectly elastic, isotropic continuum, but practical materials are not homogeneous, and behave in a non-linear elastic manner at the high stresses induced close to the crack tip. Provided the zone of stress disturbance around the crack tip is small relative to the specimen and flaw size, the elastic stresses outside this region are only slightly affected, and the stress intensity factor still provides a reasonable description of the crack tip stress field[12–14] (Fig. 6). The concepts are thus most applicable to brittle materials in which the inelastic region and the scale of inhomogeneity are small compared to the specimen and flaw dimensions.

The very high stresses in the plastic zone will not only contribute to the strain energy of the system but also give rise to an increase in the effective length of the crack. The size of the plastic zone generally increases with both a decrease in yield strength and an increase in fracture toughness. The effective crack size is, therefore, given by (Fig. 7)[9,15,16]

$$c_{\text{eff}} = \text{actual crack size} + \text{plastic zone correction factor}$$

$$= c + r_y$$

$$\text{where } r_y = \beta (K_{\text{Ic}}/\sigma_y)^2 \qquad (20)$$

where β varies between $1/2\pi$ and $1/6\pi$ depending on whether plane stress or plane strain conditions, respectively, dominate (Fig. 7).[17] The transition from plane strain to plane stress is often abrupt.

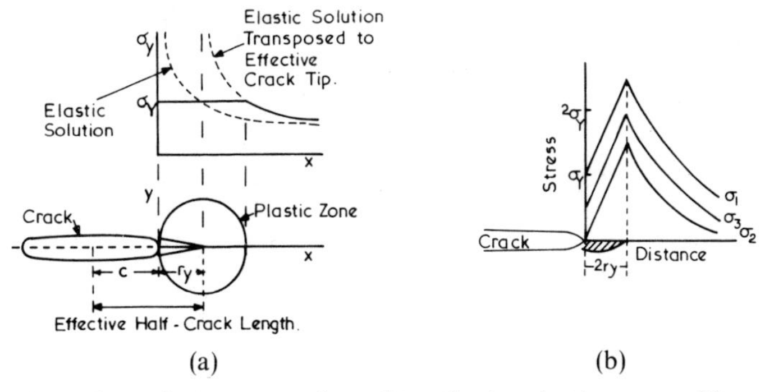

FIG. 7. Schematic representation of crack tip plastic zone with correction factor r_y; (a) in plane stress and (b) in plane strain.

6.8.5 Application of Griffith's Theory to Biaxial and Triaxial Stresses

Griffith also derived a fracture condition for biaxial states of stress from the Inglis solution of the stress distribution around an elliptical hole in a plate.[2] By considering the isotropic material to contain a large number of cracks at random, Griffith assumed fracture to occur when the maximum tensile stress at the longest crack of the most dangerous orientation reached a critical value equal to the molecular strength of the material. Griffith thus determined the relationship between the principal stresses σ_1 and σ_2 required to cause fracture and both the tensile strength, σ_0, of the material and the angle of fracture, θ (Fig. 8(a)).

The fracture criteria are given by

$$\sigma_1 = \sigma_0 \quad \text{for} \quad -3\sigma_0 \leqq \sigma_2 \leqq \sigma_1 \tag{21}$$

$$(\sigma_1 - \sigma_2)^2 + 8\sigma_0(\sigma_1 + \sigma_2) = 0, \quad \text{for} \quad \sigma_2 < -3\sigma_0 \tag{22}$$

Figure 8(b) also shows a graphical interpretation of these equations.[2,6]

The Griffith biaxial criterion has been extended to triaxial state of stress and the fracture equations then assume the same basic form as before, if σ_1 and σ_3 are considered to be the major and minor principal stresses.[6,18,19] This implies that fracture is independent of the intermediate principal stress σ_2. The fracture criterion expressed by these equations thus corresponds to Mohr's envelope at failure.

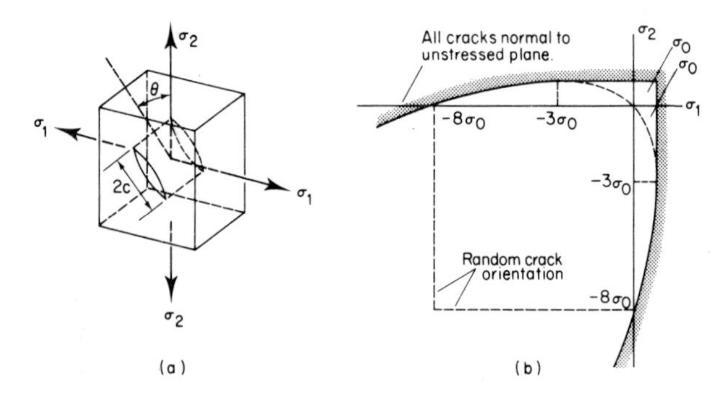

FIG. 8. (a) Griffith crack in a biaxial stress field. (b) Geometrical representation of Griffith's fracture criterion under biaxial stress.

The theory of crack propagation under triaxial stresses[18] shows that there is a parabolic relationship between the shear stress and the normal stress, acting on the plane containing the Griffith crack which propagates. This is in accordance with some reported experimental observations on concrete, sedimentary rocks, cast iron and coal, showing that brittle fracture depends on the hydrostatic as well as the deviatoric stress components although other tests also show a linear Mohr envelope in compression. Griffith's simple cracked model thus provides a physical explanation of the type of Mohr envelope observed in some concrete-like materials.

Although the tensile strength predicted by Griffith's theory has generally been found to be in reasonable agreement with some test results, in triaxial tests on concretes, rocks and coal, the predicted values have been found to be greater than those obtained by direct measurement. These discrepancies have been suggested to be due to closing or change of the shape of the internal cracks.

A modification of Griffith's theory by assuming that, in compression, the Griffith's cracks close and a frictional force (the coefficient of friction being different from Mohr's coefficient of internal friction) develops across the crack surface has also been proposed by McClintock and Walsh.[20] This modified theory yields a linear Mohr's envelope and appears to give improved agreement with test results. The point to remember is that Griffith's theory should strictly be valid only for the onset of crack propagation and not for the terminal fracture of the specimen; however, the Griffith crack mechanism can be used to account for the terminal failure surfaces under various conditions of loading.

6.9 STATISTICAL ASPECTS OF FRACTURE PROBLEMS

The basic flaw concept of Griffith's brittle fracture theory leads to statistical considerations of the distribution of flaws and strengths in a given specimen, and a consideration of fracture mechanics will be incomplete without a brief reference to the statistical aspects of fracture problems. Many tests have shown the phenomena of size effect and scatter on the strength of materials like concrete, rocks and coal, and this has led to the development of various statistical theories of fracture.[21]

Two basic and distinct approaches are available to the study of the statistical aspects of strength of materials, namely, the weakest link theory and the classical bundle concept. Both these two theories are mathematical idealisations and in reality the real characteristics of materials fall in between these two idealisations.

The primary problem in all statistical studies is the choice of a suitable and appropriate probability distribution function. The relevance of any individual distribution function for universal application cannot be deduced from experimental results alone but will have to be assessed on the physical and logical relevance of the fracture model underlying the particular distribution function. The significant aspect of progressive cracking and failure observed in concrete makes neither the weakest link nor the classical bundle concept applicable to concrete. Recently, a size-dependent strength distribution function (the Pareto distribution), which is derived by taking into account the progressive cracking of the material, has been proposed for concrete[22] and the theoretical and experimentally observed strength distributions show to compare well.

6.10 THE CRACK OPENING DISPLACEMENT (COD) APPROACH TO FRACTURE MECHANICS

The two basic requirements in obtaining fracture toughness data are to have contained plasticity and plane strain conditions. When low strength materials (such as mild steel) are tested these requirements need large test pieces. Further, large amounts of plastic deformation occur in such tests prior to fracture which invalidate the use of linear elastic fracture mechanics. The local stresses and strains at the crack tip cannot then be calculated from the applied stress nor can the crack tip stress and strain field be characterised by a single parameter. The amount of crack opening prior to crack extension can then be used as a parameter to characterise the crack tip region.

The concept of crack opening displacement can be readily appreciated in a notched beam test. As the applied stress increases, plastic yielding at the notch tip results in separation of the notch surfaces without extension of the notch. This separation of the two notch faces is termed as the crack opening displacement (*COD*). Cleavage fracture will be initiated when the opening of the tip of the crack reaches a critical value, and this specific value of the *COD* obtained at the crack tip at the instance of fracture is considered to be a measure of the toughness or notch ductility of the material.[23,24] The *COD* is related to the extent of plastic straining in the plastic zone.

It is possible to relate the crack opening displacement δ to the crack extension force G.[25,26] For conditions of plane stress and for $\sigma/\sigma_y \ll 1$, which corresponds to the requirement for a plastic zone which is small compared to the crack length,

$$G = \lambda \sigma_y \delta \tag{23}$$

where σ is the applied stress and σ_y is the yield stress of the material and λ is a constant. For plane strain conditions,

$$COD = \frac{K^2(1-\nu^2)}{\sigma_y \lambda E} \tag{24}$$

The value of λ has been calculated by various investigators, and varies from 0·97–2·14, while experimental results show $\lambda = 1$.[24–26] Finite element analyses, that might be expected to give close results to the experimental value, show least agreement. These disagreements between theory and experiment are due to differences in the definition of *COD* and substantial differences in the value of δ at initiation and total instability. Difficulties in theoretical modelling of the crack tip region would also account for these differences.

Assuming $\lambda = 1$,

$$K_{\mathrm{Ic}} = \left[\frac{E\sigma_{\mathrm{y}}\delta_{\mathrm{c}}}{1-\nu^2}\right]^{1/2}. \tag{25}$$

The advantage of *COD* as a fracture criterion is that it can be easily measured in notched specimens that yield before fracture. It can also be calculated by the methods of elastic–plastic stress and strain analysis, at stages of deformation beyond those at which the values of K and G can be calculated. The *COD* criterion has been found to be relatively insensitive to geometrical variables such as crack length, specimen shape and loading configuration. However, the critical value of *COD* applies only to the initiation of further cracking, and does not characterise the point of total instability. A full experimental verification of the relationship between G_{c} and δ_{c} is at present lacking.

Methods of measuring *COD* necessitate equipment capable of recording the crack tip opening during loading. A paddle-type gauge, known as a *COD*-meter, has been widely used, but appreciable errors can be introduced by small variations in instrument operation. Recent work shows that the measurement of plastic strain or bend angle incurred during testing or the use of a chip gauge provides a convenient and simple alternative method of determining *COD*.[27]

6.11 THE *J*-INTEGRAL FRACTURE CRITERION

More recently, an alternative method of assessing the fracture toughness of a cracked body, which cannot be regarded as linearly elastic, has been developed. The *J*-integral defined by Rice,[28] has been proposed as a fracture criterion for conditions involving both elastic and plastic deformation.[29,30] This criterion extends linear elastic fracture mechanics concepts, where only small scale plasticity is permissible, to cases where extensive plastic deformation occurs. The method relies on determining the change in potential energy, when a crack is extended by an amount da, in a manner analogous to that of the strain energy release rate G, in the linear elastic condition. The change in potential energy can be replaced by a path-independent line integral around the notch tip. The theory is developed for a non-linear elastic body and the J-integral, in a physical sense, characterises the stress–strain conditions existing near a crack tip in an elastic–plastic solid.

6.11.1 The J-integral

The J-integral is defined as a path independent energy line integral on any curve Γ surrounding the notch tip, such that J is given by[28]

$$J=\int_{\Gamma}\left(W\,\mathrm{d}y-\mathbf{T}\frac{\partial \mathbf{u}}{\partial x}\mathrm{d}s\right) \tag{26}$$

where the curve is traversed in the anticlockwise direction (Fig. 9). W is the strain energy density, $\mathbf{T}$ is the traction vector on the path Γ, $\mathbf{u}$ is the

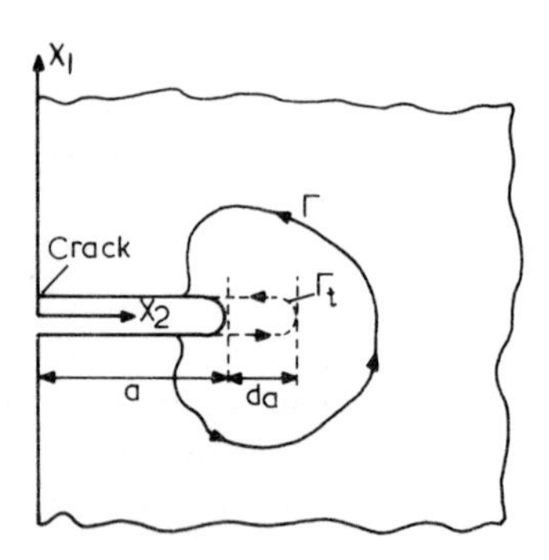

FIG. 9. Arbitrary line integral contour Γ surrounding a crack tip.

displacement vector and ds is an increment of arc length along the contour Γ beginning along the bottom surface of the crack and ending on the upper surface. It can be shown that the difference (J_1-J_2) for two curves Γ_1 and Γ_2 is zero and that the J-integral is path independent. It has been further shown that the J-integral may be interpreted as the potential energy difference between two identically loaded two-dimensional bodies having neighbouring crack sizes a and $a + \mathrm{d}a$.

$$J = -\frac{\partial U}{\partial a} \tag{27}$$

where U is the potential energy per unit thickness and a is the crack length.

Using the above energy interpretation, J-integral can be used as a fracture criterion.[29,30] The parameter J is thus a generalisation of the elastic strain energy release rate G to include non-linear behaviour. In the linear case J is the crack extension force, $\mathrm{d}a$ is interpreted as an increment of crack extension and $\mathrm{d}U$ is interpreted as the change in potential energy of the cracked specimen. These interpretations are not valid for the non-linear case.

The critical value of J is taken at the point of crack initiation and called J_{Ic}. J_{Ic} will then be a material constant for the initiation of crack extension, whether the specimen used for its determination is linear elastic or non-linear. J_{Ic} is therefore related to the linear elastic fracture mechanics parameters

$$J_{\mathrm{Ic}} = G_{\mathrm{Ic}} = K_{\mathrm{Ic}}^2 \frac{(1-\nu^2)}{E} \tag{28}$$

In the presence of significant plasticity, this energetic interpretation of J is not valid, although several investigators indicate that J_{Ic} may still adequately characterise the local crack tip condition for fracture initiation. Further, J_{Ic} reflects the value of J only at the onset of first crack extension, and there are considerable doubts as to the validity of the use of J as a characterising parameter during crack growth.

6.11.2 Experimental Determination of J_{Ic}

The most straightforward experimental method of evaluating J is based on the energy interpretation of the J-integral as the rate of change of potential energy with crack length:

$$J = -\frac{\mathrm{d}U}{\mathrm{d}a} \tag{27}$$

where U is given by

$$U = \iint_A W \mathrm{d}x\mathrm{d}y - \int_\Gamma \mathbf{T}\,\mathbf{u}\,\mathrm{d}s \tag{28a}$$

or

$$\mathrm{d}U = \mathrm{d}a \iint_A W \mathrm{d}x\mathrm{d}y - \int_\Gamma \mathbf{T}\,\mathbf{u}\,\mathrm{d}s \tag{28b}$$

$\mathrm{d}U$ is the difference in potential energy for unit thickness between two identically loaded bodies of area A which are similar in every respect except for an incremental difference in crack length $\mathrm{d}a$.

The above energy interpretation suggests that the compliance testing technique of linear elastic fracture mechanics is directly extensible through eqn. (27) to elasto-plastic materials. However, care must be taken to make the energy interpretation at a constant value of displacement or a large error could be incorporated into the J calculation. Approximate analyses can also be used since determination of J requires only overall changes in compliance.

The experimental technique developed by Begley and Landes[29,30] to measure J in non-linear materials is shown in Fig. 10. This method utilises a compliance technique and the fact that for specimens of differing crack lengths a, loaded to a constant displacement δ, i.e.

$$J = -\left(\frac{\mathrm{d}U}{\mathrm{d}a}\right)\delta = \text{const.}$$

the potential energy U is simply the area under the load–displacement diagram. The load–displacement response for various crack lengths (Fig. 10(b)) can be obtained from three-point bending tests. Graphical integration of the area under the curves in Fig. 10(b) gives the potential energy U against crack length a for various deflections (Fig. 10(c)). Then graphical differentiation of the constant displacement curves in Fig. 10(c) gives $J = -(\mathrm{d}U/\mathrm{d}a)_{\delta=\text{const}}$, thus yielding the compliance calibration curves shown in Fig. 10(d).

This multi-specimen evaluation procedure is clearly tedious and time-consuming, and the subsequent analysis required can give rise to significant errors. Bucci *et al.*[31] have given a numerical technique which is broadly similar to the experimental compliance procedure of Begley and Landes.[29,30]

There is a growing body of experimental evidence (on metals) which supports the critical value of Rice's integral J_{Ic} as a criterion for the initiation of crack extension in elastic–plastic fracture.[29,30] Small specimens which exhibit large amounts of plasticity before fracture have been shown to have a J_{Ic} value identical to G_{Ic} value measured in larger specimens which fractured in the linear elastic range. On the other hand, a $J(a)$–δ calibration can be formed from load–displacement

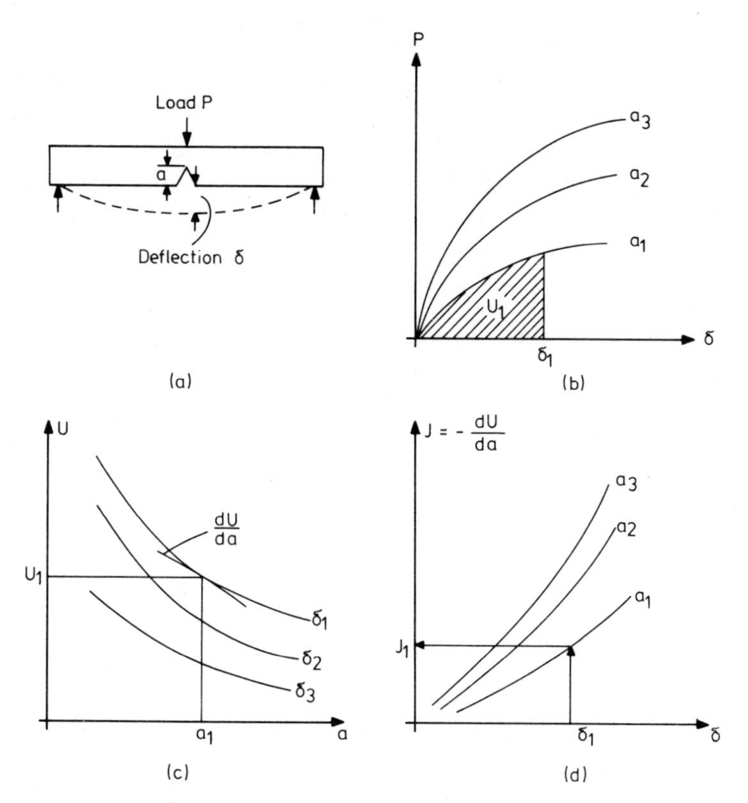

FIG. 10. Graphical method for determining $J(a)$–δ relationship. (a) Three-point bend test. (b) Load–displacement response for various crack lengths. (c) Potential energy–crack length relationship. (d) Compliance calibration curves.

records generated analytically from the plane-stress solutions of Bucci *et al.*,[31] and the estimated values of J_{Ic} using this method agreed quite well with experimental results of Begley and Landes.[29,30]

Tests on metals show that J is independent of specimen geometry and initial crack length. In a generally yielded specimen, J is directly proportional to the displacement of the loaded points, $\mathbf{u}$. In bending, $\mathbf{u}$ is proportional to the angle of bend θ, and so J is also proportional to θ. In notched bars, for a given specimen geometry, θ is proportional to the notch root displacement. It is then highly probable that J is directly proportional to COD for a given specimen geometry, but this relationship needs to be established.

6.11.3 Approximate Methods of Evaluation of the J-integral

The evaluation procedure for J originally suggested by Begley and Landes[29,30] for fracture toughness specimens is complex and needs several specimens of varying crack lengths. Various estimation procedures to simplify the process have, therefore, been proposed. One such method arises from the theoretical relationship for elastic/perfectly plastic materials between J and the energy U absorbed by a specimen subject to bending.

Rice *et al.*[32] have shown that the J-integral can be directly evaluated from single load–point displacement records for a series of crack toughness specimens having the common feature that their only significant length dimension is the uncracked ligament ($a/w \geqq 0{\cdot}6$, where a = crack length and w = width of test piece).

For a cracked specimen in pure bending,

$$J = \frac{2}{b} \int_0^{\theta_{\text{crack}}} M \, \mathrm{d}\theta_{\text{crack}} \tag{29}$$

where M = bending moment per unit thickness

θ_{crack} = angle change due to the crack

b = length of uncracked ligament.

The integral is simply the area under the M–θ_{crack} curve, i.e. the work done in loading, with the deformations with no crack present eliminated from the calculations.

For the special case of bending loads on the ligament of a deeply cracked bar,

$$J = \frac{2}{Bb} \int_0^{\delta_{\text{crack}}} P \, \mathrm{d}\delta_{\text{crack}} = \frac{2A}{Bb} \tag{30}$$

where A is the area under the P–δ curve at the displacement of interest, B is specimen thickness and b is the uncracked ligament. Here the notch depths must at least be sufficient so that plasticity encountered is confined to the uncracked ligament region ahead of the crack. Again, elastic displacements with no crack would have to be eliminated in evaluating J.

The measurement of J_{Ic} is a new technique and one of the present difficulties is that the question of where to take the measurement point of J_{Ic} is not precisely defined. The J value calculated from eqn. (30) does not give the exact J value for an advancing crack, and hence J should always be overestimated as a result of A being overestimated.[33] This overestimation would be specimen size dependent, and test results show that small specimens tend to give a slightly higher J_{Ic} value. Much work therefore needs to be done to establish standardised testing procedures[33] and to establish fully the viability of the approach and the dependence of the critical value on variables. Of course, J is not a meaningful parameter after the initiation of crack extension.

Another possible source of discrepancy in J evaluation is that in deriving the total energy U absorbed by a specimen from load–displacement diagrams (eqn. 30), allowance should be made for extraneous energy components such as those arising from testing machine compliance effects and load point indentations. The energy component of the uncracked specimen should also be subtracted from the total energy. However, more recent analysis[34] shows that the requirement for compatibility with K in the linear elastic regime implies that J analysis will generally require the retention of the unnotched beam energy. It has also been pointed out that there is an error in the derivation of eqn. (30).

6.12 APPLICATION OF FRACTURE MECHANICS TO CONCRETE

6.12.1 Fracture Behaviour of Concrete Systems

Materials such as cement pastes, mortars and concretes, and all cement-based composites are basically discontinuous, anisotropic, heterogeneous, multiphase systems. Such materials contain interfacial bond microcracks and other inherent flaws arising from volume changes and other effects during fabrication of the materials. It is these bond microcracks and interfacial discontinuities which create through their geometry the nuclei for potential crack propagation and fracture.

When such materials are subjected to external loading, the existing microcracks progressively increase and grow. The process of progressive discrete microcracking results in non-linear stress–strain behaviour, and a semi-ductile mode of failure in contrast to the ideal brittle or Griffith material in which the onset of crack growth is synonymous with fracture. The strength, stiffness and mode of failure of such materials are also affected by the size of the specimen considered.[22,35]

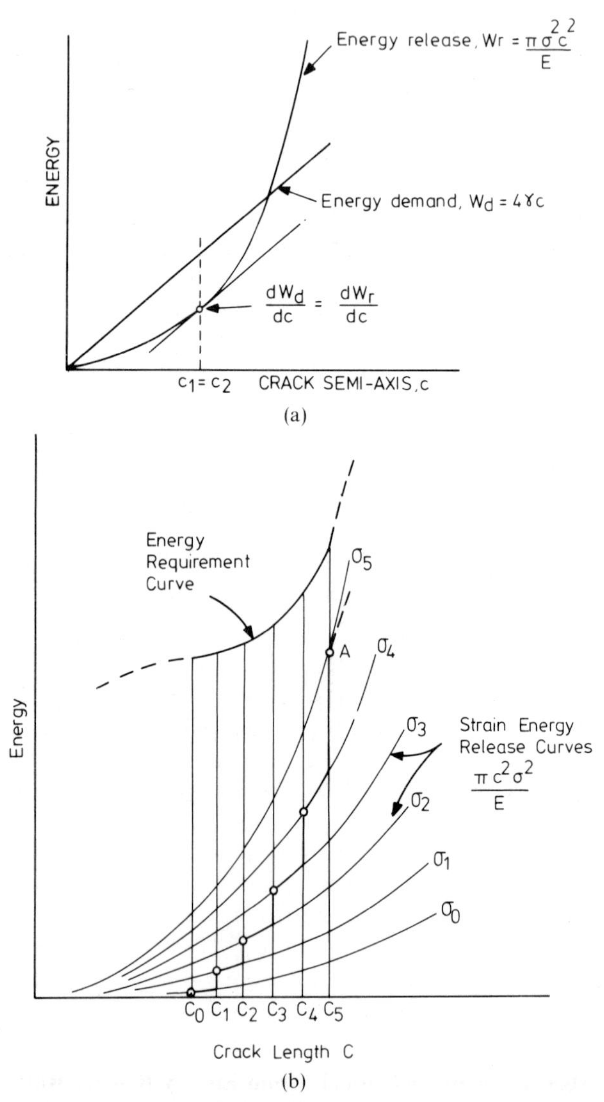

FIG. 11. (a) The ideal Griffith material. (b) Material showing stable crack growth.

There has been considerable argument about the applicab conventional Griffith concept to concrete materials, and w concrete fracture follows the weakest link concept or the class bundle concept.[36] There is a considerable volume of experimenta evidence to show that there are three basic characteristics which differentiate concrete materials from other materials. First, many tests show that the newly formed surface area in concrete materials is many times larger than the effective fracture area.

Secondly, the heterogeneity of the material ensures that cracking itself is a heterogeneous process. Even the cement paste matrix, which is probably the nearest to an elastic material in the concrete system, is itself not homogeneous, and suffers considerable microcracking prior to fracture. Five stages of crack growth can generally be distinguished in the concrete system: initiation of the crack, slow stable crack growth, crack arrest, a critical condition and unstable crack propagation. Slow satellite cracking is thus an essential feature of the fracture behaviour of concrete, and there is considerable stress transfer across interfacial microcracking.

Lastly, because of the enormous fracture surface area and the heterogeneous nature of the fracture process, the energy dissipating mechanism in concrete materials is not merely confined to the surface energy[37] (Figs. 11(a) and (b)). There is as yet no experimental or theoretical quantification of the various energy dissipating mechanisms involved in concrete fracture. However, several applications of linear elastic fracture mechanics have been made to cement paste, mortar and concrete. The results obtained by different investigators show a wide range of experimental values for the critical rate of release of elastic strain energy and the stress intensity factor, and probably confirm that forms of energy dissipation other than surface energy are involved and not accounted for in the calculated values.

In the following, the available evidence on the application of fracture mechanics to concrete materials is reported and discussed.

6.13 FRACTURE TOUGHNESS MEASUREMENTS IN CONCRETE MATERIALS

6.13.1 Measurements of Critical Strain Energy Release Rate

It has been shown earlier that the basic fracture mechanics theory can be developed in terms of an energy criterion or a stress intensity factor.

release rate has been shown to be given by eqn. (12):

$$G = \prod c\sigma^2/E$$

ld depend on the material, the specimen geometry and the growth of the crack. The critical strain energy release rate, G_c, then be interpreted as a material parameter and can be measured in a laboratory with sharply notched test specimens. Only the applied force and crack length need to be measured to compute G. The applied load at the instability point is easily obtained from an autographic load–displacement record. The appropriate crack length is more difficult to measure. The critical crack length c_c should include the equivalent length of slow crack growth in addition to the notch depth. This length of slow crack growth may be found by a compliance technique or by an ink staining technique or high speed photography.

A more reliable method of measuring G_c is by determining the compliance of the specimen at different stages in the growth of the crack. This eliminates the need for the difficult experimental measurements of critical stress and critical crack length at the onset of fast crack propagation. It can be shown that

$$G = \tfrac{1}{2}P^2(\mathrm{d}C/\mathrm{d}A) \tag{31}$$

where P is the applied load and C, the specimen compliance. Measurements of G_c have been made for glass, glassy polymers, adhesives, ceramics and concrete.

6.13.2 Fracture Toughness in Compression

Unlike those in a tensile stress field, cracks in a compression stress field are stable for a major part of the loading cycle and contribute to progressive crack propagation. There is, however, little experimental data on the determination of fracture toughness parameters in compression. Glucklich[38] has shown that corresponding to eqn. (6) for plain strain conditions,

$$\sigma_{\mathrm{comp}} = \left(\frac{8E\gamma}{\pi b(1-\nu^2)}\right)^{1/2} \tag{32}$$

where $2b$ is the minor axis of the crack, Glucklich has also suggested two different expressions for the critical strain energy release rate G_c.[39,40]

$$G_c = \frac{\pi b \sigma^2}{2E} \tag{33}$$

and

$$G_c = \frac{\pi c \sigma^2 \sin^2 \phi \cos^2 \phi (1 - v^2)}{E} \tag{34}$$

where ϕ is the inclination of the crack to the line of action of the applied stress.

Table 2 shows values of G_c in compression obtained by Knox in perspex plates[41] and Desayi on mortar and concrete prisms.[42] These data show large variations.

6.13.3 Fracture Toughness in Tension

Values of G_c have been computed by various investigators from flexural and direct tension tests.[43–55] The essential data from these studies are summarised in Table 3, which also includes values of G_c for other construction materials for comparison purposes. These studies show that G_c is influenced by a number of parameters such as size and geometry of test specimen, quality of the cement paste or mortar matrix, type and volume of aggregate, nature of aggregate–matrix bond, allowance for crack growth, the value of Young's modulus and the type of loading. The higher the value of G_c, the more difficult it becomes to achieve brittle fracture.

In G_c measurements, it is important to consider the slow crack propagation which occurs prior to rapid crack propagation,[43] and this implies that the velocity of crack propagation must also be taken into account.[37] Neglecting slow crack growth can lead to considerable underestimation in the computed values of G_c.[47] The moisture conditions of testing are also significant and tests by Glucklich and Korin[51] show that drying of cement mortar specimens from saturation to almost oven dryness increases G_c gradually, after an initial decrease.

6.13.4 Fracture Energy Measurements

When applying fracture toughness measurements care must be taken to distinguish measurements which give 'work of fracture', by observing

TABLE 2

STRAIN ENERGY RELEASE RATE IN COMPRESSION

Investigator	*Test geometry*	*Material system*	*Remarks*	*Experimental results* (G_c, kN/m)
Knox[41]	Perspex plates, 152 × 152 × 199 mm	Concrete	Crack growth considered Crack angle 20°–35°	14·2–352
Desayi[42]	Prisms, 152 × 152 × 305 mm	Mortar	No allowance for crack growth	0·1352–0·3470[a] 0·0427–0·1732[b]
		Concrete		0·1350–0·4740[a] 0·0536–0·2317[b]

[a]based on eqn. (33)
[b]based on eqn. (34)

the total energy required to fracture a notched specimen, and measurements which yield G_c. In a perfectly brittle material, measurements of the surface energy γ from 'suitable experiments' would correspond to determination of G_c. In materials which are not perfectly brittle, measurements of γ yield values greater than the true surface energy. Equivalence with G_c can then be assumed only if the effective γ is measured at the onset of rapid crack propagation. In any case, computations of fracture energy and fracture toughness from areas of load–displacement curves can be grossly misleading and contain significant errors because of the testing machine stiffness characteristics on the energy absorbed during fracture. Further, machine specimen interaction can produce extraneous fracture patterns and mask the true behaviour of a material. Considerable discrepancies can thus occur in the measured and computed values of γ if this care is not exercised.[43,48,56] (Table 4.)

There is considerable evidence from the results of tests on brittle ceramic materials and composites that fracture energies computed from notched beam tests, work of fracture method and double cantilever beam tests show considerable variation depending upon the type of test. Further, fracture energies derived from work of fracture method and notched beam tests themselves show considerable scatter, and definite dependence on the notch depth– specimen thickness ratio from about 0·2–0·9. It, appears, therefore, that all such data can only be used for qualitative comparison and cannot be treated as absolute values. The work of fracture method is strain rate dependent and invariably gives fracture energies totally unrelated to fracture energies obtained from fracture toughness type of tests.

It should also be added that surface energy measurements are influenced by (1) intrusion of impurities on the fractured surfaces, (2) presence of moisture, (3) the geometry of the cleavage surface, and (4) temperature.

Using the fracture energy reported by Cooper and Figg[57] and assuming plane stress conditions, the expected flaw sizes, for net tensile fracture stress of $5\,N/m^2$ in hardened cement paste with a Young's modulus of $14\,kN/mm^2$, likely to be present in cement paste which would propagate in an unstable manner are 7 mm (dry) and 6 mm (wet). However, these estimates of flaw sizes should be treated with caution with due regard to their applicability.

Although the energy approach has generally been abandoned in favour of the stress intensity factor approach, G_c, defined as fracture

TABLE 3

STRAIN ENERGY RELEASE RATE IN TENSION

Investigator	*Test geometry*	*Material system*	*Remarks*	*Experimental results* (G_c, kN/m)
Kaplan[43]	Notched beam, 4-point bending, 76 × 102 × 406 mm and 152 × 152 × 508 mm	Mortar Concrete	No correction for slow crack growth	0·0173–0·0285 0·0140–0·0294
	Notched beam, 3-point bending, 76 × 102 × 406 mm and 152 × 152 × 508 mm	Mortar Concrete		0·0109–0·0268 0·0067–0·0275
Romualdi and Batson[44]	Plate specimen with centre crack, 610 × 813 × 63 mm	Mortar	No correction for slow crack growth	0·0053–0·0123
Glucklich[45]	Notched beam, 4-point bending, 51 × 102 × 1067 mm	Mortar	Correction for slow crack growth	unnotched: 0·0201 notched: 0·0193
Lott and Kesler[a46]	Notched beam, 4-point bending, 102 × 102 × 305 mm	Mortar Concrete	No correction for slow crack growth	0·0035 0·0051
Welch and Haisman[47]	Notched beam, 4-point bending, 102 × 102 × 508 mm	Paste Mortar Concrete	Correction for slow crack growth '*e* + *E*(dynamic)	0·0245 0·0385 0·0187–0·0363
		Concrete	No crack growth '*e* + *E*(dynamic)	0·0063–0·0154

Moavenzadeh and Kuguel[48]	Notched beam, 3-point bending, 25 × 25 × 305 mm	Paste Mortar	No correction for slow crack growth	0·0035–0·0050 0·0245–0·0043
	Cracked (line notch) beam, 3-point bending 25 × 25 × 305 mm	Concrete		0·0084–0·0096
Naus and Lott[49]	Cracked (line notch) beam, 4-point bending, 102 × 102 × 305 mm	Concrete	No correction for slow crack growth	0·0072
Okada and Koyanagi[50]	Notched beam, 4-point bending, 47 × 100 × 388 mm	Paste Mortar Concrete	No correction for slow crack growth	0·0078 0·0098–0·0157 0·0118–0·0147
Glucklich and Korin[51]	Notched beam, 4-point bending, 12 × 25 × 280 mm	Mortar	Corrected for slow crack growth Variable moisture content	0·0098–0·0490
Mindess, Lawrence and Kesler[52]	Notched beam, 4-point bending, 76 × 76 × 381 mm	Paste Concrete Steel fibre concrete (0·23%–2·0%) Glass fibre concrete (1·0%)	No correction for crack growth	0·0088–0·0154 0·0172–0·0175 0·0114–0·0434 0·0194–0·0219
Knox[41]	Notched beam, bending, —	Concrete	—	0·0175

TABLE 3 (*continued*)

STRAIN ENERGY RELEASE RATE IN TENSION

Investigator	*Test geometry*	*Material system*	*Remarks*	*Experimental results* (G_c, kN/m)
Barr and Bear[53]	Circumferentially notched round bar, 4-point bending (CNRBB)	Mortar	—	0·0648–0·1174
Mazars[54]	Plate specimen with centre crack and single-edge crack, 600 × 340 × 80 mm	Concrete	Variable crack length	0·0102
	Notched beam, 3-point bending, 150 × 100 × 1000 mm		Finite element analysis	
George[55]	Notched beam, 3-point bending, 76 × 76 × 286 mm	Soil cement	No correction for slow crack growth	0·0035–0·0166

Hahn, Kanninen and Rosenfield[81]	—	Plain carbon steels ductile fracture	500–900
		High strength steels ductile fracture	5–130
		Low to medium strength steels brittle fracture	0·6–60
		Epoxy resin	0·22
		Doulgas Fir, parallel to grain	0·03
		Glass	0·002–0·008
		Polymethyl methacrylate	0·5

[a]Values of G_c derived from K_{Ic}

TABLE 4

FRACTURE ENERGIES OF VARIOUS MATERIALS

Investigator	*Material*	*Test conditions*	*Experimental fracture surface energy* (J/m^2)
Brunauer, Kantro and Weise[56]	Tobermorite	Heat of dissolution method	0·39
Moavenzadeh and Kuguel[48]	Hydrated cement paste (28 d)	25 × 25 × 300 mm; water cement ratio = 0·5, 3-point bending test	2·35–4·22
	Mortar (7 d)		2·14–4·75
	Concrete (28 d)		3·50–6·25
Cooper and Figg[57]	Hydrated cement paste	25 × 50 × 200 mm; water–cement ratio = 0·5, slow bend test	14·90 (dry) 12·40 (wet)
Kaplan[43]	Mortar 1–2·4	75 × 100 × 400 mm, and 150 × 150 × 500 mm, 3- and 4-point bending tests	19·30 (measured) 1·61 (calculated)

Harris, Varlow and Ellis[58]	Mortar	Notch beam, 3-point bending	
	1–3·0	Dry, variable age	20–23
		Wet, variable age	25–40
Auskern and Horn[59]	Paste	13 × 13 × 76 mm, 3-point bending test	8·7–9·5
	Polymer impregnated paste		66·0–84·0
Beaumont and Aleszka[60]	Fibre concrete	25 × 25 × 125 mm, 3-point bending	1·75
	Polymer impregnated fibre concrete		20·00
—	Silica glass		5–10
	Polymethylmethacrylate		350–600
	Mild steel	Dry, variable age	2970–3200
		Wet, variable age	3800–4050

toughness, has recently been analytically related to compressive strength and the flexural moment of the modulus of rupture test.[61] The values of G_c computed from existing direct tension and flexural tests show reasonable agreement between the results. This approach has been used to evaluate the shear at diagonal tension cracking in both flexurally cracked and flexurally uncracked reinforced beams, and the torque at diagonal tension cracking, in a reinforced concrete beam subjected to torsion.

6.13.5 Measurements of Stress Intensity Factor

One of the limitations of direct measurements of G_c is that it does not involve a knowledge of the stress distribution in the vicinity of the crack. Further, the phenomena of surface energy and plastic work accompanying fracture development are not yet fully understood, so that the stress intensity approach is now generally preferred to the energy approach. The K_c parameter is often a more useful representation of a material for engineering applications, and G_c is more useful when fracture mechanics is applied to material development.

The basis for the application of linear elastic stress field fracture mechanics is that a value of K_c or K_{Ic}, determined from the unstable fracture of suitable test specimens, will be identical to the K value at fracture instability for a crack in a structure in service. The K value as a function of load and specimen geometry for test specimens (called K calibration) may be determined by theoretical elastic stress analysis[13] or by experimental compliance measurements.

The experimental methods of fracture toughness determination are well established.[14,17] The general procedure is to adopt a test specimen, of suitable geometry and dimensions, into which a crack of suitable size is introduced. The specimen is then loaded at a slow rate so that the crack is incrementally extended, and some means provided to measure the crack extension.

Since K_{Ic} is considered to be a material property, there are minimum dimension requirements to guard against general yield of the specimen. The specimen dimensions are determined by the ratio of toughness to the yield strength of the material (Fig. 6). The required degree of sharpness of the notch or slit is also important. In general, a natural crack should be initiated at the root of the notch or slit. The largest source of error in K_c (and G_c) measurements is the uncertainty about the exact crack length at the moment of instability. Several methods of measuring crack length have been used, such as penetrating liquids (ink staining), cinematography, optical microscopy, electric potential, acoustic emission, and displacement gauges. More recently, ultrasonics have been used for detecting the onset of crack extension and for monitoring continued crack growth in fracture toughness tests.

Although a wide range of specimen geometries is available for fracture mechanics measurements, in concrete fracture toughness testing, there are no standard specimen dimensions or methods of testing, and many of the reported tests do not conform to the requirements of K_{Ic} determination. In spite of the lack of standard specifications for specimen geometry and test methods, a considerable amount of research has been carried out to determine the fracture parameters K_{Ic} (and G_{Ic}) of hardened cement paste, mortar and concrete. The test methods used to determine these parameters include 3-point bending and 4-point bending of notched beams, plate specimens with centre crack, double cantilever beam (DCB), compact tension (CT) specimen, double torsion plate specimen and double-edge-cracked tests (Figs. 12–17). The influence of a wide range of parameters on fracture toughness has been studied, and the major test results are shown in Tables 5 and 6.

From the reported test data, the following conclusions can be drawn. Factors influencing mix design also generally affect fracture toughness. Water–cement ratio, air content, curing regime, fine aggregate content, and the type and volume fraction of coarse aggregate all influence fracture toughness. The two major factors affecting fracture toughness appear to be the matrix strength and the type and volume concentration of coarse aggregate inclusions which act as crack arresters.

The fracture toughness values of aggregates (e.g. marble and quartz) are about an order of magnitude larger than those of the cement paste matrix.[71] Addition of aggregate therefore increases fracture toughness, but it also results in a progressive increase in toughness with crack growth (Fig. 18)[64] although crack initiation probably occurs at a K_{Ic} value corresponding to that for the paste. The higher the proportion of aggregate, generally the larger is the increase in toughness. Volume fraction, rather than grading, of aggregate appears to be more important. There is some evidence that K_{Ic} varies approximately linearly with aggregate volume,[64] and there is also some evidence that toughness increases with maximum size of the coarse aggregate

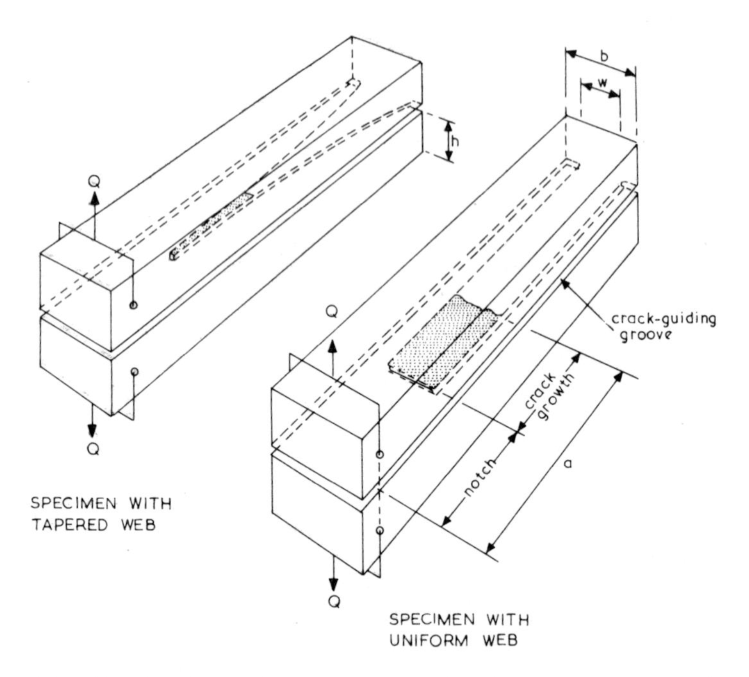

FIG. 12. The double cantilever beam test for fracture toughness. General arrangement of (DCB) specimens.

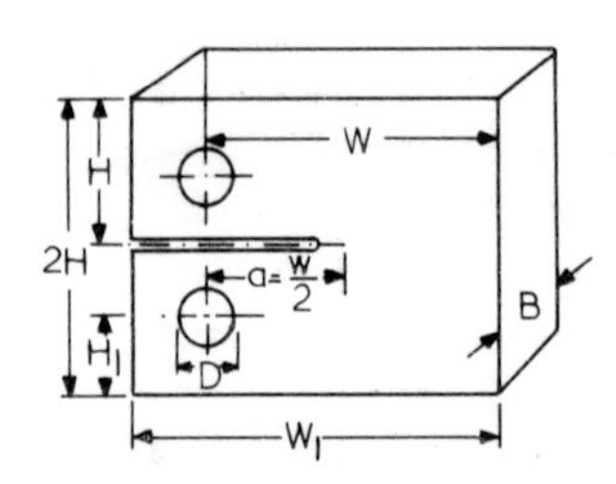

FIG. 13. Compact Tension (CT) specimen: wedge opening loading (WOL). $W = 2{\cdot}0B$, $D = 0{\cdot}5B$, $a = 1{\cdot}0B$, $W_1 = 2{\cdot}5B$, $H = 1{\cdot}2B$, and $H_1 = 0{\cdot}65B$.

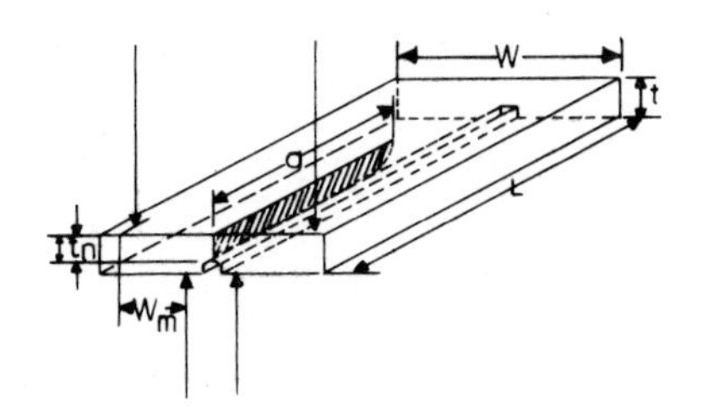

FIG. 14. Double torsion test: plate specimen.

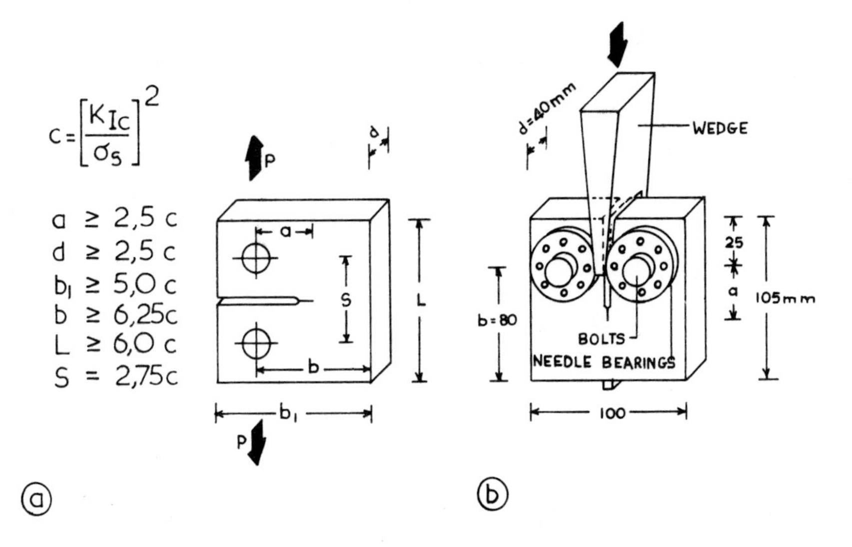

FIG. 15. Compact tension (CT) specimen. (a) Specimen according to ASTM, Plane-strain Fracture Toughness of Metallic Materials, Designation E99/72. (b) Wedge loaded specimen.

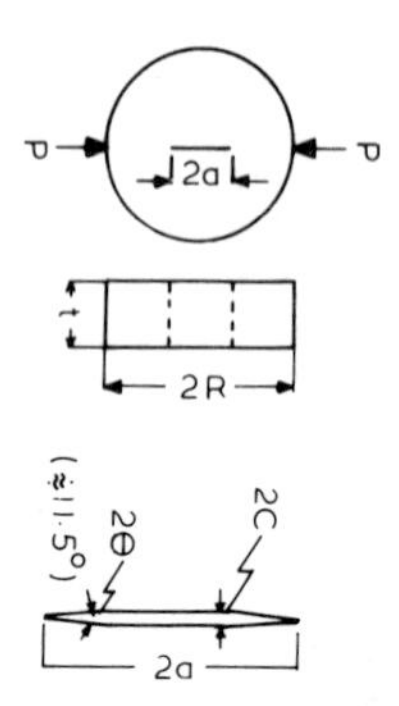

FIG. 16. Diametral compression test specimen.

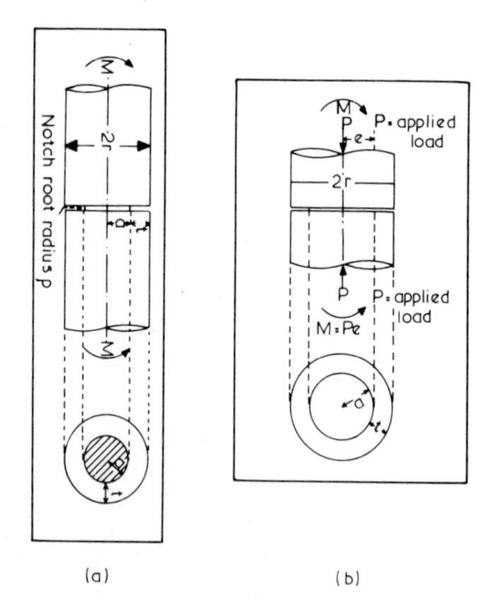

FIG. 17. (a) Circumferentially notched round bar under bending (CNRBB). (b) Circumferentially notched round bar under eccentric loading (CNRBEL).

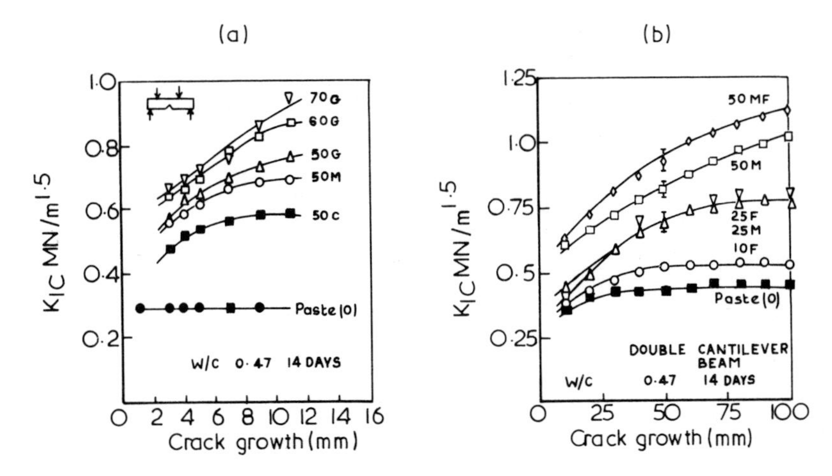

FIG. 18. Variation of fracture toughness with crack growth of cement paste, mortar and concrete. (a) Notched beam, 4-point bending test. (b) Double cantilever beam test. Aggregate sizes: F = fine (52–100 mesh), M = medium (14–25 mesh), MF = mixture of equal proportions of M and F, C = coarse (10–5 mm) and G = graded (10–100 mesh). (By permission of Pergamon Press Ltd.[64])

particles.[72] When considering different types of coarse aggregates, the aggregate–matrix bond may be the more influential factor affecting fracture toughness, since the fracture toughness of aggregate–matrix interfaces is considerably lower than that of the hardened cement paste.[71]

Many tests have ignored the effect of crack growth on K_{Ic} (Table 5). Hardened cement paste is generally brittle, and it is difficult to restrict and control crack growth, particularly in flexural tests. K_{Ic} would, therefore, be expected to be independent of crack growth in cement paste. Both notched beam and DCB tests show that this is so; wedge loaded compact tension tests, however, show that fracture toughness decreases with crack growth and eventually stabilises to a constant value when one major crack develops to fracture.[64, 71]

Khrapkov *et al.*[72] also found that K_{Ic} at unstable crack growth is less than that at stable crack growth. For concrete, they obtained values of 0·75–0·85 MN/m$^{3/2}$ at the former stage compared to a value of 1·90 MN/m$^{3/2}$ at stable crack growth. The explanation probably lies in that at stable crack growth there is extensive satellite microcracking at the pseudoplastic zone of the crack tip. A higher energy is required for

TABLE 5

CRITICAL STRESS INTENSITY FACTORS FROM FRACTURE TOUGHNESS TESTS

Investigator	*Test geometry*	*Material system*	*Remarks*	*Experimental results* (K_{Ic}, $MN/m^{3/2}$)
Kaplan[a,43]	Notched beam, 4-point bending, 76 × 102 × 406 mm and 152 × 152 × 508 mm	Mortar Concrete	No correction for slow crack growth	0·64–0·78 0·55–0·92
	Notched beam, 3-point bending, 76 × 102 × 406 mm and 152 × 152 × 508 mm	Mortar Concrete		0·67–0·89 0·57–1·15
Romualdi and Batson[44]	Plate specimen with centre crack, 610 × 813 × 63 mm	Mortar	No correction for slow crack growth	0·34–0·52
Lott and Kesler[46]	Notched beam, 4-point bending, 102 × 102 × 305 mm	Mortar Concrete	No correction for slow crack growth	0·29–0·33 0·34–0·40
Welch and Haisman[47]	Notched beam, 4-point bending, 102 × 102 × 508 mm	Paste Mortar Concrete	Correction for slow crack growth $+E$ (dynamic)	0·85 1·08 1·78–1·28
		Concrete	No crack growth $+E$ (dynamic)	0·45–0·83

Moavenzadeh and Kuguel[48]	Notched beam, 3-point bending, 25 × 25 × 305 mm	Paste Mortar	No correction for slow crack growth	0·13–0·17 0·14–0·15
	Cracked (line notch) beam, 3-point bending, 25 × 25 × 305 mm	Concrete (water–cement ratio = 0·5; age of test 3–28 days)		0·23–0·26
Naus and Lott[49]	51 × 51 × 356 mm (paste and mortar)	Paste		0·31–0·45
	Cracked (line notch) beam, 4-point bending	Mortar	No correction for slow crack growth	0·21–0·57
	102 × 102 × 305 mm	Concrete		0·37–0·77
Kesler, Naus and Lott[62]	Plate specimen with through-thickness flaw, wedge loaded, 51 × 305 × W (457–914) mm	Paste Mortar Concrete	With and without correction for slow crack growth	0·08–0·26 0·30–1·26 0·34–1·43
Harris, Varlow and Ellis[58]	Notched beam, 3-point bending	Mortar	Dry, variable age Wet, variable age	0·40–0·29 0·41–0·44
			No correction for slow crack growth	
Walsh[63]	Notched beam, 3-point bending, 76 mm × 1·5d × 5d (d = 152 and 254 mm)	Concrete	—	0·54–1·07
Okada and Koyanagi[50]	Notched beam, 4-point bending, 47 × 100 × 388 mm	Paste Mortar Concrete	No correction for slow crack growth	0·22 0·29–0·39 0·33 – 0·46

TABLE 5 (*continued*)

CRITICAL STRESS INTENSITY FACTORS FROM FRACTURE TOUGHNESS TESTS

Investigator	*Test geometry*	*Material system*	*Remarks*	*Experimental results* (K_{Ic}, $MN/m^{3/2}$)
Brown and Pomeroy[64]	Cracked (line notch) beam, 4-point bending, $38 \times 38 \times 250$ mm	Paste Mortar and concrete	Crack growth considered	0·30 0·45–0·95
	Double cantilever beam (DCB) $50 \times 100 \times 350$ mm (Fig. 12)	Paste Mortar		0·30–0·45 0·35–1·10
Carmichael and Jerram[65]	Compact tension (CT) specimen, 25·4 mm thick (wedge opening loading) (Fig. 13)	Mortar	—	0·32–0·48
Nadeau, Mindess and Hay[66]	Double torsion plate specimen, $229 \times 76 \times 13$ mm (Fig. 14)	Paste	Machined notch sharp crack	0·34 0·29
	Notched beam, 3-point bending $13 \times 13 \times 38$ mm	Paste		0·32
Mindess, Nadeau and Hay[67]	Double torsion plate specimen, $229 \times 76 \times 10$ mm	Paste	Variable curing conditions	0·31–0·37
	Notched beam, 3-point loading, $10 \times 10 \times 37$ mm	Paste		0·33–0·48
Mindess and Nadeau[68]	Notched beam, 3-point bending, $B \times 51 \times 203$ mm ($B = 45$–254 mm)	Mortar Concrete	Variable width of crack front	0·47 0·76

Higgins and Bailey[69]	Cracked beam (line notch), 3-point bending, 25 × 14 × 90 mm	Paste	No correction for slow crack growth (variable notch width)	0·36–0·46 0·40–0·65
	Cracked specimens, direct tension (double-edge-cracked specimen)	Paste	Variable specimen depth	0·32–0·65
			No correction for slow crack growth	0·39–0·49
Kitagawa and Suyama[70]	Notched beam, 3-point bending, 40 × 80 × 320 (span) mm,	Mortar	No correction for crack growth	0·27
	30 × 60 × 240 (span) mm, and 20 × 40 × 160 (span) mm	Concrete		0·34
Hillemeier and Hilsdorf[71]	Wedge loaded compact-tension specimen, 100 × 105 × 40 mm	Paste	Correction for slow crack growth	0·49–0·31
		Aggregate	Marble Quartz	2·0 3·3
		Aggregate-Matrix interface		0·10–0·12
		Paste	Polymer modified matrix	0·24
Khrapkov, Trapesnikov, Geinats, Pashchenko and Pak[72]	Notched cylinders, 100 × 150 (diameter) mm	Concrete	Variable age	0·15–0·61
	Notched cylinders, 400 × 400 (diameter) mm	Concrete	Variable aggregate size	0·56–1·01

TABLE 5 (*continued*)

CRITICAL STRESS INTENSITY FACTORS FROM FRACTURE TOUGHNESS TESTS

Investigator	*Test geometry*	*Material system*	*Remarks*	*Experimental results* (K_{Ic}, $MN/m^{3/2}$)
Mindess, Lawrence and Kesler[52]	Notched beams, 4-point bending, 76 × 76 × 381 mm	Paste	No correction for slow crack growth	0·50–0·66
		Concrete		0·87–0·88
		Steel fibre concrete (0·23%–2·0%)		0·68–1·32
		Glass fibre concrete (1·0%)		0·81–0·86
Harris, Varlow and Ellis[58]	Notched beam, 3-point bending	Fibre (mild steel) concrete	Dry, variable age	0·61–0·63
			Wet, variable age	0·74–0·83
			No correction for slow crack growth	
Swamy (unpublished date)	Notched beams, 4-point bending, 100 × 100 × 500 mm	Steel fibre concrete (2%; 0·40 × 37·5 mm)	Cured in internal environment	1·0–9·5
		Polypropylene fibre concrete (fibrillated, 2%; 50 mm long)	Corrected for slow crack growth	0·8–7·5
		Glass fibre concrete (AR, 2% chopped 12·5 mm long)	K_{ic} with increasing crack growth	1·0–5·0
George[55]	Notched beam, 3-point bending 76 × 76 × 286 mm	Soil cement	No correction for slow crack growth	0·09–0·14

Hahn, Kanninen Rosenfield[81]	Plain carbon steel, ductile fracture	≈300–400
	High strength steels, ductile fracture	30–150
	Low to medium strength steels, brittle fracture	10–100
	Epoxy resin	0·80
	Douglas Fir, parallel to grain	0·30
	Douglas Fir, normal to grain	0·40
	Glass	0·3–0·6

[a]Values of K_{Ic} computed from G_c.

TABLE 6

CRITICAL STRESS INTENSITY FACTORS FOR FRACTURE TOUGHNESS TESTS

Investigator	*Test geometry*	*Material system*	*Remarks*	*Experimental results* (K_{Ic}, $MN/m^{3/2}$)
Kitagawa, Kim and Suyama[73]	Notched ring, diametral compression test,	Mortar	No correction for for crack growth	0·27–0·29
	100 × 33 mm and 150 × 50 mm (Fig. 16)	Concrete		0·30–0·36
Barr and Bear[53,74]	Circumferentially notched round bar, 4-point bending (CNRBB) 41 × 150 mm and 54 × 150 mm	Mortar	No correction for slow crack growth	0·51–0·63 0·74–1·00 0·71–1·02
	Circumferentially notched round bar, eccentric loading (CNRBEL), 54 mm diameter (Fig. 17)	Mortar	Variable notch depth	0·51–0·86

multiple crack growth. As cracking proceeds and unstable crack growth is reached, not all the cracks propagate at the same rate and final fracture is initiated by only one major crack and hence the K_{Ic} value is reduced. It may well be that these contradictions are also the result of the different type of tests used, and can only be clarified by further research.

The addition of polymer dispersions and other admixtures also appears to reduce the fracture toughness of the hardened cement paste matrix.[72]

For hardened cement pastes, although fracture toughness increases in the early ages, it remains practically constant after about 28 days. It would, then, appear that fracture toughness of cement paste is dependent not merely on capillary porosity but also on the structure and composition of the solid constituents.[64,69] However, with concrete, fracture toughness is reported to increase with age.[72]

The crack geometry appears to have a clear effect on the fracture toughness parameters. The degree of sharpness, or the lack of it, will influence the ease or difficulty with which the crack is initiated at the crack tip. Tests have been reported, however, that the type of notch has no significant effect if slow crack growth is considered.[47] The width of the crack or slit, i.e. the size of the flaw relative to the size of the microcracking zone has a significant influence on fracture toughness (Fig. 19).[62,69] There is thus a distinct size effect on fracture toughness and many tests show that K_{Ic} varies with specimen size.[43,62,63,69,70]

Minor differences in fracture toughness have also been observed between sawn and cast-in notches, but these appear to be not so significant. Fracture toughness appears to be independent of the length of the crack front in notched beam tests;[68] in plate tests with a central flaw, on the other hand, the length of the crack front appears to influence fracture toughness.[62,70] Fracture toughness also appears to be relatively insensitive to rate of loading.[53,69]

Associated with the specimen size and crack geometry is the question of notch sensitivity of cementitious systems. Notch sensitivity depends very much on the type of test; notch beam tests show that hardened cement paste is notch sensitive,[69,71,75] whereas notch plate tests show the opposite. These tests show that as the notch depth increases, the strength of the notched specimen is reduced by a factor of four; considering cementitious systems alone, the hardened paste is clearly notch sensitive. However, if compared with other materials such as glass, where even the smallest of flaws could reduce strength by factors

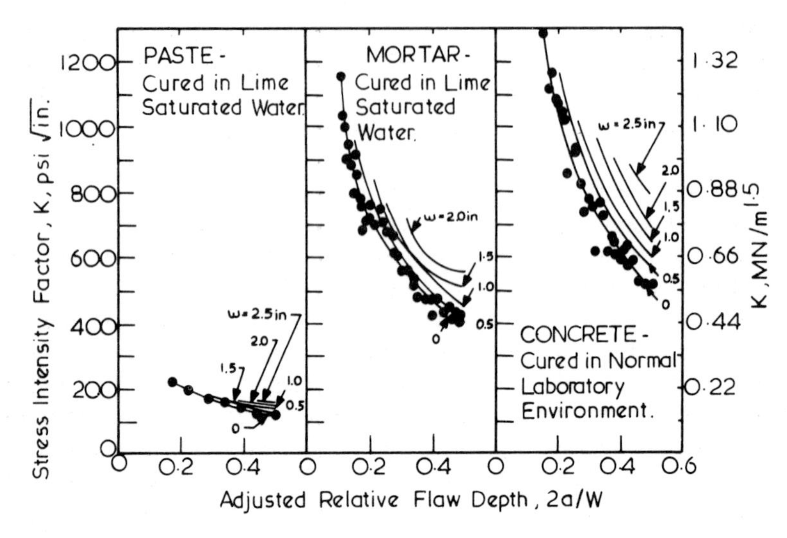

FIG. 19. Variation of stress intensity factor with relative flaw depth in relation to assumed microcracking zone size. (Plate specimens, crack length 2a, W = width of specimen, ω = increase in total crack growth.) (By permission of The Society of Materials Science, Japan.[62])

of the order of 100, the hardened cement paste could be considered to be relatively notch insensitive.

Although it is well established that size effects in mortar and concrete influence mean strength, scatter of results and mode of failure,[35] in terms of fracture toughness mortar and concrete, may be considered to be relatively notch insensitive compared to the hardened cement paste[43,49,75] probably because of the inelastic deformation or microcracking at the root of the notch. However, here again the perspective should not be lost. For laboratory size specimens, concrete may be relatively notch insensitive; but for larger systems like cladding panels and structural members, the material may be very much more susceptible to crack propagation.[22,35,63]

Mindess *et al.* have recently applied the method suggested by Rice *et al.*[32] to evaluate J from load–displacement diagrams of notched beams under 4-point bending. These tests show that J_{Ic} is a promising fracture criterion for cement paste, plain concrete and fibre-reinforced concrete[52] while both K_{Ic} and G_{Ic} are almost certainly not, except for cement paste. These tests also showed that neither G_{Ic} nor K_{Ic}

adequately reflected the changes in concrete behaviour due to fibre additions, whereas J_{Ic} showed significant increases with fibre additions (Table 7). It should, however, be remembered that the reported values of J_{Ic} may well be underestimated.[34]

6.13.6 Quasi-static Fracture Toughness Testing

In materials where cracks run in a slow, controllable manner, so that negligible generation of kinetic energy occurs during cracking, Gurney and Hunt[76] have suggested a quasi-static method of fracture toughness determination. The quantity used in this method as a measure of fracture toughness is the local work required to spread the crack expressed per unit of nominal crack area. This quantity is represented by R† and is equal to G_c and K_c^2/E. The method was initially applied to linear elastic materials and extended to cracking where irreversible deformation is confined to a boundary layer near the crack surface.

The method was originally developed for materials like polymethylmethacrylate (perspex) in which cracks can be made to propagate slowly. The specimen type used by Gurney and Hunt and the typical load–displacement graph obtained autographically are shown in Fig. 20. The area OA_1A_2 then represents the energy required to increase the crack area from A_1 to A_2 and the mean value of R is given by

$$R = \frac{\text{Area } OA_1A_2}{A_2 - A_1} \tag{35}$$

It can be shown that (eqn. (31))

$$R = \tfrac{1}{2}P^2(\mathrm{d}C/\mathrm{d}A) = G$$

so that by using the compliance technique, the need to measure the crack area can be eliminated.

This method has also been extended to quasi-static cracking of materials with high fracture toughness and low yield stress, i.e. in situations where general yielding precedes crack propagation. The size parameter which controls the cracking–yielding transition is ER/σ_y^2 where E is the Young's modulus, R fracture toughness and σ_y, yield

†It has been suggested that strictly speaking, R measures fracture resistance rather than fracture toughness, K_{Ic} or G_c, which relate to the onset of unstable crack propagation under essentially plane strain conditions.

TABLE 7

THE J-INTEGRAL FRACTURE TOUGHNESS

Investigator	*Test geometry*	*Material system*	*Remarks*	*Experimental results* (J_{Ic}, *kN/m*)
Mindess, Lawrence Kesler[52]	Notched beams, 4-point bending, 76 × 76 × 381 mm	Paste	No correction for slow crack growth	0·0110–0·0152
		Concrete		0·0400–0·0422
		Steel-fibre concrete (0·23 %–2·0 %)		0·0305–0·8298
		Glass-fibre concrete (1·0 %)		0·1354–0·2389

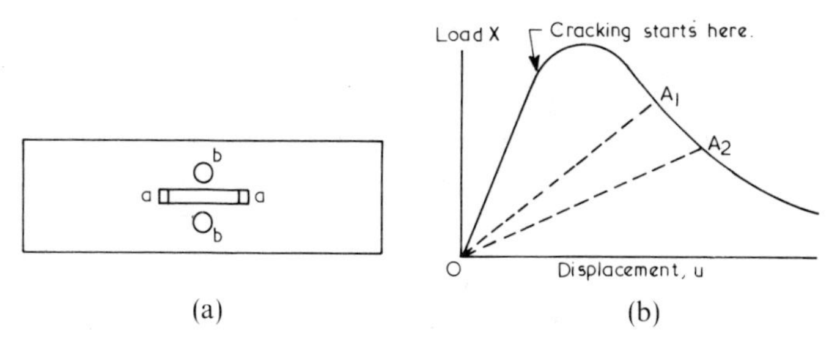

FIG. 20. (a) Perspex test specimen for quasi-static testing, slot aa, loading holes bb. (b) Typical load–displacement diagram. A_1 and A_2 correspond to known crack areas.

stress. A simple and versatile laboratory-size test rig which enables the large structure fracture mode of materials with high values of ER/σ_y^2 (>50 mm) to be determined using small-scale test pieces in conventional testing machines has recently been developed.[77]

The irreversible work area method of determining fracture toughness has been found particularly applicable to fibre-reinforced composites and environmental fracture toughness testing in which both the environmental medium (liquids or gases) and temperature can be varied. Improved specimen geometries have also been developed such as the double-ended specimen for glass-fibre-reinforced plastics and the single-ended or double cantilever specimen for environmental testing.[78] The technique can, however, be readily used for most materials which, in a suitable specimen form, exhibit stable crack growth when deformed in a stiff testing machine under normal laboratory conditions. Thus, the method has been used to determine the fracture toughness of glass-fibre-reinforced cement using crack line loaded single-edge-crack specimens[79] and the work of fracture of asbestos cements using grooved double cantilever beam specimens.

6.14 CONCLUSIONS

The data presented in this chapter clearly show that the application of fracture mechanics to concrete materials and concrete structures is very much in its infancy, and that much of this application is owing to developments that have taken place in other material systems. Experience with metals shows that fracture mechanics concepts can be powerful tools in predicting critical crack lengths and flaw sizes, in determining the safe period of operation of a structure under a given applied stress, and in estimating the life of a structure. Although application of fracture mechanics concepts to concrete lags behind in almost all these aspects, there are welcome signs of the use of these concepts to cracking in concrete structures under various stress systems, and to fibre composites.[52,61,65] Khrapkov *et al.*[72] have reported the application of fracture mechanics to the analysis of massive concrete dams operating in cold temperatures.

There is no doubt that there is the need to know more about the fracture process in concrete and how this ultimately relates to mechanical behaviour and strength both in laboratory tests and in service. The state of knowledge of fracture processes in concrete is clearly not so advanced as it is for other engineering materials like metals, glass, ceramics and rocks. The state of knowledge appears to lack reliable quantification, and many questions remain unanswered. With respect to concrete materials at any rate, much of the basic fracture mechanics concepts have remained to be of a purely theoretical nature because in real engineering materials, the critical strain energy release rate lacks a direct physical meaning and surface energies cannot be easily measured. The problem is made more difficult by the lack of a rational and logical mathematical model to relate fracture processes to other physical data such as bond strength and effective flaw size. However, vast experience has been accumulated in fracture mechanics testing and measurements for metal systems, and more recently for ceramic materials and polymers, and this fund of information must be applied to the development of concepts and techniques suitable for concrete systems.[80,81]

The stress intensity approach, on the other hand, offers a more rational and logical concept to view the strength behaviour of concrete and concrete structures in terms of their microstructure. It is in this direction that much of the future work has to be directed.

REFERENCES

1. INGLIS, C. E. (1913). *Trans. Inst. Naval Architects*, **55,** 219.
2. GRIFFITH, A. A. (1920). *Phil. Trans. Roy. Soc. London, Ser. A*, **221,** 163; Proc. First Int. Cong. App. Mech. (Delft), 1924, 55.
3. OROWAN, E. (1945–6). *Trans. Inst. Engrs Shipbuilders Scot.*, **39,** 165.
4. SACK, R. A. (1946). *Proc. Phys. Soc.*, **58,** 729.
5. SNEDDON, I. N. (1946). *Proc. Roy. Soc. London, Ser. A*, **187,** 229.

6. OROWAN, E. (1948–9). *Reports on Progress in Physics*, **12,** 185.
7. OROWAN, E. (1952). *Fatigue and Fracture of Metals*. John Wiley and Sons Inc., New York, p. 139.
8. IRWIN, G. R. (1957). *Proc. Ninth Internal, Cong. App. Mech., Brussels*, **8,** 245.
9 IRWIN, G. R. (1958). *Encyclopaedia of Physics*, **6,** 551.
10. RIVLIN, R. S. and THOMAS, A. G. (1955). *J. Polymer Sci.*, **18,** 189.
11. MURRELL, S. A. F. (1963). *Proc. 5th Rock Mech. Symp.* (Ed. by C. Fairhurst). Pergamon Press, Oxford, p. 563.
12. IRWIN, G. R. (1957). *J. Appl. Mech.*, **24** (3), 361.
13. PARIS, P. C. and SIH, G. C. (1965). In *Fracture Toughness Testing, ASTM STP 381*, Am. Soc. Test. Mat., Philadelphia, Pa., USA., p. 30.
14. BROWN, W. F. and SRAWLEY, J. E. (1966). In *Plane Strain Crack Toughness Testing of High Strength Metallic Materials, ASTM STP 410*, Am. Soc. Test. Mat., Philadelphia, Pa., USA, p. 1.
15. IRWIN, G. R. (1961). *Proc. Seventh Sagamore Res. Conf. Ordnance Materials 1960*, US Office of Technical Services, Washington DC, p. IV. 63.
16. MCCLINTOCK, F. A. and HULT, J. A. H. (1957). *Proc. Ninth Internat. Cong. App. Mech.*, Brussels, **8,** 51.
17. SRAWLEY, J. E. and BROWN, W. F. (1965). In *Fracture Toughness Testing ASTM-STP 381*, Am. Soc. Test. Mat., Philadelphia, Pa., USA, p. 133.
18. MURRELL, S. A. F. (1964). *Brit. J. Appl. Phys.*, **15,** 1195.
19. HOCK, E. and BIENIAWSKI, Z. T. (1966). *Proc. 1st Congress Internat. Society on Rock Mech., Lisbon*, **1,** 243.
20. MCCLINTOCK, F. A. and WALSH, J. B. (1962). *Proc., 4th U.S. Nat. Cong. of App. Mech.*, Vol. 2, Am. Soc. Mech. Engrs., p. 1015.
21. FREUDENTHAL, A. M. (1968). In *Fracture: Vol 2, Mathematical Fundamentals* (Ed. by H. Liebowitz). Academic Press, London.
22. KAMESWARA RAO, C. V. S. and SWAMY, R. N. (1974). *Cement and Conc. Res.*, **4,** 669.
23. WELLS, A. A. (1963). *Brit. Welding J.*, **10,** 563.
24. WELLS, A. A. (1961). *Proc. Crack Propagation Symp., Cranfield*, **1,** 210.
25. BURDEKIN, F. M. and STONE, D. E. W. (1966). *J. Strain Analysis*, **1** (2), 145.
26. ROBINSON, J. N. and TETELMAN, A. S. (1974). In *Fracture Toughness and Slow-stable Cracking, ASTM STP 559*, Am. Soc. Test. Mat., Philadelphia, Pa., USA, p. 139.
27. ELLIOTT, D. and MAY, M. J. (1968). *BISRA Report MG/C/51/68*, The Inter-group Laboratories of the British Steel Corporation, London, p. 1.
28. RICE, J. R. (1968). *J. App. Mech.*, **35,** 379.
29. BEGLEY, J. A. and LANDES, J. D. (1972). In *Fracture Toughness, ASTM STP 514*, Am. Soc. Test. Mat., Philadelphia, Pa., USA, p. 1.
30. LANDES, J. D. and BEGLEY, J. A. (1972). In *Fracture Toughness, ASTM STP 514*, Am. Soc. Test. Mat., Philadelphia, Pa., USA, p. 24.
31. BUCCI, R. J., PARIS, P. C., LANDES, J. D. and RICE, R. J. (1972). In *Fracture Toughness, ASTM STP 514*, Am. Soc. Test. Mat., Philadelphia, Pa., USA, p. 40.
32. RICE, J. R., PARIS, P. C. and MERKLE, J. G. (1973). In *Flow Growth and Fracture Toughness Testing, ASTM STP 536*, Am. Soc. Test. Mat., Philadelphia, Pa., USA, p. 231.

33. LANDES, J. D. and BEGLEY, J. A. (1974). In *Fracture Analysis, ASTM STP 560*, Am. Soc. Test. Mat., Philadelphia, Pa., USA, p. 170.
34. CHIPPERFIELD, C. G. (1978). *J. of Testing and Evaluation*, **6,** 253.
35. SWAMY, R. N. and KAMESWARA RAO, C. V. S. (1973). *Cement and Conc. Res.*, **3,** 413.
36. SWAMY, R. N. and KAMESWARA RAO, C. V. S. (1977). *Proc. Conference on Fibre Reinforced Materials: Design and Engineering Applications.* Institution of Civil Engineers, London, p. 87.
37. SWAMY, R. N. (1968). *Proc. Internat. Conference on the Structure of Concrete, London, 1965.* Cement and Concrete Association, p. 212.
38. GLUCKLICH, J. (1962). *Theoretical and App. Mech. Report No. 215, Uni. of Illinois*, 1.
39. GLUCKLICH, J. (1963). *Proc. Am. Soc. Civ. Engrs., J. Eng. Mech. Div.*, **89,** 127.
40. GLUCKLICH, J. (1968). *Proc. Internat. Conference on the Structure of Concrete, London*, 1965, Cement and Concrete Association, p. 176.
41. KNOX, W. R. A. (1970). In *Fracture Toughness of High Strength Materials: Theory and Practice, Publication 20.* Iron and Steel Institute, London, p. 158.
42. DESAYI, P. (1977). *RILEM Materials and Structures*, **10,** 139.
43. KAPLAN, M. F. (1961). *J. Am. Conc. Inst.*, **58,** 591.
44. ROMUALDI, J. P. and BATSON, G. B. (1963). *Proc. Am. Soc. Civ. Engrs., J. Eng. Mech. Div.*, **89,** 147.
45. GLUCKLICH, J. (1965). *Proc. 1st Internat. Conf. on Fracture*, **2,** 1343.
46. LOTT, J. L. and KESLER, C. E. (1966). Symp. Structure of Portland Cement Paste and Concrete, *Highway Res. Board, Special Report 90*, 204.
47. WELCH, G. B. and HAISMAN, B. (1969). *RILEM Materials and Structures*, **2,** 171.
48. MOAVENZADEH, F. and KUGUEL, R. (1969). *J. of Materials*, **4,** 497.
49. NAUS, D. J. and LOTT, J. L. (1969). *J. Am. Conc. Inst.*, **66,** 481.
50. OKADA, K. and KOYANAGI, W. (1972). *Proc. Internat. Conf. Mech. Behaviour of Materials, Kyoto, 1971*, **4,** 72.
51. GLUCKLICH, J. and KORIN U. (1975). *J. Am. Ceramic Soc.*, **58,** 517
52. MINDESS, S., LAWRENCE, F. V. and KESLER, C. E. (1977). *Cement and Concrete Res.*, **7,** 731.
53. BARR, B. and BEAR, T. (1976). *Concrete*, **10,** 25.
54. MAZARS, J. (1977). *Fourth Internat. Conf. on Fracture*, **3B,** 1205.
55. GEORGE, K. P. (1970). *Proc. Am. Soc. Civ. Engrs., J. Soil Mech. and Foundations, Div.*, **96,** 991.
56. BRUNAUER, S., KANTRO, D. L. and WEISE, C. H. (1959). *Can. J. Chem.*, **37,** 714.
57. COOPER, G. A. and FIGG, J. (1972). *J. British Ceramic Soc.*, **71,** 1.
58. HARRIS, B., VARLOW, J. and ELLIS, C. D. (1972). *Cement and Concrete Res.*, **2,** 447.
59. AUSKERN, A. and HORN, W. (1974). *Cement and Concrete Res.*, **4,** 785.
60. BEAUMONT, P. W. R. and ALESZKA, J. C. (1978). J. Mat. Science, **13,** 1749.
61. HAWKINS, N. M., WYSS, A. N. and MATTOCK, A. H. (1977). *Proc. Am. Soc. Civ. Engrs., J. Struct. Div.*, **103,** 1015.

62. KESLER, C. E., NAUS, D. J. and LOTT, J. L. (1972). *Proc. Internat. Conf. on Mechanical Behaviour of Materials, Kyoto, 1971*, **4,** 113.
63. WALSH, P. F. (1972). *Indian Concrete Journal*, **46,** 469; *Eng. Fract. Mech.*, **4,** 533 and, *Proc. Am. Soc. Civ. Engrs., J. Eng. Mech. Div.*, **98,** 1611.
64. BROWN, J. H. and POMEROY, C. D. (1973). *Cement and Concrete Res.*, **3,** 475.
65. CARMICHAEL, G. D. T. and JERRAM, K. (1973). *Cement and Concrete Res.*, **3,** 459.
66. NADEAU, J. S., Mindess, S., and HAY, J. M. (1974). *J. Am. Ceramic Soc.*, **57,** 51.
67. MINDESS, S., NADEAU, J. S., and HAY, J. M. (1974). *Cement and Concrete Res.*, **4,** 953.
68. MINDESS, S. and NADEAU, J. S. (1976). *Cement and Concrete Res.*, **6,** 529.
69. HIGGINS, D. D. and BAILEY, J. E. (1976). *J. Mat. Science*, **11,** 1995.
70. KITAGAWA, K. and SUYAMA, M. (1976). *Proc., 19th Japan Congress on Materials Research*, 156.
71. HILLEMEIER, B. and HILSDORF, H. K. (1977). *Cement and Concrete Res.*, **7,** 523.
72. KHRAPKOV, A. A., TRAPESNIKOV, L. P., GEINATS, G. S., PASHCHENKO, V. I. and PAK, A. P. (1977). *Fourth Internat. Conf. on Fracture*, **3B,** 1211.
73. KITAGAWA, H., KIM, S. and SUYAMA, M. (1976). *Proc. 19th Japan Congress on Materials Research*, 160.
74. BARR, B. and BEAR, T. (1977). *Concrete*, **11,** 30.
75. SHAH, S. P. and MCGARRY, F. J. (1971). *Proc. Am. Soc. Civ. Engrs., J. Eng. Mech. Div.*, **97,** 1663.
76. GURNEY, C. and HUNT, J. (1967). *Proc. Roy. Soc. London, Ser. A*, **229,** 508.
77. GURNEY, C., MAI, Y. W. and OWEN, R. C. (1974). *Proc. Roy. Soc. London, Ser. A*, **340,** 213.
78. ASHWELL, D. G. and HANCOCK, M. G. (1977). *Proc. Conference on Fibre-Reinforced Materials, Design and Engineering Applications*, Institution of Civil Engineers, London, p. 97.
79. PATTERSON, W. A. and CHAN, H. C. (1975). *Composites*, **6,** 102.
80. KENNY, P. and CAMPBELL, J. D. (1967). *Progress in Materials Science*, **3,** 137.
81. HAHN, G. J., KANNINEN, M. F. and ROSENFIELD, A. R. (1972). *Annual Review of Materials Science*, **2,** 381.

Chapter 7

THE DIPHASE CONCEPT, WITH PARTICULAR REFERENCE TO CONCRETE†

N. CLAYTON and F. J. GRIMER

Building Research Establishment, Building Research Station, Watford, UK

SUMMARY

The paradoxical behaviour of a range of solids, including concrete, under certain unusual applied stress conditions is firstly described, followed by the attempt made by the original discoverer of the behaviour to fit it into a conventional theoretical framework. The authors then propose that a greater insight into material behaviour, paradoxical or otherwise, can be achieved by adopting a diphase approach, the most important consequence of which is that materials owe their strength to an external source, in contrast to the traditional explanation of internal bonding.

Experiments, on concrete specimens in particular, which demonstrate the reasonableness of this viewpoint are presented. It must be made clear, however, that experimental results, as such, can never provide coercive proof of the need to reject existing concepts; rather, this can only come about by a general recognition of the aesthetic and overall practical advantages of adopting a fresh viewpoint.

†The original experiments described in this paper were carried out by the authors as employees of the Building Research Establishment, Garston, Watford. The views expressed are those of the authors, not necessarily those of the Establishment.

7.1 THE PARADOX

It is useful to begin by describing a simple test carried out on a standard concrete cylinder. The specimen is inserted into a specially designed apparatus which permits water pressure to be applied to its bare curved surface (Figs. 1 and 2). That is, to use conventional

FIG. 1. Water pressure experiment apparatus.

terminology, the concrete is subjected to axisymmetrically applied compressive stress using water as the loading medium without the use of an intervening membrane. The result of this test is surprising, both in terms of the ultimate applied stress and the mode of failure. The loading is compressive and uniform, hence, using conventional knowledge of concrete, it is expected that the measured stress at fracture would be high and that the fracture would be by crushing or multiple cracking. In fact, the ultimate stress is low (less than one-fifth of the crushing strength†) and the specimen breaks across a single plane transverse to the axis (Fig. 3). That is, the fractured specimen looks for all the world as though it has been pulled apart by its ends—subjected to applied tension, it seems, not compression at all!

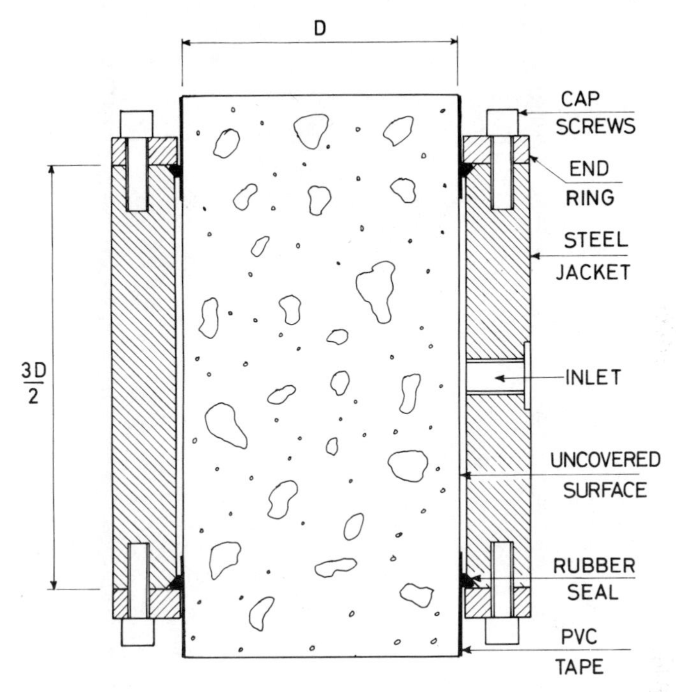

FIG. 2. Water pressure experiment details.

It was stated that the result is surprising; this should perhaps be qualified by excluding those engineers who have, in the past, done research in relation to concrete gravity dams, for example Serge Leliavsky Bey,[1] who investigated the effect of uplift in dams due to the permeation of water. His tests were somewhat more involved but he arrived at a similar result. In fact, since he applied the water pressure over a longer period of time, the failure pressure was even lower; not

†On the few occasions when the term 'crushing strength' is used, it means the ultimate applied stress obtained by loading a standard concrete cylinder axially in a conventional compression testing machine.

significantly different from the ultimate stress of the concrete under uniaxial tension. Since this phenomenon was known 30 years ago, it must consequently be explained why it is considered necessary to

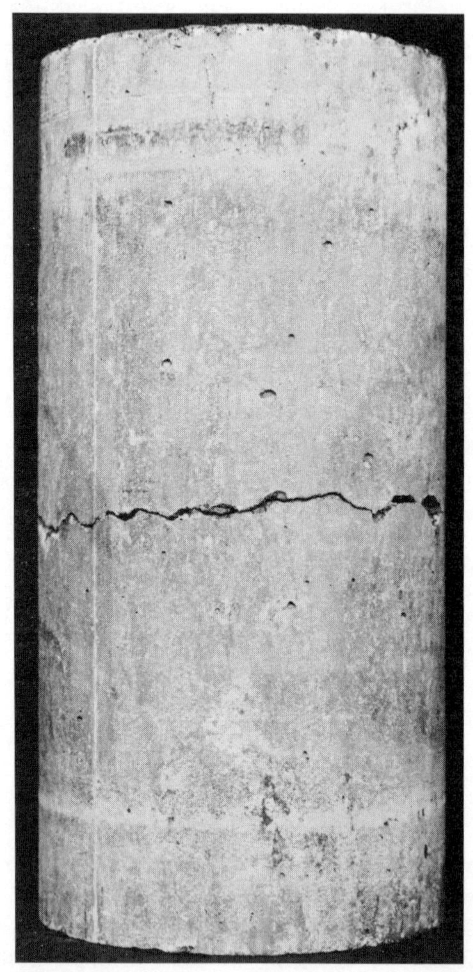

FIG. 3. Fracture of concrete cylinder due to water pressure.

describe it here. There are three main reasons. First, the authors have found that engineers who are familiar with conventional testing are usually not aware of this behaviour. Secondly, the existing strength of materials theory is not capable of predicting the behaviour. And thirdly, the real significance of the behaviour in relation to our knowledge of materials in general is not appreciated. It will become apparent that these three points are related to one another.

Since the behaviour of the concrete cylinder subjected to water pressure is not compatible with the conventional notion of concrete being strong in compression, there are two alternative ways of proceeding. Either it is interpreted in other terms thus making it a special case, or it is accepted as being indicative of the deficiency of the conventional theory which must consequently be replaced. That is, either carry out a patching up job or start all over again. On the evidence of just this one example of anomalous behaviour, it is obviously easier to provide an *ad hoc* explanation. For example, the following rationalisation would do: as the concrete is permeable to water, the specimen is pushed apart from within due to internal pore pressures. So it may now, upon reconsideration, be stated that concrete is strong in compression unless the loading medium enters the concrete, in which case the specimen fails in tension. In other words, subjecting concrete to water pressure is an exceptional type of loading and is not compression at all, but actually tension. The original theory that concrete is strong in compression is then left intact, and that is what really matters—it cannot be expected that deeply held views will change simply because a new piece of data comes along. However, it will be shown that, by taking a broader viewpoint, a greater insight into the nature of compressive loading may be found and that this enables the anomalous behaviour of concrete described above to be seen in an entirely new light.

What would happen if fluid pressure were applied to the curved surface of a solid cylinder of material other than concrete? It was as long ago as 1912 that such tests were carried out by the American physicist, Percy Bridgman, who found that, for a range of brittle and ductile materials, the specimen behaves just as though its ends have been pulled apart.[2] That is, for a brittle substance like glass, fracture takes the form of a single cleavage plane perpendicular to the cylinder axis (similar to concrete) and, for a ductile substance like mild steel, the specimen necks and then separates with a 'cup and cone' fracture surface as in the uniaxial tension test. The generality of this phenomenon is striking indeed. It is a result which cries out to be noticed. Indeed, from a conventional theoretical viewpoint, its existence is highly embarrassing. Bridgman, himself, recognised the difficulty only too well:

> 'The paradoxical thing about this rupture is that,..., there is no force along the fibre across which rupture occurs.'[3]

For a long time, Bridgman was vexed by this paradox. Because of this and the discovery of other anomalies, he felt that a general understanding of material behaviour under applied stress could never be achieved. He made the following pessimistic statements:

> 'It is not difficult in any special case to invent considerations that appeal to our intuitions as adequate explanations of the observed rupture,... the difficulty comes in bringing all possible cases under one point of view.... For myself, I am exceedingly sceptical as to whether there is any such thing as a genuine criterion of rupture.'[3]

However, a few years later, he published the beginnings of an explanation[4] and, in so doing, made an important step towards gaining a real insight into the nature of materials. He argued as follows.

Application of axisymmetric pressure to a cylindrical specimen (or, similarly, biaxial stress applied to a rectangular prism) is equivalent to a hydrostatic pressure applied to the specimen (triaxial stress) plus an applied axial tensile stress of the same value (Fig. 4). He further argued that since 'hydrostatic stress does not essentially modify qualitative behaviour', this term in the equation disappears and the required result, that the axisymmetric pressure (or biaxial stress) is equivalent to applying an axial tension, remains. This led Bridgman to find out what quantitative effect hydrostatic pressure has on the strength of materials, by carrying out conventional tests inside a small 'pressurized testing laboratory'.[5]

Now let this argument be applied to the specific case of concrete. It is necessary to explain why, when water pressure is applied to the curved surface of a concrete cylinder, the specimen breaks as though it had been subjected to axial tension. Imagine, first of all, the concrete cylinder being placed into a high pressure water environment; for example at the bottom of the sea at a depth of a few hundred metres. Now if the conventional uniaxial tension test is carried out (the practical difficulties involved in this test are no doubt even greater on the sea bed!), the fracture would be the same as if the test had been carried out in the laboratory at atmospheric pressure. This is what Bridgman meant by saying that applying hydrostatic pressure does not change the qualitative behaviour of the material. The fracture achieved in these conditions is consequently understandable. And since, from

Bridgman's argument, applying water pressure axisymmetrically to the specimen is equivalent to this imaginary test (assume it is arranged so that the specimen fractures just as the applied tensile stress reaches the hydrostatic pressure value), it appears that the required explanation of the anomalous result has been found. Unfortunately, whilst the reason

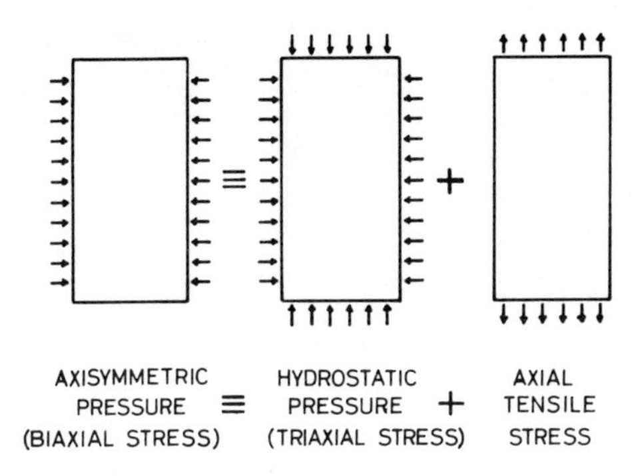

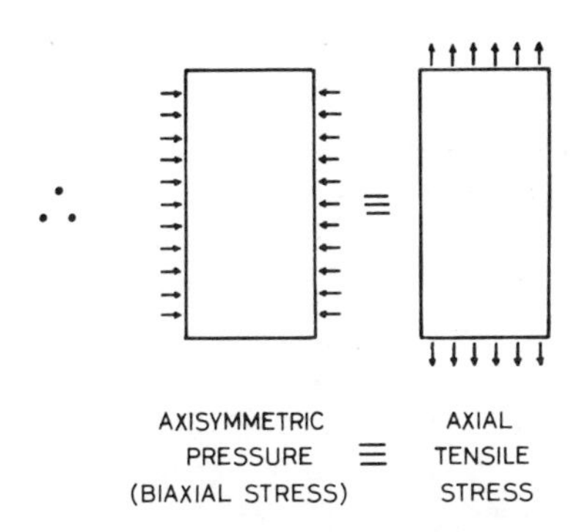

FIG. 4. Bridgman's explanation.

for the equivalence of axisymmetrically applied water pressure to hydrostatic pressure plus axial tension may be clear in the schematic representation of Fig. 4, in physical terms it is not so obvious. That is, it seems strange why water pressure applied to the curved surface of a concrete cylinder should be the same as carrying out a uniaxial tension test in a high pressure water environment, even though the mathematics shows that this is perfectly reasonable. This difficulty is because our language does not tie up with the reality of the behaviour which we are studying. The mathematics expresses the fact that the wrong terms are being used. As Paul Weiss has said:

> 'Scientific language, as the tool for the articulation of scientific knowledge, must be kept in step with the advance of knowledge.... Instead of adapting our accustomed tools of expression to the widening scope of new insights into nature, we often, unwittingly, try to constrain nature, or rather our views of it, into the narrow box of our traditional ways of representing it. In other words, habits of language tend to force patterns of thought into rather rigid molds.'[6]

Bridgman's argument depends upon seeing the applied tension in the imaginary experiment as counteracting the effect of the axial component of hydrostatic pressure. It is this crucial point which the mathematics of the situation brings out, but which our language tends to conceal. An experiment will now be described which demonstrates this particular aspect extremely well.

A cylinder of concrete has its ends covered by two circular steel caps and then put into a vessel filled with water at a high pressure. Its value is set at the pressure which would be required to fracture the specimen if it were applied to its curved surface only, say 5 N/mm^2. The end caps are consequently forced onto the ends of the specimen with a stress of 5 N/mm^2; a rubber seal at each end of the specimen prevents entrance of water pressure into the space between each cap and the end of the concrete. *There is no bonding agent between the cap and the specimen.* A small diameter steel rod is connected to the centre of each end cap and one projects out of the top and the other out of the bottom of the pressure vessel. The actual apparatus design is shown in section in Fig. 5. The complete apparatus is put into a tension testing machine, the projecting rods being fixed into the grips as shown in Fig. 6. The procedure of the conventional uniaxial tension test is then carried out. It is found that the specimen fractures transverse to the axis when the stress applied by the machine is 5 N/mm^2. It is most important to note that, but for the presence of the water pressure in the vessel, the caps would be pulled off the ends of the specimen without fracture taking place (the frictional resistance between the end seals and the specimen is negligible). The tension could not be applied by the machine but for the axial component of the water pressure which holds the caps onto the ends of the specimen. If this is not understood, the whole point of the experiment is missed.

This demonstration is a physical representation of the argument presented in mathematical terms shown in Fig. 4. It demonstrates

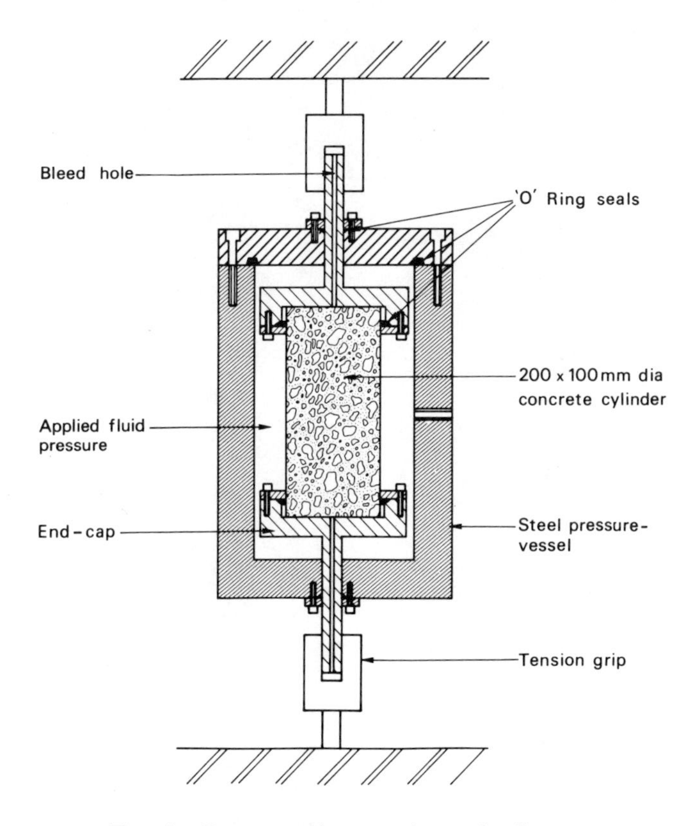

FIG. 5. Demonstration experiment details.

clearly why applying water pressure to the curved surface of a concrete cylinder produces the same result as if the specimen's ends were pulled apart. It is necessary to explain this in full.

Consider the viewpoint of an observer who is not aware of the presence of the water pressure in the vessel. He sees the ends of the specimen as being pulled apart and therefore thinks the demonstration is an ordinary uniaxial tension test. However, we, the experimenters, know that the specimen itself is not being subjected to applied tension

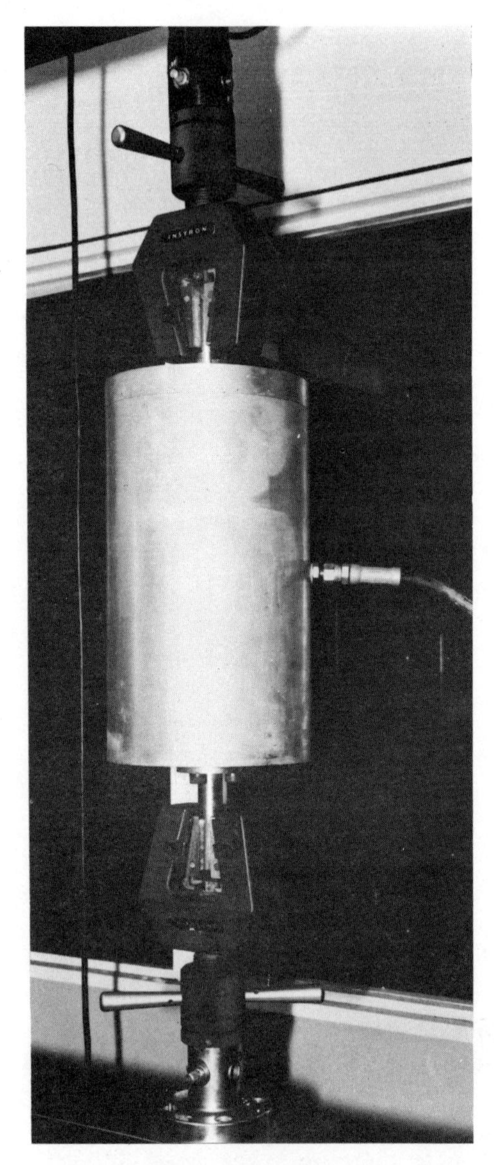

FIG. 6. Demonstration experiment apparatus.

when the rods connected to the end caps are pulled apart. Rather, the axial compressive stress, due to the water pressure acting on the end caps, is being reduced. We, the experimenters, recognise that what appears upon cursory examination by the observer to be a uniaxial tension test may also be seen as an indirect way of subjecting the specimen to water pressure on its curved surface only.

Which view is correct? Is the experiment a uniaxial tension test or an axisymmetric water pressure test? To answer one or the other would be as ridiculous as saying that the person in the well known drawing shown in Fig. 7 is in reality a pretty young lady with her face looking

FIG. 7. Ambiguous drawing.

away or, alternatively, that the drawing is of an ugly old woman staring slightly downwards. Either view is permissible—that is the point of the illustration. Similarly, the point of the experiment described above is to demonstrate that there is no necessary reason for explaining the axisymmetric water pressure test in terms of uniaxial tension or, for that matter, the other way round. Just because it may be preferable to look at pretty young ladies does not mean that the ugly old woman does not exist in the picture. Similarly, just because the tension test may be more familiar does not mean that the behaviour of concrete subjected to water pressure on its curved surface is not as basic in its own right.

The truth is that both views are only partial: they are incomplete in themselves. And since the views are so different, it is not possible to hold them simultaneously in such a way as to give a complete picture. In order to understand why one form of loading produces the same fracture as loading a specimen in an apparently vastly different manner, it is necessary to question the corresponding viewpoints which lead to two apparently contrasting situations when the specimen, judging from its behaviour, does not discriminate. It is the viewpoints which must be transformed so as to correspond, more closely, with the manner in which the material itself discerns applied stresses. Unfortunately, as with all prejudices, these viewpoints are deeply rooted; so much so that they are not usually questioned even when they lead to contradictions. As mentioned earlier, Bridgman concluded at one stage that the behaviour of materials is beyond reason, instead of scrutinising his own preconceptions. In order to arrive at a meaningful description of the material behaviour under discussion, a unifying concept is required which will lead to a new means of expression into which our previous biased viewpoints may be translated. Its power must be assessed by its ability to describe previously unrelated empirical facts by the use of the same terms.

The key to achieving the new understanding which is now being advocated, is the hydrostatic pressure introduced by Bridgman, mathematically, as shown in Fig. 4 and which was demonstrated, physically, in the experiment shown in Figs 5 and 6. There is an important difference, however, between its use in each case. Bridgman's aim was to change the reference datum of stress by imagining the specimen as subjected to hydrostatic pressure, so that the concept of tension could be used to explain what would otherwise remain, to him, a paradoxical result. In short, since there is no tension actually applied when a cylinder of material is subjected to fluid pressure on its curved surface, he manufactured one by performing a clever mathematical transformation. In contrast, the intention here was to show, in physical terms, that what one observer sees as an applied tension can, from a different viewpoint, be seen as the removal of a pre-existing external pressure. Bridgman did not realise that the hydrostatic pressure might have a physical significance. That is, a material *in its natural state* could be recognised as being subjected to hydrostatic pressure from a source which, at present, we do not perceive but which is nonetheless real. Applied tension is then seen, not as an entity in itself, but as the reduction of this pre-existing external pressure.

7.2 THE DIPHASE APPROACH

A theoretical model will now be described, which explicitly indicates how the presence of a high pressure environment can have a real function in relation to strength of materials.[7] The concepts upon which this approach is based arose partially out of the recognition that certain principles of soil mechanics can be used to gain an insight into the nature of strength of materials in general. The essential nature of clays, silts and sands depends upon the fact that the soil particles are held in a state of triaxial compressive stress due to the presence of the interstitial pore water. By analogy, a material may be regarded at each level of consideration as comprising just two phases which are given the general names 'Solid' and 'Fluid.† The Solid phase provides the form of the material and is best conceptualised as a finite number of inert, homogeneous particles, each having an indefinitely high strength and stiffness. The scale of these particles depends upon the level at which the material is being considered and can range from that of the specimen itself down to atoms and below. The Fluid phase is the active constituent and provides the pressure which holds the particles of the Solid phase together. In contrast, however, with the conventional view of soils, the Fluid pressure acts externally on the Solid phase. That is, the Solid phase is held together from the outside and this is the manner in which the diphase model represents how a material gains its strength.

The conceptual switch involved in adopting this viewpoint is not easy, but for the moment this problem will be put aside in order to remark upon how it corresponds with Bridgman's mathematical formulation and then to describe the general outlook upon which the diphase approach is based.

In order to formulate a pattern into which could be fitted the fracture of materials under multi-axial stress, including the particular case of axisymmetrically applied stress described earlier, Bridgman recognised that a three-dimensional representation had to be used.[4] That is, all applied stress systems including the uniaxial and biaxial cases have to be represented in terms of triaxial stress states. By adopting the diphase approach, this conclusion is a natural outcome since a material is seen as being in a triaxial state of stress, due to the

†The diphase system terms are given capital letters so as to avoid confusion with the particular usage of the words 'solid' and 'fluid' in ordinary language.

L

action of the Fluid phase, before and during the application of any additional stress.

The important general basis of the diphase approach is that the model uses two complementary concepts which are mutually exclusive. These are applicable at all levels of consideration, hence providing the requisite variety to represent the complexity of the actual subject being studied. The choice of the concepts is critical since they are used to describe the essential nature of the subject. Each of the concepts only has a meaning in relation to its complement and therefore neither one can be used on its own. This type of conceptual approach is not unprecedented. It has been used in apparently widely differing branches of science as a tool for dealing with organised complexity. A few examples must be quoted to illustrate the meaning of this paragraph.

In the science of management and effective organisation, Stafford Beer has stated in his book, *Designing Freedom*, that a concept which is intuitively thought to be absolute has, in reality, no meaning if it is not seen in contrast to its implied complement:

> 'the concept of freedom is not meaningful for any person except within measurable variety constraints'[8]

The ideal of absolute freedom is as self defeating as the ideal of absolute constraint: both lead to the death of the very system which is in need of being liberated/controlled. The model based upon the mutually dependent concepts of freedom and constraint is applicable at all recursive levels of the social system, where it is to be noted that:

> 'the level of recursion is simply the focus of attention at which we contemplate any viable system, and that one level is contained within the next.'[8]

In epistemology, Michael Polanyi made a signficant breakthrough† by recognising that in every acquisition of human knowledge, whether theoretical or practical, there is a mutual dependence of our 'subsidiary awareness' of the particulars involved and our 'focal awareness' of the coherent entity under study.[9] This applies at all levels of man's understanding of nature. In relation to human perception, our observation of an object depends upon the subliminal sensing of its parts together with explicit recognition of the meaning or function of the object. In relation to scientific knowledge, a theoretical discovery

†In view of the number of epistemological problems, both ancient and modern, which Polanyi's approach solves, this is perhaps an understatement.

depends upon an intuitive appreciation of the relevant clues together with an overall understanding of the significance of the theory. The two aspects, the implicit and the explicit, are inseparably linked with each other at every level of consideration.

The last example is chosen from nearer home. In the study of structural stability, Graham Armer has explicitly made use of the fact that on every scale the aspect of the structural system which is being considered must be seen relative to its particular environment and that the converse applies in specifying the nature of the relevant aspect of the environment, the latter relation being less obvious.[10] How a structural element functions depends upon the geometry and magnitude of loading (not upon the engineer's preconception of how it should function) which is, conversely, dependent upon the type and position of the element. On the scale of the building itself, its architectural features are determined by the service for which it is built and will consequently fail, as effectively as if it had been knocked down element by element, if this need is not fulfilled (e.g. Centre Point, London—at present unoccupied due to the high cost of its office space). Conversely, the use to which the building can be put relies upon its particular form (e.g. Centre Point cannot be used to house families). At the level of populations of buildings, their number and degree of uniformity is determined by the economic and political climate which, conversely, depends upon the availability of particular building methods. To sum up, stability is seen in terms of the multi-level structure–environment system.

The description of the diphase model for the study of materials can now be continued with the above conceptual basis in mind. Its use in an engineering context will be illustrated by referring to tests on concrete specimens.

The diphase view explicitly acknowledges that all materials are differentiated.[1] The highest level of differentiation is that of the specimen itself which is clearly distinguishable from its environment. Within the specimen, the differentiation continues at all levels of discrimination so as to produce a hierarchy of Solid–Fluid systems. The highest level may conveniently be numbered the first and at this and every other level, the Fluid is external to the corresponding Solid. Hence, the first level Fluid is the immediate environment of the first level Solid, the second level Fluid is the immediate environment of the second level Solid, and so on. The first hierarchical level therefore consists simply of the specimen and its environment as shown in Fig. 8(a).

The first level Solid can be subdivided into many particles plus an internal Fluid phase as shown in Fig. 8(b). Each individual particle is the second level Solid and surrounding each one is the second level Fluid; the very same Fluid which is internal to the first level Solid. The second level Fluid is consequently both external and internal depending upon the level of Solid to which reference is being made. This obviously applies also at every subsequent level of the hierarchical system.

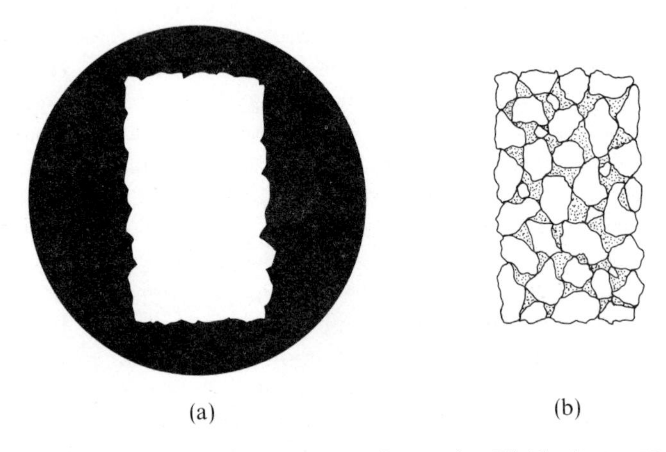

FIG. 8. The diphase model (a) ■—first order Fluid phase; □—first order Solid phase; (b) ▩—second order Fluid phase; □—second order Solid phase.

Complementarily, every Solid phase has a dual role depending upon the level of Fluid to which reference is being made. In relation to its external Fluid (i.e. at its own level) it exists as an undifferentiated whole, whereas in relation to its internal Fluid (i.e. at the level below) it exists as a number of separate particles.

A material owes its existence to the Fluid phase acting externally upon the Solid phase. In engineering terms, strength is caused by the pressure of the Fluid which pushes onto the Solid at its own level. Viewed internally, the particles of the Solid are held together so that a material's strength is gained in a similar manner to a heap of sand or stones which is put under triaxial compression (see Appendix 1). As long ago as 1885, Osborne Reynolds unintentionally demonstrated the meaningfulness of this view by means of a simple experiment in which a thin rubber bag filled with sand particles turned into a rigid body when it was partially evacuated of air, so that the atmospheric pressure held the particles together from the outside.[11] Recently, this effect was used in a more sophisticated form to develop a material, made explicitly of two phases, which is flexible at higher temperatures but which solidifies on cooling.[12]

Of course, a material which is formed due to the relatively low value of external pressure provided by the atmosphere (0·1 N/mm^2), only has a minute strength compared with most ordinary solids. If, as the authors firmly believe, the first level Fluid has meaning in reality, there must exist quasi-atmospheres which can exert pressures of very much higher magnitudes than ordinary atmospheric pressure. The relevance of this conclusion to the science of physics is very important and it is an aspect which the authors have investigated with great interest. However, since it involves cosmological issues which are not of immediate concern in the present context, it is necessary to adopt a pragmatic attitude towards this controversial issue. That is, it will be described how adopting the viewpoint that the Solid is held together from the outside provides a meaningful insight into the behaviour of materials subjected to applied stress.

In passing, the following point may be noted. It should not be assumed that the Fluid 'hyperatmosphere' mentioned above is homogeneous, any more than the ordinary atmosphere is. One reason for using the term 'first level Fluid' (instead of the word 'aether' for example) is that attention is not restricted to one highly specific quasi-atmosphere, whether conceptual or real. The complexity of this Fluid must not be underestimated. Like all fluids, it may contain currents and turbulence on various scales and will therefore not always act isotropically.

The first level Fluid must be seen as exerting a very high pressure indeed. If this Fluid were the only influence, a material's strength would be much greater than it is. Acting internally to the first level Solid, and therefore opposed to the first level Fluid, is the second level Fluid. These external and internal Fluids act against one another across the Solid phase boundary, and it is the difference between them which determines the strength of the material. Alternatively, the Solid phase may be seen as providing a semi-permeable membrane producing a pressure drop between the outside and the inside. It is the effectiveness of the Solid phase in its ability to cause a partial reduction, within its boundary, of the ambient external Fluid pressure which determines the strength of the material (this is why the form of the Solid is so important in relation to material behaviour under applied stress).

There are consequently two equally valid views of the same system depending upon which of the two alternative viewpoints is adopted. In relation to study of the resistance of materials to directly applied stress, however, the former viewpoint is the more appropriate, since it leads straight to a useful fracture criterion.

Fracture of a material occurs when the resultant compressive stress on the Solid phase has been removed. At this point there is no pressure holding the Solid together and the material therefore has no strength. Fracture is consequently the point at which the material has been reduced to equilibrium; in all other states there is a condition of non-equilibrium, this being the requirement for a material to have strength. To repeat; the natural state of a material is when there is pressure on the Solid: the criterion for fracture is that the pressure on the Solid is zero. Evidence which demonstrates the validity of this reasoning, obtained from experiments on concrete specimens and data on the compressibility of water, has already been presented elsewhere.[7] It can now be shown how this viewpoint leads to an understanding of the fracture of specimens under several different types and configurations of directly applied stress.

By recognising the stress on the Solid phase as being due to a pressure difference, it is clear that its removal, and consequently fracture of the material, can be achieved by two general methods:

(1) Decreasing the external stress on the Solid phase
(2) Increasing the internal stress on the Solid phase

An example of the first method is what is conventionally termed applied 'tension'. As far as the material itself is concerned, however, there is no such entity as 'tension', only a reduction of the pre-existing compressive stress. Indeed the adjective 'compressive' is itself redundant because all stresses in the diphase system are compressive. Pressure and stress are synonymous. The concept of 'tension' is necessary for the anthropocentric viewpoint only; the material recognises only compression or the lack of it (a direct analogy exists in relation to temperature where, in everyday language, the concept of 'cold' is very useful; the science of thermodynamics, however, recognises only heat or the lack of it). So as to further clarify this point, reference is now made to a slightly modified version of the previously described experiment shown in Figs. 5 and 6. This time a gas, nitrogen, is used as the applied fluid environment and its pressure is set at, say, $4\,N/mm^2$. The set-up is shown in Fig. 9 where the apparatus is essentially the same as before with the projecting rods, connected to the specimen end caps, being fixed into the grips of a tension testing machine which proceeds to pull them apart (the pressure vessel shown has perspex sides, enabling the specimen to be seen during the test). The specimen fractures transverse to the axis as shown, when the applied stress reaches the uniaxial 'tensile strength' of the specimen (3 N/mm^2). Two points should be noted. First, this test is quantitatively, as well as qualitatively, indistinguishable from an ordinary uniaxial 'tension' test. Hence, as far as the strength of the high level material is concerned, the presence of the gas pressure environment has no effect. This fluid has effectively become part of the existing Fluid phase of the material and the partial removal of its external action on the Solid phase in the axial direction causes the specimen to fracture. Similarly, in a conventional 'tension' test, the external stress on the Solid phase in the axial direction due to the external Fluid phase, is reduced, thereby causing fracture.†

The second point to notice about the demonstration test is that the specimen is still being subjected to axial compression from the external action of the gas pressure in the vessel at the point of fracture, its value in the experiment shown in Fig. 9 being 1 N/mm^2 (4 – 3 N/mm^2). The gas pressure used in this experiment, which is shown as 4 N/mm^2, can be set much higher and the specimen will still fracture at the same value of applied 'tension', i.e. 3 N/mm^2. Evidently, therefore, the axial compression on the specimen at fracture from the external action of the gas pressure can be made indefinitely high. This demonstrates that in an ordinary 'tension' test, the effect of the external Fluid phase pressure acting on the Solid is only partially reduced, fracture occurring when its value is lowered just enough so as to equal that of the internal pressure on the Solid.

Now consider another way in which fracture of a specimen can be caused due to decreasing the external stress on the Solid. This type of fracture occurs in certain circumstances when a previously applied uniaxial compressive stress of a high value is subsequently removed. For this phenomenon to occur, the applied compressive stress must not cause failure of the specimen but it must irreversibly alter the form of

†An additional aid to achieving the conceptual switch required in appreciating this explanation may be gained by thinking of the action of pulling a rubber sucker off a smooth surface. The apparent bond between the sucker and the surface is broken by applying a pull large enough to counteract the external action of the atmospheric pressure.

FIG. 9. Modified demonstration experiment.

the Solid so as to remove the pre-existing stress in that direction due to the action of the Fluid phase. In these circumstances, the applied uniaxial compressive stress is the only agent holding the Solid phase together in that direction and its removal consequently causes the specimen to fracture. Referring again to Bridgman:

> 'Although such conditions are not common in every day experience, they are nevertheless not infrequent. All through my high pressure experimenting I have been continually bothered by fractures that occurred during *release* of pressure.'[13]

He also mentions tests carried out by David Griggs,[14] a colleague of his at Harvard, who studied the behaviour of rocks. He found that subjecting a specimen of limestone to high confining pressure enables an abnormally large uniaxial compressive load to be applied without crushing but with considerable plastic deformation of the specimen. This fulfils the conditions specified above. Bridgman commented:

> 'Griggs found in many cases that if the experiment were terminated by release of compressive load and then release of pressure before the occurrence of compressive rupture, but after considerable plastic flow had taken place, the specimen was found, on removal from the apparatus, to be ruptured into discs on planes perpendicular to the original direction of compression. Although this sort of fracture was observed many times, Griggs for some reason did not describe it in his paper.'[13]

It is reasonable to suppose that the same behaviour will occur with concrete, but due to the practical difficulties of operating with the high confining pressure which would be necessary (several hundred N/mm^2) it has not been found possible to carry out the experiment. If other researchers obtained the result whilst working from a conventional theoretical standpoint, it is likely to be dismissed as an unaccountable experimental error due to its highly paradoxical nature, and therefore it may well be unreported.

The second general method of achieving fracture of a material, as stated earlier, is to increase the internal stress on the Solid phase. This involves applying compressive stress to the specimen in such a way as to achieve transmission of pressure to the internal Fluid phase. The experiment described at the beginning of this paper provides a pertinent example; that is, applying fluid pressure to the curved surface of a solid cylinder of material is an effective way of increasing the internal pressure on the Solid. In the direction of the applied fluid pressure, the change of internal Fluid pressure is counteracted by the change of external Fluid pressure, thus maintaining the stress on the Solid. In the axial direction of the cylindrical specimen, however, there is no change of external Fluid pressure and hence the stress on the Solid is reduced due to the increase of internal Fluid pressure. As the applied fluid pressure increases, the stress on the Solid in the axial direction is steadily further reduced and when this reaches zero, fracture occurs. The mode of fracture is the same as if the specimen had been subjected to an applied axial 'tension' because the direction in which the stress on the Solid is removed is the same in both cases. That is, as far as the first level Solid is concerned, the two tests are identical.

In the particular example of a concrete cylinder subjected to water pressure on its curved surface, the applied stress at fracture is nearly twice† the value of applied 'tensile' stress which would be needed to achieve the same fracture. This is because the water pressure is not as effective in increasing the internal pressure as the 'tension' is in decreasing the external pressure on the Solid phase. If, however, gas is used instead of water to apply the pressure to the curved surface of the concrete cylinder, the applied pressure at fracture is reduced because this fluid is able to transmit pressure much more effectively than water to the internal Fluid phase, and becomes equal† to the applied 'tensile' stress which would achieve the same fracture. (This fact has been used to introduce a new simple means of determining the 'tensile strength' of concrete,[16] the conventional test being notoriously difficult to carry out in practice.) In contrast, if the loading medium is less effective than water in raising the internal pressure on the Solid, the applied stress at fracture will be higher. The authors have found that using small ball bearings surrounding the curved concrete surface, as the loading medium, gives an applied stress at fracture of two or three times the value obtained using water pressure, whilst the mode of fracture remains the same (Fig. 10). Similar evidence has been obtained by Langan and Garas[17] who found that subjecting a concrete cylinder to wire winding around its curved surface produced the same mode of fracture at an applied stress of approximately half the crushing strength of the concrete. If an almost rigid surface is used to apply the load, only a small proportion of the applied stress is communicated to the

†These relationships obviously depend upon the specific conditions of testing: details are given in Ref. 15.

internal Fluid phase and consequently the apparent strength of the specimen is even higher. Richart, Brandtzaeg and Brown[18] have shown that when the load is applied via a close fitting metal sleeve surrounding the curved surface of a concrete cylinder, the applied stress

FIG. 10. Fracture of concrete cylinder due to ball bearing pressure.

at fracture is as great as the crushing strength. Again, the specimen fractures across a single plane perpendicular to the axis.

It is apparent from the evidence presented in the above paragraph that a whole range of ultimate applied stress values may be obtained by using different loading media, whilst achieving exactly the same type of fracture of the concrete specimen. That is, the same qualitative behaviour can be achieved under quantitatively different conditions. This is because the level of the material at which fracture occurs is the same in all cases and therefore involves the same Solid–Fluid system, but since there are different methods of increasing the internal stress on the Solid, some more direct than others, the external loading conditions which achieve fracture may vary considerably.

It is arbitrary whether a certain common method of loading is incorporated into the nature of concrete itself: if it is, the term 'compressive strength' becomes a meaningful expression. However, this built-in assumption must be made explicit in order to avoid the danger of concluding that concrete is in some way inherently strong in compression. No property is inherent: this is one of the main functions of adopting the diphase system which represents strength as being due to both internal and external components.

7.3 THE NON-ESSENTIAL TENSION

Throughout, this paper has been mainly concerned with one particular type of fracture of a specimen; that is to say the one shown in Fig. 3. This is because it is the simplest type of fracture which can occur under directly applied stress; and a start has to be made somewhere. Despite efforts in this paper to prove to the contrary, there are no doubt many people who will refer to it as a 'tensile fracture' and who will mistakenly interpret the evidence presented as showing that fracture is always due to 'tension' on one scale or another (in opposition, for example, to those who maintain that all fracture is due to shear). In order to finally demonstrate that this type of fracture has no necessary connection whatever with the purely anthropocentric concept of 'tension', reference must be made to the following triaxial experiment.

A concrete cylinder is loaded with fluid pressure (using gas or water) applied to its bare curved surface in the same manner as described earlier, whilst simultaneously its ends are subjected to compression by means of ordinary steel platens as shown in Figs. 11 and 12. Similar loading conditions have been studied by other researchers investigating the effect of a water pressure environment on the behaviour of concrete.[19] In those experiments, the water pressure was fixed at a particular value and the applied axial compressive stress necessary to cause failure by crushing of the specimen was measured. In this experiment, on the other hand, the two loads are increased simultaneously with the rate of increase of the applied fluid pressure higher than the rate of increase of the applied axial stress. Under these conditions, the specimen fractures perpendicular to the axis, as shown in Fig. 3; that is, perpendicular to the direction of applied axial

FIG. 11. Triaxial experiment apparatus.

compressive stress. This observation does not refute, however, the possible claim that the fracture is due to an 'indirect tension' which would be signified by an axial elongation of the specimen. In other words, although there is no obvious 'tensile stress' present, the phenomenon may be explicable by the presence of a 'tensile strain'. In

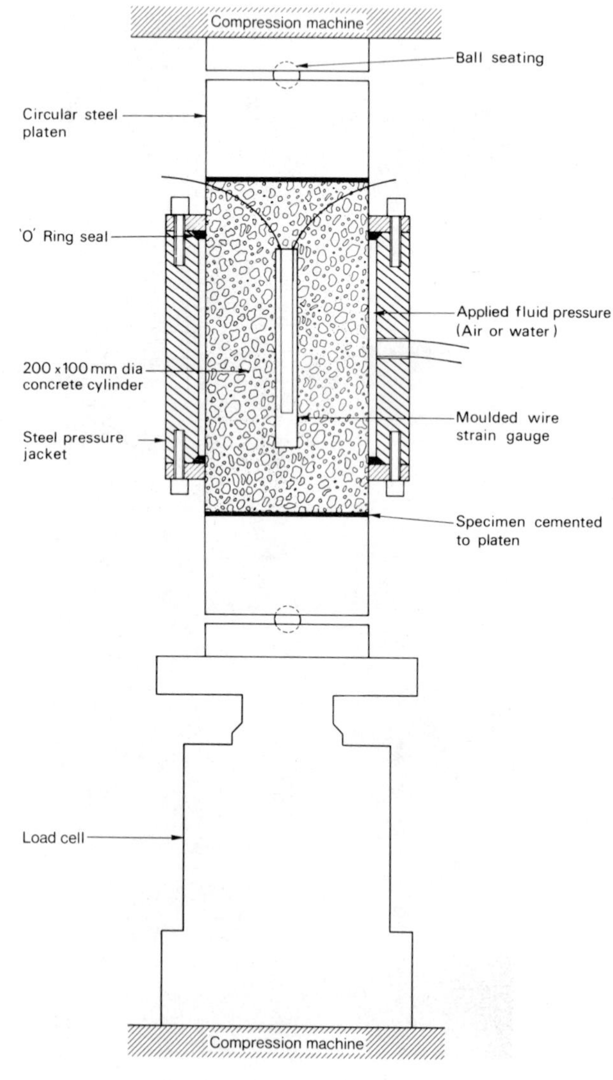

FIG. 12. Triaxial experiment details.

order to investigate this possibility, the strain in the axial direction was measured by means of a resistance gauge cast into the centre of the specimen, as shown. It was found that when the fluid pressure was raised at two or more times the rate of increase of the applied axial compressive stress, the specimen did indeed elongate in the axial direction. But upon lowering the ratio of fluid pressure to applied axial compressive stress, the value of the elongation at fracture was reduced and could be made zero. Upon further lowering of the applied stress ratio, whilst still keeping it above unity, the axial deformation of the specimen at fracture was no longer an elongation but a contraction. That is, when the fluid pressure was increased at a fractionally greater rate than the axial compressive stress, the specimen fractured across a plane perpendicular to an applied *compressive strain.* A programme of tests using both water and gas as the loading fluid has corroborated this observation (see Appendix 2). In certain cases, fracture occurred when the axial compression was over 100 microstrain. Clearly, 'tension' is a redundant concept in relation to fracture.

This experiment also demonstrates that strain has no necessary connection with fracture nor with the specimen's state of stress. For example, the measured axial compression of over 100 microstrain which was produced by applying an axial compressive stress of 15 N/mm^2 and a water pressure of 21 N/mm^2, could equally well be produced in an ordinary uniaxial compression test at an applied stress of about 4 N/mm^2, in which case the specimen would not be on the point of fracture. Of course, in certain circumstances, for a given specimen and applied loading conditions, there will be a definite relationship between stress and strain, but this will change, sometimes drastically, when the conditions are altered. A particular stress–strain relationship gives information about the system produced by the interaction of the specimen with the applied loading: it does not describe an inherent characteristic of the material.

7.4 FINAL COMMENTS

The value of the diphase conceptual approach lies in its ability to draw attention to important features which remain implicit in the conventional study of strength of materials. To begin with, the explicit use of two complementary concepts mirrors the use of common engineering terms which are successful only when applied in pairs. For example, stress and strain; brittle and ductile; elastic and plastic; homogeneous and heterogeneous; rigidity and flexibility; stability and adaptability; and continuous and discrete. Any one of these terms loses its meaning if applied absolutely without reference to its implied complement; for example, in order to make the statement that a material is homogeneous, either implicit or explicit reference must be made to a particular size of specimen and this is, of course, the scale at which the material is heterogeneous (in relation to its environment). Conversely, in order to state that a material is heterogeneous, reference must be made to the scale at which the material is homogeneous, that is, at the subparticle level. The other pairs of examples given in the list are also mutually dependent concepts.

In the diphase system, the Solid represents the readily perceived part of a material. The behaviour of materials cannot be understood, however, by reference to the Solid alone. By recognising the Fluid phase as being the equally important constituent of a material at every hierarchical level of consideration, the reductionist fallacy, in which all phenomena are seen as being due to the indeterminate interplay of ultimate particles, is avoided. Ludwig von Bertalanffy has made this point forcibly in relation to science in general:

> 'Reality, in the modern conception, appears as a tremendous hierarchical order of organized entities, leading, in a superposition of many levels, from physical and chemical to biological and sociological systems. Unity of Science is granted, not by a utopian reduction of all sciences to physics and chemistry, but by the structural uniformities of the different levels of reality.'[20]

The action of the internal Fluid phase is manifested by the formation of voids and crack growth inside a specimen. In contrast to the customary view, however, the diphase approach recognises voids and cracks to be substantial, and their spread and coalescence as visible evidence of the internal Fluid. Most importantly, the internal Fluid is under pressure and this presents the possibility of interpreting internal crack growth in an analogous manner to how caviation in liquids occurs due to the formation of vapour pressure bubbles.[21,22] The Fluid phase also makes the dynamic nature of a material explicit. Properties such as creep and fatigue obviously relate to Fluid flow, not just within the specimen, but to and from the environment. This is, therefore, an open system approach in which the concept of the 'steady state'[23] is used instead of the customary view of a material as being in overall equilibrium.

The diphase system approach draws attention to the fact that the recorded characteristics of a specimen depend as much upon its environment as its internal form. Whether the ultimate source of a material's strength is the external Fluid pressure, as the authors propose, will probably remain a controversial issue for some time, but there is no doubt that the nature of the loading environment is of major importance in relation to material behaviour. This was clearly demonstrated earlier when it was described how the same fracture of a concrete cylinder can be achieved under widely ranging ultimate axisymmetric stress values by simply changing the loading medium. Another pertinent example is provided in the study made by Hannant *et al.* on the cylinder splitting ('Brazilian') test for the indirect tensile strength determination of concrete.[24] They showed that most of the measured strength was provided by the action of the testing machine via the packing strips. Their tests involved comparing the results obtained from pre-split cylinders (made by casting each cylinder in two halves, the plane of separation being along the length and diameter corresponding to the usual fracture plane) with those from ordinary cylinders, and found that the pre-split cylinder tests gave over half the load carrying capacity of the ordinary ones! These examples show that strength depends upon the nature of loading. The dependence of the recorded strain behaviour of a concrete specimen on the loading environment was demonstrated by the triaxial tests described earlier, in which the axial strain when fracture occurred varied from tensile to compressive depending upon the ratio of the fluid pressure to the axial compression. Consequently, neither strength nor stiffness are inherent characteristics; they are determined by the interaction of the loading environment with the specimen.

It must be emphasised that this view differs from the customary opinion in that, although it is well known that the method of loading can affect the recorded strength and stiffness, it is still widely thought that concrete has inherent structural characteristics which are somehow independent of the loading environment. For example, it has long been known that the 'compressive strength' of concrete is an elusive quantity, yet this is customarily put down to practical difficulties of testing which prevent the 'real' value emerging. The diphase system approach takes a different view. When pressure is applied to a specimen, that is, to the first order Solid, the strength value obtained will depend upon how the pressure is shared between the second order Solid and Fluid phases. If most of the pressure is put onto the Solid, a high strength is recorded; a low strength if most of the pressure is put onto the Fluid.† There is no logical necessity for choosing one form of loading rather than the other.

A meaningful study of concrete behaviour can only be made in terms of specific structural contexts; in other words, study must be of the composite load–specimen system, as opposed to the customary approach which attempts to study each aspect of the system in isolation. By adopting this method for the study of structural elements, the emphasis moves away from 'material properties' and is placed upon the interaction of the relevant system components. As a consequence, the qualitative aspects of the system become the focus of interest and quantities such as strength and stiffness are seen as providing information-free data.

The final aspect of the diphase approach, to be mentioned here, is in connection with the concept of 'tension'. As explained earlier, 'tension' is represented in diphase terms as the reduction of a pre-existing pressure. Now, whether this view is accepted or not, the fact remains that the type of fracture caused by an applied 'tension' can equally be produced by reducing a previously applied compressive stress. Experiments described in this chapter have demonstrated this, in addition to the evidence referred to previously by Bridgman[13] on his own research and that of Griggs.[14] This is yet another example of material behaviour which is counter-intuitive and therefore must be taken account of explicitly by theory. Unfortunately, inconvenient facts of this nature involve such a stretching of the existing theoretical framework which, due to its resistance to fundamental change, is forced into dividing itself again and again in order to cope. This compartmentalism of materials science means that prediction in new situations is almost impossible and real discovery is prevented. To be blunt, materials science at present suffers from galloping empiricism, an ailment all too common in this technological age. We are flooded with

†The 'tensile strength' of concrete is also as sensitive to how the load is applied. Usually, most of the applied 'tensile' stress is removed from the Solid, thus giving a low strength. It is interesting to speculate, however, on the possibility of achieving a high strength by reducing the external Fluid pressure directly. The 'tensile strength' recorded would then approach the value normally associated with the 'compressive strength' of concrete.

data, but there is a pitiful shortage of information. The unfortunate truth is that the cost of information gained at one level is change at another. Acquisition of data, in itself, is not enough. As Stafford Beer has stated in another context:

> 'We can generate data indefinitely; we can exchange data for ever; we can store data, retrieve data, and file them away. All this is great fun; maybe useful, maybe lucrative. But we have to ask why. The purpose is regulation. And that means translating data into information. *Information is what changes us.* My purpose too is to effect change—to impart information, not data.'[25]

REFERENCES

1. Leliavsky Bey, S. (1947). *Trans. ASCE*, **112,** 444.
2. Bridgman, P. W. (1912). *Phil. Mag.*, **24,** 63.
3. Bridgman, P. W. (1931). *The Physics of High Pressure.* G. Bell & Sons, London, pp. 92–93.
4. Bridgman, P. W. (1939). *Mech. Eng.*, February, 107.
5. Bridgman, P. W. (1949). *Research*, **2,** 550.
6. Weiss, P. A. (1971). *Hierarchically Organized Systems in Theory and in Practice.* Hafner, New York, p. vi.
7. Grimer, F. J. and Hewitt, R. E. (1971). In *Proc. of Southampton 1969 Civ. Eng. Mats. Conf., Structure, Solid Mechanics and Engineering Design* (Ed. by Te'eni, M.). Wiley-Interscience, London, pp. 681–691.
8. Beer, S. (1974). *Designing Freedom.* Wiley, London, pp. 82 and 58–59.
9. Polanyi, M. (1966). *Philosophy*, **41,** 1.
10. Armer, G. S. T., Building Research Station. Private Communication.
11. Reynolds, O., *Phil. Mag. S.5. 1885*, **20,** 469.
12. Smith and Nephew Ltd. British Patent No. 1 456 831, Nov. 1976.
13. Bridgman, P. W. (1938). *J. Appl. Phys.*, **9,** 517.
14. Griggs, D. T. (1936). *J. Geol.*, **44,** 541.
15. Clayton, N. (1978). *Mag. Concrete Res.*, **30,** 26.
16. Building Research Establishment (1976). *BRE News*, **37,** 15.
17. Langan, D. and Garas, F. K. (1971). In *Proc. of Southampton 1969 Civ. Eng. Mats. Conf., Structure, Solid Mechanics and Engineering Design* (Ed. by Te'eni, M.). Wiley-Interscience, London, pp. 1061–1088.
18. Richart, F. E., Brandtzaeg, A. and Brown, R. L. (1928). *Univ. Illinois Engineering Experiment Station Bulletin No. 185.*
19. Akroyd, T. N. W. (1961). *Mag. Concrete Res.*, **13,** 111.
20. von Bertalanffy, L. (1973). *General System Theory.* Penguin University Books, Harmondsworth, pp. 87–88.

21. APFEL, R. E. (1972). *Sci. Am.*, Dec., 58.
22. FISHER, J. C. (1949). *Sci. Monthly*, **68,** 415.
23. BURTON, A. C. (1939). *J. Cellular Comp. Physiol.*, **14,** 327.
24. HANNANT, D. J., BUCKLEY, K. J. and CROFT, J. (1973). *Mat. et Const.*, **6,** 15.
25. BEER, S. (1975). *Platform for Change.* Wiley, London, p. 223.

APPENDIX 1: INTERNAL BONDS VERSUS EXTERNAL PRESSURE

The diphase view that a material owes its strength to an external pressure obviously disagrees fundamentally with the conventional concept of strength being due to internal bonds. The following arguments are put forward in favour of the proposal that the concept of bonds is inadequate.

Bonds exist at many different levels; from those on the nuclear scale between neutrons and protons to those in the macroscopic range between visible particles. These various types are explained in one of two possible ways; by means of either a binding agent between the constituent parts or an inherent attraction of the parts to each other. The first method is simply a gluing together of the particles, albeit by an extremely strong glue. This glue can be very sophisticated as in the case of the metallic bond where an 'election cloud' is thought to hold the atoms together. However, no matter which euphemism is used, if bonding is to be explained in terms of a physical entity between the particles, then glue it must be.

From an aesthetic point of view, the picture of a world which is glued together is totally repugnant. From a rational point of view, the concept of a glue is hardly an explanation at all since any material which has the property of bonding inert particles together must be more complex than the particles themselves. A fine example of this type of rationalisation is provided by Banesh Hoffmann[1] in relation to Heisenberg's account of the forces which bind the atom's nucleus together:

> 'Heisenberg therefore ascribed them to exchange phenomena within the nucleus,.... According to the curious picture scientists have to use in thinking of these things, this exchange is a sort of rhythmic interchange of position between the particles comprising

> the nucleus. Now a neutron can become a proton by shedding an electron and a neutrino, and a proton can become a neutron by absorbing them. Thus the interchange of place between a proton and a neutron can be pictured as a sort of tossing to and fro between them of an electron and a neutrino, as in a long, fast rally in tennis. The neutron serves, and in serving becomes a proton. The original proton receives, and in receiving becomes a neutron. It at once returns the serve....'

At best, such a description provides a mnemonic, but in no circumstances can it be seen as providing any real insight or understanding of the problem. The concept of a glue similarly raises more difficulties than it removes and must, therefore, be rejected as being scientifically unproductive.

Alternatively, it may be considered that bonds are due to a mutual attraction of the particles themselves without the need for an intermediate substance. This view must be taken rather more seriously since it is quite an appealing philosophy. The notion of action at a distance, which is implied, is suggestive of spiritual interaction and a condition where unforced harmony prevails. If the pun may be forgiven, it is truly an attractive world. Unfortunately, laudable though it is, such a philosophy is hardly reconcilable with strength of materials theory; adopting non-materialistic ideas to explain material behaviour is an admission of defeat and leads to no progress whatever. No compromise is possible; either a completely non-materialistic philosophy is adopted, in which case a strength of materials theory has no place, or the concept of action at a distance is rejected altogether. The latter is the less radical solution and is the one chosen here.

APPENDIX 2: TRIAXIAL TEST PROGRAMME

As described in the test, the triaxial tests involved applying fluid pressure axisymmetrically to the bare curved surface of the cylinder simultaneously with, but at a greater rate than, the axial compressive stress applied mechanically to the ends. Two series of tests were performed, the first using water pressure and the second using nitrogen gas pressure. Each series of tests was carried out using several ratios of fluid pressure to axial stress, these being 1·0–1, 1·2–1, 1·4–1, 1·6–1, 1·8–

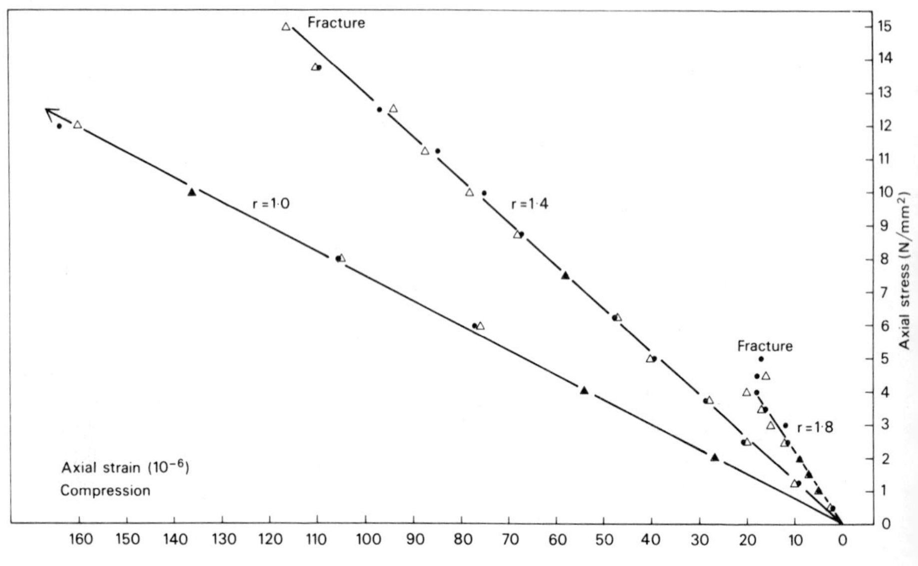

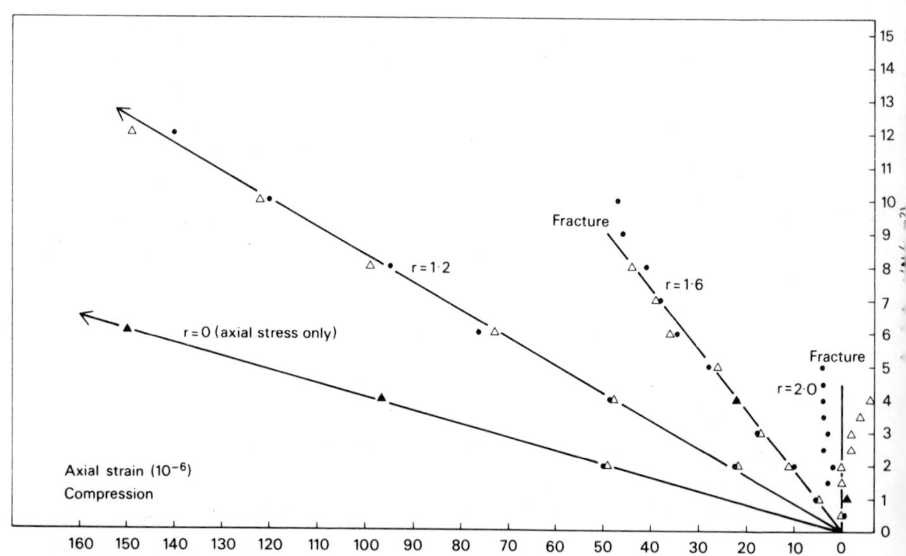

FIG. A1. Stress–strain curves for the triaxial tests using water pressure (for clarity, alternate curves are shown on each of two graphs). Each point plotted represents one reading: △—set 1 results; ●—set 2 results. r = applied stress ratio = $\frac{\text{axisymmetric water pressure}}{\text{axial stress}}$.

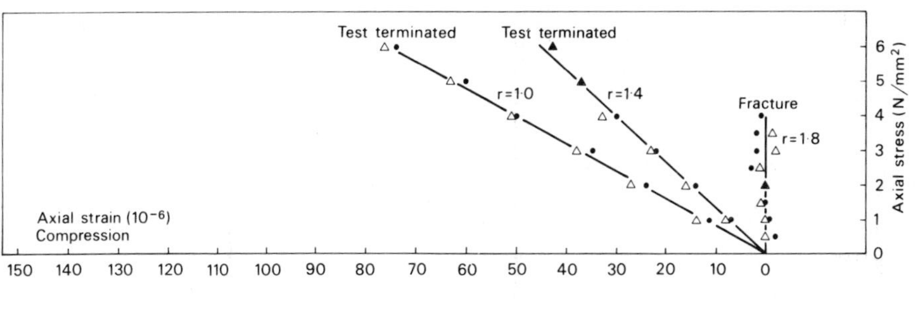

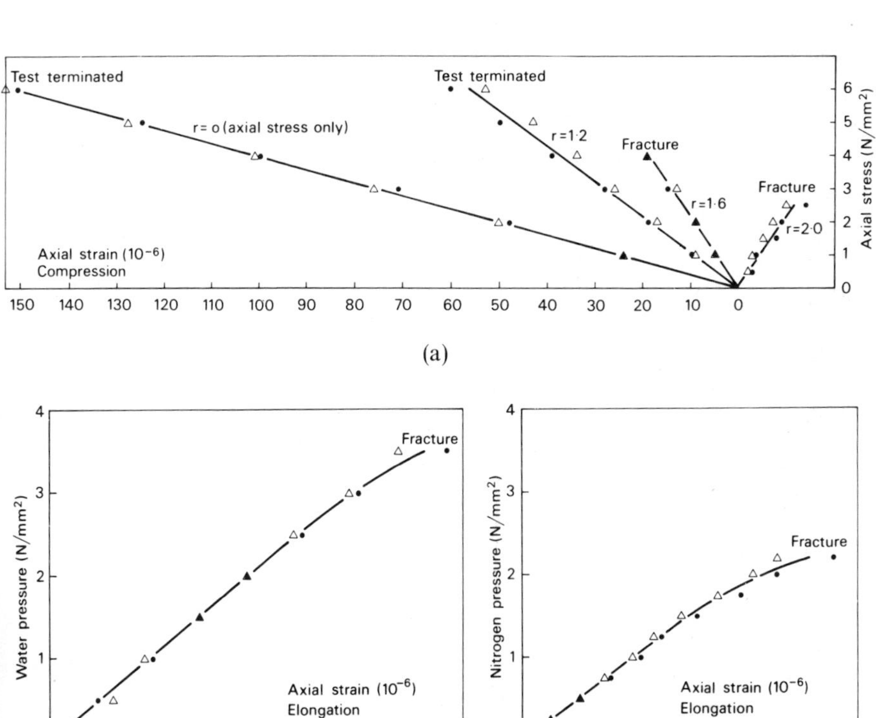

FIG. A2. (a) Stress–strain curves for the triaxial tests using nitrogen pressure (for clarity, alternate curves are shown on each of two graphs). r = applied stress ratio $= \dfrac{\text{axisymmetric nitrogen pressure}}{\text{axial stress}}$. (b) Stress–strain curves for the tests using fluid pressure only. Each point plotted represents one reading: △—set 1 results; ●—set 2 results.

1 and 2·0–1. In addition to the triaxial tests, as a comparison, the axial strain was measured on specimens subjected to axial compressive stress only, and also on specimens subjected to fluid pressure only. Hence, for each series of tests, there were a total of 8 loading conditions. Two specimens were tested at every loading condition for each series and no specimens were tested more than once. A separate batch of concrete for each complete set of loading conditions was cast (see Ref. 15 for details of the mix, etc.), making 4 batches in all, each consisting of 8 specimens.

The results are expressed graphically in Figs. A1 and A2 where, for each of the two series of tests, the axial strain is plotted at each increment of axial compressive stress. Each corresponding value of applied fluid pressure can be found by multiplying the axial stress by the applied stress ratio (r) given for each line plotted. In addition, the graphs of fluid pressure versus axial strain are given for the tests carried out with no applied axial stress. In some cases the tests had to be terminated without fracture occurring, due to the limit of the apparatus. Lines with an arrow head indicate the tests were terminated at higher values which are not plotted. All the plots for which fracture occurred are shown complete.

REFERENCE

1. HOFFMANN, B. (1959). *The Strange Story of Quantum.* Penguin Books, Harmondsworth, p. 187.

INDEX